Advanced AI Technologies for Improving Education Systems

Editor:
Amit Kumar Tyagi
National Forensic Sciences University, Gandhinagar, Gujarat, India

CRC Press
Taylor & Francis Group
Boca Raton London New York

CRC Press is an imprint of the
Taylor & Francis Group, an **informa** business
A SCIENCE PUBLISHERS BOOK

First edition published 2026
by CRC Press
2385 NW Executive Center Drive, Suite 320, Boca Raton FL 33431

and by CRC Press
4 Park Square, Milton Park, Abingdon, Oxon, OX14 4RN

CRC Press is an imprint of Taylor & Francis Group, LLC

Library of Congress Cataloging-in-Publication Data (applied for)

ISBN: 978-1-041-07061-0 (hbk)
ISBN: 978-1-041-07210-2 (pbk)
ISBN: 978-1-003-63936-7 (ebk)

DOI: 10.1201/9781003639367

Typeset in Times New Roman
by Prime Publishing Services

Preface

Education has always been the major contributor in human progress, yet traditional teaching and learning methods often struggle to keep pace with the rapidly evolving needs of modern society. But today with the integration of Artificial Intelligence (AI) into education is no longer a futuristic concept but an emerging reality that is reshaping the way knowledge is delivered, assessed, and personalized. This book is written to explore this transformative journey, offering readers an in-depth understanding of how cutting-edge AI tools can enhance education at every level.

The book begins by introducing the foundations of AI as they relate to education, i.e., from intelligent tutoring systems and adaptive learning platforms to natural language processing and generative AI that can create personalized content. It examines the potential of machine learning algorithms to predict student performance, recommend tailored learning paths, and identify at-risk learners early. These innovations are helping educators shift from a "one-size-fits-all" approach to a more individualized and inclusive learning experience.

Another major focus of this work is the use of AI for improving administrative efficiency. Automating routine tasks such as grading, attendance, and feedback allows educators to devote more time to meaningful interactions with students. Note that AI-driven analytics provide valuable insights into curriculum effectiveness, student engagement patterns, and institutional performance, empowering data-driven decision-making.

While the promise of AI is significant, this book also recognizes the challenges that accompany its adoption. To note, ethical issues such as data privacy, algorithmic bias, and the digital divide must be addressed to ensure equitable and responsible use of AI in education. Throughout the text, readers will find balanced discussions of both the opportunities and the risks associated with AI-driven education systems.

This book is designed for almost to all people who are involve in learning (directly/indirect) and who are eager to explore innovative approaches to teaching and learning. Hence with combining theoretical insights with real-world applications, it provides a roadmap for leveraging AI to build smarter, more inclusive, and future-ready education systems. We hope that this work inspires educators and stakeholders to embrace AI not as a replacement for human teaching, but as a powerful ally that enhances creativity, critical thinking, and lifelong learning. The future of education is intelligent, adaptive, and student-centered and AI is a key enabler of that transformation.

If you like the book, please recommend it to your students/colleagues.

Contents

List of Contributors

Aditya Krishnan
School of Computer Science and Engineering, Vellore Institute of Technology, Tamil Nadu, India.

Anirudh AVS
School of Computer Science and Engineering, Vellore Institute of Technology, Tamil Nadu, India.

Anu Pius
Department of Science and Humanities, St.Joseph's College of Engineering and Technology, Kerala, India. https://orcid.org/0009-0007-8577-5997

Aparna Nair M.K.
Department of Electronics and Communication Engineering, Government College of Engineering, Kannur, Kerala; and APJ Abdul Kalam Technological University, Kerala. https://orcid.org/0000-0002-0575-0850

Awotude Joseph Bamidele
Faculty of Information and Communication Sciences, Department of Computer Science, University of Ilorin, Nigeria. https://orcid.org/0000-0002-1020-4432

B. Niveditha
Department of Computer Science and Engineering, Marri Laxman Reddy Institute of Technology and Management, Telangana, India.
https://orcid.org/0009-0005-5813-8082.

Devi Priya R.
Department of Computer Science and Engineering, KPR Institute of Engineering and Technology, Tamil Nadu, India.

Dilliraj Ekambaram
Department of Electronics and Communication Engineering, SRM Institute of Science and Technology, Tamil Nadu, India.
https://orcid.org/0000-0002-9245-2746

Dudekula Siddaiah
Department of Computer Science, Pulla Reddy Institute of Computer Science, Telangana, India.

Elangovan G.
Department of Data Science and Business Systems, SRM Institute of Science and Technology, Tamil Nadu, India. http://orcid.org/0000-0003-2385-2129

Evelyn Damilola Temitayo
Department of Computer Science, University of Ilorin, Nigeria.
https://orcid.org/0009-0006-5483-2891

I. Gusti Bagus Yosia Wiryakusuma
Department of International Business Management, Universitas Ciputra, Indonesia. https://orcid.org/0000-0001-5845-8661

Idowu Dauda Oladipo
Department of Computer Science, University of Ilorin, Nigeria.
https://orcid.org/0000-0002-1472-2237

Ir. Bambang Sugiyono Agus Purwono
Mechanical Engineering Department, State Polytechnic of Malang, Indonesia.
https://orcid.org/0000-0002-5982-8983

Jeevanantham Vellaichamy
Department of Computer Science and Engineering, Vel Tech Rangarajan Dr. Sagunthala R&D Institute of Science and Technology, Tamil Nadu, India.
https://orcid.org/0000-0003-3787-4855

Joseph Bamidele Awotunde
Department of Computer Science, University of Ilorin, Nigeria.
https://orcid.org/0000-0002-1020-4432

Kanimuthu A.R.M.
Department of Computer Science and Engineering, KPR Institute of Engineering and Technology, Tamil Nadu, India.

Kapil Dev Tyagi
Department of ECE, Jaypee Institute of Information Technology, Uttar Pradesh, India. https://orcid.org/0000-0002-4596-2321

Karimunnisa Shaik
Department of Information Technology, Marri Laxman Reddy Institute of Technology and Management, Telangana, India.
https://orcid.org/0009-0007-8160-252X

Karmel A
School of Computer Science and Engineering, Vellore Institute of Technology, Tamil Nadu, India.

Keshav Dahal
School of Computing, Engineering and Physical Sciences, University of the West of Scotland, UK

Lakkineni Swathi
Ford Motor Company, MI, USA. https://orcid.org/0009-0006-2769-1421

M. Ramprasath
Department of Data Science and Business Systems, SRM Institute of Science and Technology, Tamil Nadu, India. http://orcid.org/0000-0002-2667-9184

M.V. Krishna Reddy
Department of Computer Science and Engineering, Chaitanya Bharathi Institute of Technology, Hyderabad, India. https://orcid.org/0000-0002-7247-6580

Manjushree Nayak
Amity School of Engineering & Technology, Amity University Chhattisgarh, India. https://orcid.org/0000-0001-6383-780X

Mohan Singh
Department of ECE, G.L. Bajaj Institute of Technology and Management, Uttar Pradesh, India. https://orcid.org/0000-0002-2973-3676

Muyideen Abdulraheem
Department of Computer Science, University of Ilorin, Nigeria. https://orcid.org/0000-0003-4568-7566

Nadana Ravishankar
Department of Data Science and Business systems, SRM Institute of Science and Technology, Chennai, India. https://orcid.org/0000-0002-9854-5150

Navaneethakrishnan M.V.
Department of Computer Science and Engineering, KPR Institute of Engineering and Technology, Tamil Nadu, India.

Nemanja Zdravkovic
Faculty of information Technology, Metropolitan University, Serbia. https://orcid.org/0000-0002-2631-6308

Nishath J.P.
Department of Computer Science and Engineering, KPR Institute of Engineering and Technology, Tamil Nadu, India.

Oluwakemi Christiana Abikoye
Department of Computer Science, University of Ilorin, Nigeria. https://orcid.org/0000-0002-8912-6333

Paul Olujide Adebayo
Department of Computer Science, University of Ilorin, Nigeria. https://orcid.org/0000-0002-4088-0634

Prakash Duraisamy
Department of Computer Science and Engineering, University of Wisconsin-Green Bay, USA.

Prerna Mishra
Department of CSE, IIIT Nagpur, India. https://orcid.org/0000-0003-2895-6146

R. Felista Sugirtha Lizy
Department of Computer Science, A.P.C. Mahalaxmi College for Women, Tamil Nadu, India. https://orcid.org/0000-0003-3648-0205

R. Felista Sugirtha Lizy
Department of Computer Science, A.P.C. Mahalaxmi College for Women, Tamil Nadu, India. https://orcid.org/0000-0003-3648-0205

R. Ravinder Reddy
Department of Computer Science and Engineering, Chaitanya Bharathi Institute of Technology, Hyderabad, India. https://orcid.org/0000-0001-6488-9935

R. Suya Padhra Haridha
Department of Physics, V. O. Chidambaram College, Thoothukudi- Tamil Nadu, India, 628008. https://orcid.org/0000-0002-9988-0946

Rajalakshmi Nagarnaidu Rajaperumal
Department of Computer Science and Engineering, Vel Tech Rangarajan Dr. Sagunthala R&D Institute of Science and Technology, Tamil Nadu, India. https://orcid.org/0000-0001-9705-0917

Rajesh Manoharan
Department of Computer Science and Engineering, Aarupadai Veedu Institute of Technology, Vinayaka Mission's Research Foundation (DU), Tamil Nadu, India. https://orcid.org/0000-0003-0003-473X

Ravi Uyyala
Department of Computer Science and Engineering, Chaitanya Bharathi Institute of Technology, Hyderabad, India. https://orcid.org/0000-0002-6913-7529

S. Durga Devi
Department of Computer Science and Engineering, Chaitanya Bharathi Institute of Technology, Hyderabad, India. https://orcid.org/0000-0002-5163-8886

S. Sowjanya Chintalapati
Department of CSE, Sreyas Institute of Engineering and Technology, Hyderabad, India. https://orcid.org/0009-0005-5651-5438

S.N. Sithi Shamila
Department of Information Technology, Wavoo Wajeeha Women's College of Arts and Science, Tamil Nadu, India. https://orcid.org/0000-0003-0439-2534

Saravanan Krishnan
Department of Computer Science and Engineering, Anna University, Tamil Nadu, India. https://orcid.org/0000-0001-9517-9792

Saravanan Srinivasan
Department of Computer Science and Engineering, SRM Institute of Science and Technology, Tamil Nadu, India. https://orcid.org/0000-0001-7884-3532

Satyanarayana Nimmala
CVR College of Engineering, Hyderabad, India.
https://orcid.org/0000-0002-9433-1489

Seema Singh
Department of Education, Lloyd Institute of Management & Technology, Uttar Pradesh, India. https://orcid.org/0009-0009-4776-2241

Shweta Patel
Amity Business School, Amity University Chhattisgarh, India.
https://orcid.org/0000-0002-5241-9984

Usman-Hamza Fatimah Enehezei
Department of Computer Science, University of Ilorin, Nigeria.
https://orcid.org/0009-0004-1115-4555

V. Kavitha
Department of Data Science and Business Systems, SRM Institute of Science and Technology, Tamil Nadu, India. https://orcid.org/0000-0002-2728-5646

Vaibhav Bhushan Tyagi
Department of ECE, ISBAT University, Uganda
https://orcid.org/0000-0001-8153-3607

Vengatesan Krishnasamy
Department of Artificial Intelligence and Data Science, Sanjivani University, Maharashtra, India. https://orcid.org/0000-0002-8896-4925

Vijayakumar Ponnusamy
Department of Electronics and Communication Engineering, SRM Institute of Science and Technology, Tamil Nadu, India.
https://orcid.org/0000-0002-3929-8495

Vineeth V.K.
Department of Mechanical and Industrial Engineering, Texas A&M University, USA. https://orcid.org/0000-0003-1378-0619

CHAPTER 1

All are Ori Revolutionizing Education: AI-driven Smart Facilities for Modern Schools

Vengatesan Krishnasamy[1*], *Rajesh Manoharan*[2], *Nadana Ravishankar*[3] and *Anu Pius*[4]

1. Introduction

School operations, student learning, and academic achievement are all being transformed by the integration of artificial intelligence (AI) into educational systems. In light of the great technical advances made in Last 10 years, smart facilities driven by AI have become a game-changing force in contemporary education (Makinde et al., 2024). Efficiency, security, and individualized education are all taken to new heights by these establishments. By combining analytics driven by AI, automation, and the Internet of Things (IoT), these smart systems aim to make schools run more smoothly, get students more involved, and create a safer, more adaptable learning environment (Kavitha et al., 2024). As educators and lawmakers recognize the need to update educational environments, AI-powered solutions are taking front stage in shaping future schools. The idea that schools require updating is gaining more and more support.

1.1 The Applications of AI in the Academic Setting

From its humble beginnings as an ancillary tool in the realm of education, AI has steadily grown into a crucial part of how institutions operate. One of the first

1 Department of Artificial Intelligence and Data Science, Sanjivani University, Maharashtra, India. https://orcid.org/0000-0002-8896-4925

2 Department of Computer Science and Engineering, Aarupadai Veedu Institute of Technology, Vinayaka Mission's Research Foundation (DU), Tamil Nadu, India. https://orcid.org/0000-0003-0003-473X

3 Department of Data Science and Business Systems, SRM Institute of Science and Technology, Chennai, India. https://orcid.org/0000-0002-9854-5150

4 Department of Science and Humanities, St.Joseph's College of Engineering and Technology, Kerala, India. https://orcid.org/0009-0007-8577-5997

* Corresponding author: vengicse2005@gmail.com

DOI: 10.1201/9781003639367-1

applications of AI was in adaptive learning systems, which adapted lessons to each student's unique needs (Makinde et al., 2024). Conversely, AI now serves as a whole facility manager, overseeing everything from smart security systems to automated administrative workflows and energy-efficient building designs. All throughout the globe, schools are implementing AI systems to streamline their routine tasks, which in turn allows for better use of resources and an easier learning environment for everyone involved.

Utilizing technologies powered by AI in educational institutions signifies not just a fresh technological development but also a sea change. More and more, modern educational institutions are moving away from antiquated administrative methods and luxuries in favour of intelligent environments that can dynamically adjust to the needs of both students and instructors (Mishra, 2025). This shift allows schools to ditch the one-size-fits-all model and instead create an ecosystem where AI improves every part of education, from personalizing lessons for each student to predicting when school buildings will need repairs.

1.2 Intelligent Buildings Driven by AI: Their Advantages and Essential Characteristics

The ability of smart facilities driven by AI to enhance the security of educational institutions is one of its most striking features (Adiguzel et al., 2023). AI-powered surveillance systems can detect potential dangers, track students' whereabouts, and verify their adherence to school rules in real time. Facial recognition software, biometric access controls, and AI-powered anomaly detection might all work together to make schools safer places to study. It is now possible to detect unauthorized access and eliminate threats with the use of these technologies (Palanivel, 2020).

It is also crucial to think about automating administrative tasks. Using AI-powered chatbots and virtual assistants, school administrators may streamline communication with parents and students, as well as handle queries and schedule meetings (Swargiary, 2024). This alleviation of administrative burdens allows faculty and staff to focus more on the learning and growth components of education. In addition, data-driven choices, including those on resource allocation, student performance evaluation, and maintenance scheduling, can be aided by predictive analytics in school administration.

One other manner in which AI is revolutionizing education is through intelligent learning environments. The advent of AI-powered interactive whiteboards, personalized AI teachers, and real-time language translation technology has led to more inclusive and engaging learning sessions (Zeeshan et al., 2022). By analysing student replies, adapting educational materials to those responses, and providing timely feedback, these technologies make sure that no student falls behind because of variances in learning style or speed.

1.3 Sustainability in Energy Consumption and Green School Practices

The development of energy-efficient school settings is a key component of the modern educational infrastructure's growing focus on sustainability, which in turn relies heavily on AI. AI-driven intelligent energy management systems improve heating, cooling, and lighting in educational institutions by analysing energy consumption trends (Srinivasa et al., 2022). By cutting operating costs and carbon emissions as much as feasible, this aids educational institutions in reaching their worldwide sustainability objectives.

The optimal condition of educational facilities is also ensured by predictive maintenance made possible by AI. Machine learning has the potential to foresee potential issues with machinery by constantly scanning electrical, plumbing, and heating, ventilation, and air conditioning (HVAC) networks (Alsbou & Alsaraireh, 2024). Both downtime and maintenance costs decrease as a result of this. The improvements in this area not only make school buildings last longer, but they also help create a more sustainable and affordable educational system.

1.4 Thinking about Difficult and Ethical Considerations

There are many advantages to smart buildings driven by AI, but there are also some challenges in installing these facilities. Schools are increasingly using monitoring technology based on machine learning and biometric recognition, which has raised concerns about data privacy and security (Kamalov et al., 2023). Achieving and maintaining trust in AI-driven learning environments hinges on the protection and ethical use of student data (Shafiq, 2025). Not all schools, especially those in economically deprived regions, have the financial or technological means to successfully implement AI-driven solutions, which is a major barrier due to the digital divide.

There is also the ethical consideration of whether or not AI has biases that might unintentionally disadvantage some student groups. Unfair treatment of students or an inaccurate portrayal of their abilities can arise when machine learning models are trained on biased datasets, which can exacerbate pre-existing inequalities (Alam, 2022). To tackle these concerns, we need to establish strict AI governance frameworks, make AI algorithms publicly available, and check their performance regularly to ensure that AI-driven educational solutions are fair and inclusive (Badshah et al., 2023).

With their ability to optimize administrative operations, boost energy efficiency, personalize educational experiences, and increase safety, intelligent facilities driven by AI are undeniably transforming modern education. Despite ongoing constraints such data privacy concerns, ethical dilemmas, and resource inequities, the potential benefits of AI in education much outweigh these problems. The development of inclusive and ethical AI frameworks requires close collaboration among lawmakers, educators, and developers. The advancements in AI will continue to benefit many, but this will make sure that all children may enjoy them. In order to provide a more efficient, safe, and adaptive learning environment, educational institutions are considering the use of smart facilities driven by AI. This might help students be ready for a future when technology and education are closely connected.

2. Modernization in AI Education

The AI education today makes AI more accessible, participative, and beneficial for students and professionals. To prepare students for the AI-driven future, schools are employing advanced technology, customized learning, and real-world applications. AI literacy and competence develop with cutting-edge technology, hands-on experiences, and multidisciplinary methods.

The shift from theoretical to practical AI education is significant. Traditional AI education used texts and lectures with minimal hands-on work. Online platforms, coding environments, and AI-powered simulations allow students to apply AI concepts. Google TensorFlow, IBM Watson, and OpenAI GPT models help students interactively develop, test, and improve AI models (Rahiman & Kodikal, 2024). This hands-on approach promotes AI application comprehension and problem-solving across sectors. AI-driven adaptive learning systems personalize AI education. These methods adjust the curriculum based on student strengths, weaknesses, and development. AI courses on Coursera, Udacity, and edX include quizzes, coding, and customized content (Dimitriadou & Lanitis, 2023). Personalization enables varied learners understand AI subjects at their own speed, reducing knowledge gaps.

Modernization incorporates AI in schools. Since AI expertise is no longer confined to computer science students, schools and universities are incorporating it into other topics. AI is used in biology, economics, healthcare, and the arts, demonstrating its versatility. Finance students utilize AI algorithms to predict market trends from historical data, while medical students use AI-driven research tools to detect illnesses from medical images. This integrative approach shows pupils how AI disrupts industries. Future AI education advancements include instructors and chatbots. AI-powered assistants answer queries and clarify complex AI topics live. AI teachers are available 24/7, making learning global. Google Socratic and educational app AI-driven chatbots use natural language processing (NLP) to explain AI subjects and teach students without teachers (Adel, 2024). This accomplishment makes AI education accessible to rural and poor students.

Modern AI education focuses on ethics and responsible AI development. As AI advances, discrimination, justice, transparency, and accountability become ethical issues. AI ethics are being taught in schools to help students comprehend their work's societal effects. Universities and online platforms educate about AI fairness, bias detection, and explainable AI to help students build ethical and socially responsible AI. Responsible AI prevents machine learning model biases and benefits all cultures. Academic, corporate, and government partnership is another important aspect of AI education modernization. Google, Microsoft, and NVIDIA provide university AI research, internships, and industry training (Qazi et al., 2024). AI education stays updated with technology and industry demands with this relationship. Students address real-world AI challenges alongside professionals via bootcamps, hackathons, and research funding. Students may use AI to develop machine learning, robotics, and NLP through these linkages.

The AI education revolution also requires AI-powered learning analytics. To improve teaching and curriculum, schools use AI to assess student involvement, performance, and learning. AI can help teachers identify and help struggling students. These analytics enable colleges assess AI courses and change curriculum based on research and industry demands. Institutions create a learning feedback loop by employing AI to improve AI education.

Future AI education will use virtual reality (VR) and augmented reality (AR) to teach ideas. VR-powered AI labs enable students to test machine learning models in realistic settings, while AR overlays explain AI. These technologies improve AI education and help students visualize complicated concepts. VR simulations explain how neural networks absorb information, providing learners a new perspective on AI models. Other revolutionary trends include open-source and community-driven AI education democratization. OpenAI, TensorFlow, and Hugging Face provide models, courses, and datasets for learners worldwide to experiment with cutting-edge AI technology (Ejjami, 2024). Students may cooperate, share, and participate in global AI challenges on Kaggle and GitHub. This open-access strategy simplifies AI learning and empowers diverse pupils.

The AI education is expanding beyond programmers and data scientists to include non-technical learners. AI courses for business, healthcare, and social good are becoming specialized for professionals who want to understand AI's impact. These courses prioritize ethics, decision-making, and AI applications over deep programming. When institutions make AI education more accessible to non-technical audiences, society benefits. AI education may educate more AI as it advances. AI-powered mentors, holographic AI teachers, and fully automated AI classrooms may change AI education. Modernizing AI education provides a dynamic, inclusive, and hands-on learning experience to educate students for the fast-changing profession. Modern AI education influences the next generation of innovators and responsible practitioners through customized learning, transdisciplinary applications, ethics, and industrial involvement.

3. AI-Driven Smart Facilities for Modern Schools

Smart facilities use AI with physical and digital infrastructure to improve school safety, efficiency, and learning. Smart technologies maximize resources, handle school operations, and tailor learning using AI (Liando & Tatipang, 2024). Digital revolution in education requires AI-driven facilities for seamless operations and a more engaging and dynamic learning environment. Figure 1.1 represents AI integration in modern schools.

The AI-powered smart facilities impact school monitoring and security. Real-time AI-powered cameras and facial recognition systems can monitor schools for unauthorized individuals and security threats. Machine learning algorithms alert school administrators and security workers to unusual activities like loitering in banned areas or unlawful access. By restricting school admission to registered students, staff, and visitors, AI-based biometric attendance systems increase security. AI-powered solutions protect students and staff from identity fraud and illegal access. AI enhances current school energy and resource allocation beyond security. Based on occupancy and weather, AI-driven smart grids and IoT devices control lighting, heating, and cooling. Intelligent systems reduce energy waste, saving money and boosting sustainability. Schools save water using AI-powered water management systems that detect leaks and limit consumption. AI optimizes educational resource use, lowering expenses and improving education.

Student learning is customized and participatory in AI-driven smart classrooms. AI-enabled smartboards can read handwriting, translate languages, and summarize lessons instantaneously. Chatbots and virtual teachers may provide students tailored

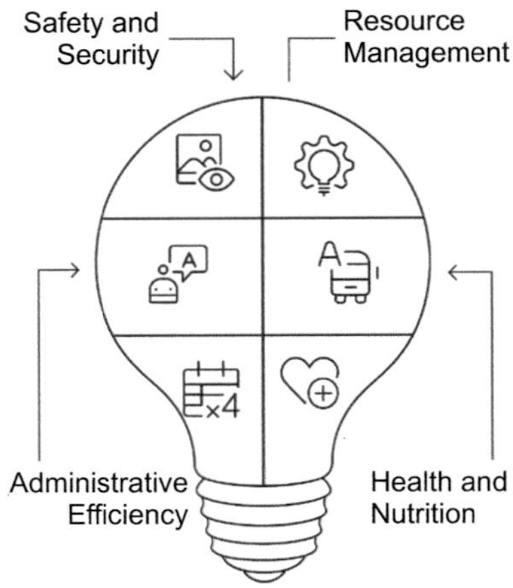

Fig. 1.1 AI in transforming modern school environments.

replies. Innovative algorithms adapt to student learning speeds, guaranteeing no student falls behind. Learning management systems (LMSs) using AI identify student strengths and weaknesses and provide study tactics (Cao et al., 2020). AI can identify learning patterns and adapt treatments to improve academic achievement for all students. School transportation is another AI-driven smart facility strength. AI fleet management solutions optimize school bus routes based on traffic and fuel economy in real time. To guarantee student safety and punctuality, parents and school administrators get real-time bus location, projected arrival times, and route deviation information. AI-powered bus monitoring systems may quickly report bullying and illegal stops. AI can save fuel and operational costs and increase school transport safety.

Artificial intelligence streamlines administrative operations in modern schools. AI-powered scheduling solutions maximize space usage and minimize conflicts in class. Machine-graded assignments and timely feedback free teachers to focus on more engaging and inventive teaching techniques. AI-powered chatbots answer student and parent queries on schedules, processes, and events. Communications improve with virtual assistants, reducing school staff labour and enhancing administration. AI transforms libraries into digital information centres. NLP helps students find books and study resources quickly with AI-driven cataloguing systems. AI recommends materials based on students' reading history and academic interests. Some libraries utilize robots to retrieve books, navigate aisles, and teach. AI offers voice-assisted search, making learning Preparing learning materials easier and quick access for visually impaired students. Libraries with AI encourage inquiry and research by making learning more inclusive and efficient.

The AI-enabled facilities help improve student health. AI detects early signs of sickness, stress, and fatigue in pupils' physical and mental health. School nurses and administrators receive real-time medical updates via smart sensors and wearables (George & Wooden, 2023). AI-powered mental health chatbots help pupils cope and refer to counsellors. Schools can prevent health issues with AI in student wellness initiatives. School cafeterias using AI improve meal planning and nutrition. Based on kids' diets, allergies, and nutritional needs, smart cafeterias serve balanced meals. AI-powered inventory management systems estimate demand and track food supply in real time, avoiding waste. Some schools sell lunches quickly with AI-powered self-checkout devices. They also reduce food waste, raise productivity, and improve student nutrition using AI.

Automation, machine learning, and data analytics will transform school AI-driven smart facilities. Advances in AI will increase security, sustainability, customized learning, and operational efficiency in schools (Akobe et al., 2025). Schools can stay adaptable and future-ready by using AI-powered predictive analytics to predict enrolment, infrastructure maintenance, and curriculum changes.

Thus, AI changes education by making schools safer, smarter, and more efficient. AI can help institutions create dynamic learning environments for students and teachers. This change improves academics and school life, preparing kids for a tech-driven future.

3.1 Impact of AI-driven Facilities for Modern Schools

Modern schools use AI to improve delivery, management, and experience. Intelligent schools employ AI to improve learning, security, administration, and efficiency. To prepare children for a technologically advanced society, AI-driven solutions can make schools more adaptable, resource-efficient, and student-centred.

Security and safety are major benefits of AI-driven facilities. AI-powered surveillance systems with facial recognition, anomaly detection, and predictive analytics monitor schools in real time. These systems can identify unauthorized users, suspicious conduct, and quickly notify administrators to take action. Only registered students, staff, and permitted visitors can enter biometrically controlled areas, boosting school security. AI-powered emergency response systems evaluate fires and medical situations and trigger safety measures, reducing response time and risk. Learning spaces are personalized and made interactive using AI. Unlike traditional schools, AI-driven classrooms adapt to students' learning styles, speeds, and needs. AI-enabled smartboards identify handwriting, convert notes to text, and translate quickly, making lectures more accessible (Selvakumar et al., 2025). AI-powered chatbots and tutors answer questions, clarify complex ideas, and suggest learning tactics. These advanced tools identify knowledge gaps and provide personalized study materials to keep students ahead.

The AI-powered facilities reduce administrative workload for educators. Based on teacher availability, topic demands, and classroom occupancy, automated scheduling systems improve schedules. AI-powered grading systems assess student work and provide immediate feedback, freeing teachers to teach and advise. For tuition, admissions, and event notifications, AI-driven chatbots can help connect students, parents, and staff. AI-powered facilities minimize administrative work, allowing teachers to teach. Educational facilities powered by AI effect energy efficiency and sustainability. AI-powered smart networks govern lighting, heating, and cooling based on occupancy and weather. These devices prevent energy waste, save utility expenses, and encourage sustainability. AI-driven leak detection and water optimization technologies help schools function sustainably. AI resource management meets global sustainability goals by saving schools money and helping the environment.

The AI-powered school transport is safer, faster, and more dependable. Tracking school buses in real time, intelligent fleet management systems estimate arrival times and optimize routes based on traffic. Live updates help school administrators and parents cooperate and protect students. AI-driven transportation solutions increase safety, punctuality, and parent and school administrator confidence (Strielkowski et al., 2024). AI-driven smart libraries have transformed student academic resource consumption. Students can quickly find books and study materials by utilizing AI-powered cataloguing systems. For customized learning, recommendation algorithms suggest reading based on students' interests and past searches. Some AI-driven libraries contain robots that search books, counsel students, and manage inventory. Disabled students can access education via voice-assisted search. Through accessibility and efficiency, AI-driven libraries allow exploration and self-directed learning.

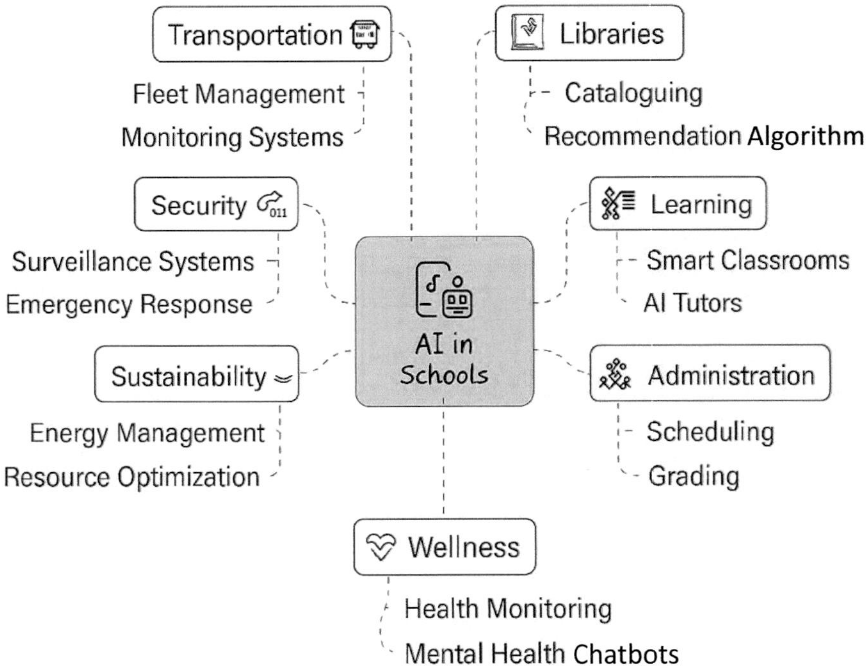

Fig. 1.2 Impact of AI-driven facilities for modern schools.

Modern schools also consider impact of AI on student wellness. Wearable sensors and AI-powered health monitoring systems assess students' physical and mental well-being in real time. These devices alert school nurses and counsellors to early illness, stress, and fatigue. Intelligent mental health chatbots assist students to cope and direct them to specialists. AI improves student wellness initiatives by reducing absenteeism and improving academic performance.

The AI-driven meal planning and nutrition management changes school cafeterias. Smart cafeterias provide balanced meals for student diets, allergies, and nutrition. AI-powered inventory management reduces food waste and cost. Some schools use AI-driven self-checkout kiosks to distribute meals faster and improve eating habits. AI improves cafeteria efficiency and sustainability by encouraging healthy eating and reducing food waste. AI-driven predictive analytics help schools forecast enrolment, infrastructure, and curriculum. These findings aid institutions' data-driven sustainability and adaption decisions. By evaluating abilities, interests, and market trends, AI-driven job guidance systems help students pick academic and professional pathways. Figure 1.2 displays impact of AI in modern schools.

Beyond the classroom, AI-driven facilities impact education and prepare students for the changing job environment. Students gain digital and problem-solving skills and benefit from AI-driven sector adaption via AI-powered learning tools. AI-enabled schools prepare students for future problems and opportunities.

4. Categories of Modernized Facilities

Educational institution efficacy and student learning are impacted by modernized school facilities. These establishments are more than simply structures; they encompass a range of environments, resources, and technologies that are designed to enhance education, security, and environmental friendliness. School buildings have undergone modernization to meet the needs of both students and teachers, who have high expectations for their educational achievements. These groups encourage innovation, long-term viability, and well-rounded student growth as a result of recent technical and pedagogical developments. Figure 1.3 shows the parts of modernized facilities.

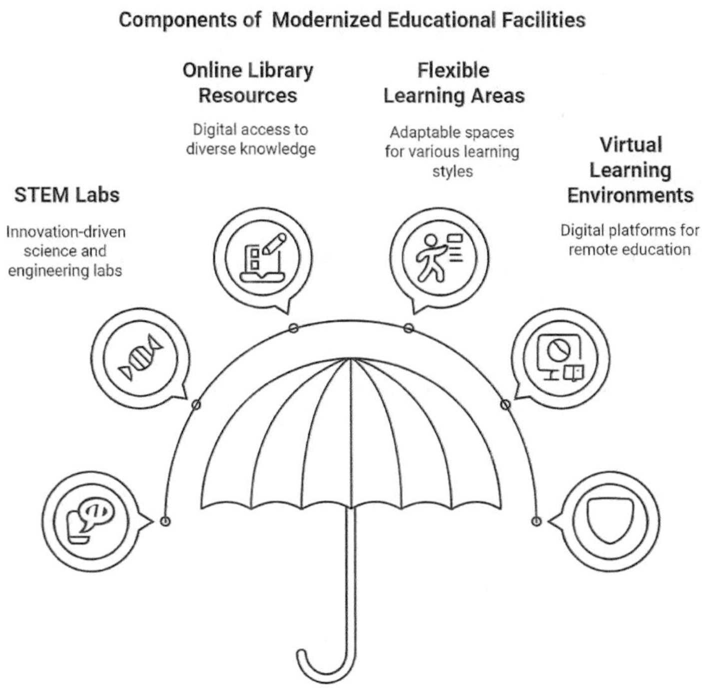

Fig. 1.3 Modernized facilities components.

4.1 Online Lecture Halls

In order to make learning more engaging and interactive, smart classrooms employ digital tools and current technology (Tiwari & Fahrudin, 2024). Smart classrooms integrate technology to make the learning environment immersive and dynamic (Swargiary, 2024). Computers, software, the internet, audio-visual equipment, digital projectors, and interactive whiteboards are standard. Traditional education is being attempted to be made more dynamic, multimedia-rich, and relevant to different learning styles in smart classrooms. It enhances learning, engagement, and results via the use of technology.

Modern education has been revolutionized by smart classrooms, which provide both students and teachers with powerful tools to enhance learning. These educational settings make use of cutting-edge technology to boost efficiency, engagement, and creativity. Students are better prepared for a world where technology and creativity are evolving at a rapid pace in smart classrooms (Raza & Alam, 2025). Following smart Technologies used in classrooms:

- *Intelligent teaching aids*: Smart classrooms employ smart boards and other forms of interactive whiteboarding. Digital material may be projected onto large touch-sensitive displays, allowing teachers to provide interactive courses. This makes learning more interesting and encourages student participation (Seldon & Abidoye, 2018).
- *A/V system*: High-definition projectors, speakers, and microphones are standard in modern smart classrooms. According to Aithal and Maiya (2023), these features enhance the use and participation of multimedia content.
- *Moodle or Blackboard*: These learning management systems allow smart classrooms to organize their courses, distribute materials, administer assessments, and facilitate collaboration between teachers and students.
- *No books*: Document cameras make it easier to incorporate a variety of learning tools by allowing instructors to showcase physical documents, textbooks, and 3D objects.
- *Devices for students*: Students in smart classrooms typically have access to laptops, tablets, and other digital tools, allowing for customized learning (Salla et al., 2025).
- *Internet access*: In order to use online resources, do research, and collaborate with peers, smart classrooms require reliable and fast internet connections (Adipat & Chotikapanich, 2024).
- *Educational apps and software*: Smart classrooms make use of a variety of educational apps and software to back up learning goals and activities.
- *Tools for collaboration*: Students in smart classrooms may work together on projects either locally or globally, thanks to the collaboration software and tools that are standard in these types of classrooms.

Figure 1.4 represents smart classroom parts. Technology for video conferencing allows for the extension of classroom instruction to non-traditional locations, such as virtual field trips and remote guest lecturers. Learning, engagement, and preparedness for the digital age could all take a leap forward in smart classrooms. Technology, instructional approaches, and teacher training are essential for a successful deployment.

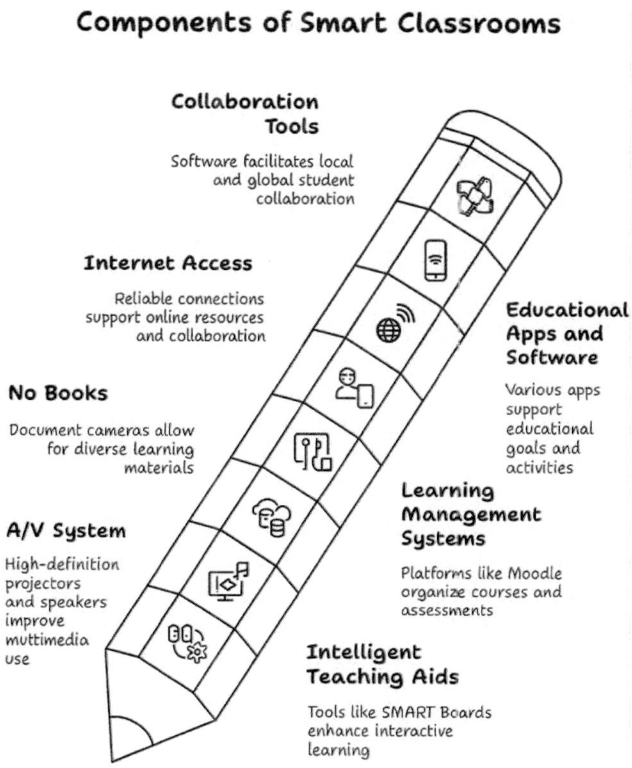

Fig. 1.4 Smart classroom components.

4.2 STEM Labs

One innovative method of teaching in the modern day is the Science, Technology, Engineering, and Mathematics (STEM) lab. These speciality classrooms inspire and equip students to investigate vital fields that drive technological advancement and innovation, according to the National Research Council (2011). The National Science Foundation (2021) asserts that STEM education's significance has grown due to the fact that it equips students with the necessary skills and knowledge to tackle complex, real-world challenges and prepares them for careers in STEM fields. Children that attend STEM labs are shaped to be innovators and problem solvers in the future.

The STEM labs allow students to actively investigate these fields. Modern scientific tools, software, 3D printers, and robot kits are available at these labs. In order to acquire skills that are valued by employers, students are motivated to think critically, work together, and find solutions to actual issues. STEM labs also provide an opportunity for students to put into practice what they have learnt in the classroom. Research by Ioannou-Sougleridi et al. (2024) found that students' problem-solving and creative abilities were enhanced through hands-on experiments and projects, which also helped them understand mathematical concepts, engineering principles, and technological developments. The STEM fields, including STEM labs, are essential in today's rapidly evolving technological landscape. They assist students in being ready for emerging technologies, environmental concerns, and international challenges.

4.3 Online Library Resources

With the advent of the digital era came a new kind of library: the digital library. The way we store, retrieve, and use knowledge is being transformed by these online data warehouses. There are vast quantities of text, images, audio, and video information stored in digital libraries, which are an evolution of conventional libraries (Saaida, 2023). Books, articles, multimedia, manuscripts, and other forms of written and oral knowledge are all part of digital libraries, which are sometimes called electronic libraries. Scientific articles, historical records, multimedia collections, and more are also housed in these online archives. These digital libraries are accessible to users through computer networks, such as the internet. The users may peruse, search, access, and retrieve material. The accessibility, searchability, and digital resource management are all enhanced by the following components (see Figure 1.5):

- *Library of text and images*: In the content repository, electronic documents are kept. Included are digital materials such as text, images, videos, and audio files. A content repository that is well organized allows students easy access to textbooks, research papers, and multimedia elements. Academics benefit from this accessibility as it broadens students' comprehension.
- *Cataloguing and metadata*: They are crucial for defining and classifying digital resources. Some examples of descriptive data are title, author, date, and keywords. Searching digital resource is made easier with cataloguing and metadata. Finding appropriate information quickly helps students save time and enhance students focus on preparation.
- *Digital library*: The digital library's user interface is where patrons may do their actual reading. It has features that allow one to search, browse, and access digital resources. To enhance the user experience, a user-friendly interface is essential. The convenience of the digital library makes it more likely that students will utilize it frequently.
- *Approaching digital library*: Systematic approaches of searching for and retrieving digital resources constitute this category. Search engines and advanced querying capabilities are commonplace on these sites. With the help of modern search and retrieval tools, students may quickly locate the information they need. Research and information literacy are both enhanced by this.
- *Preserving digital materials*: This represent the digital content are accessible over the long term by preventing degradation, obsolescence, and data loss. Digital preservation ensures that resources are accessible to students for the long term. Educational materials should be preserved so that they can be used by students in the future.
- *Security*: To protect sensitive information and ensure that only authorized users may access them, digital libraries implement access control and security measures. To ensure the safety of sensitive information and intellectual property, access control methods are implemented in learning environments.
- *Protocols*: Students may be certain that their academic work is safe in digital libraries since these institutions follow protocols for digital material protection,

accessibility, and interoperability. In the future, digital libraries may be able to connect with other school systems, providing pupils with even more instructional materials.

- *Assistance for users and their needs*: Learn how to find and use digital resources with the help of digital libraries' reference aids, training, and support services. Student usage of the digital library is facilitated by user services such as reference and tutorials. Following this advice can help you become a better researcher and information seeker.

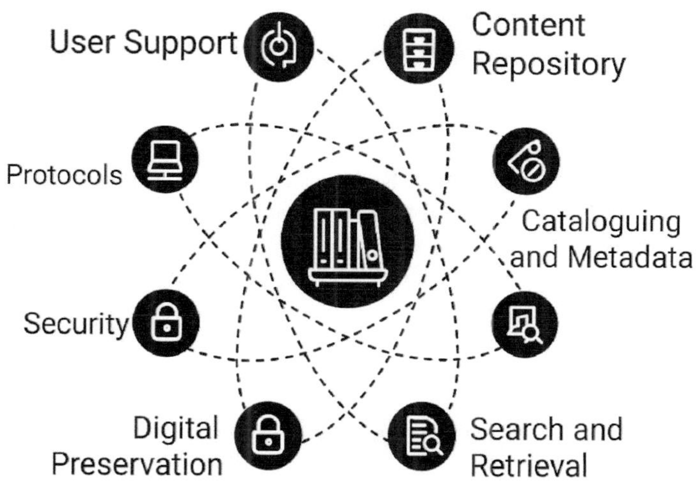

Fig. 1.5 Digital library parts.

4.4 Flexible Learning Areas

Flexible learning spaces, also known as flexible classrooms or adaptable learning environments, accommodate varied learning requirements, engage students, and enable active and dynamic teaching (Roopa et al., 2021). These rooms may be adapted to different teaching methods, learning activities, and group sizes, thereby improving learning. Aeni et al. (2024) note that modern educational institutions use adaptable and technology-rich spaces to improve learning by accommodating diverse teaching methods, fostering student engagement, and encouraging active and collaborative learning. Discusses flexible learning environments, their importance, and their potential to transform education. Flexible learning environments are replacing rows of static desks and a podium in classrooms. Fluid learning spaces enable traditional lectures, group discussions, hands-on activities, and technology-enhanced learning by providing a fluid and ever-changing setting. Saaida (2023) emphasizes adaptable learning space features as shown in Figure 1.6.

- *Collaborative learning spaces*: Facilitate group work, conversation, and projects using furniture and arrangements. Collaboration, communication, and problem-solving are fostered in collaborative learning settings. They promote collaborative learning and provide a friendly atmosphere for pupils to share ideas.

- *Active learning classrooms*: These facilities encourage student-centred instruction with mobile furniture, technology, and interactive displays. Active learning classrooms emphasize involvement over listening. They improve student comprehension, critical thinking, and retention. More students ask questions and participate in conversations.
- *Technologically sophisticated places*: These spaces have interactive whiteboards, multimedia materials, and digital learning connectivity. Technology-enabled environments enable mixed and digital learning. They teach digital literacy and prepare pupils for the digital age. Many digital resources make learning more dynamic and interesting for students.
- *Quiet and From it*: These are ideal for research, reflection, and concentration. Students can study alone in quiet settings. They benefit introverted students. Students can read, think, and self-learn.
- *Makerspaces*: Creative spaces featuring tools and materials for design, prototyping, and experimenting. Makerspaces foster creativity, innovation, and hands-on learning. They foster problem-solving, creativity, and STEM knowledge among pupils.
- *Lecture capture spaces*: These places record and stream lectures and presentations. Lecture capture rooms provide flexible content delivery. Reviewing lectures, accessing missed sessions, and studying at your own speed improves accessibility and convenience.

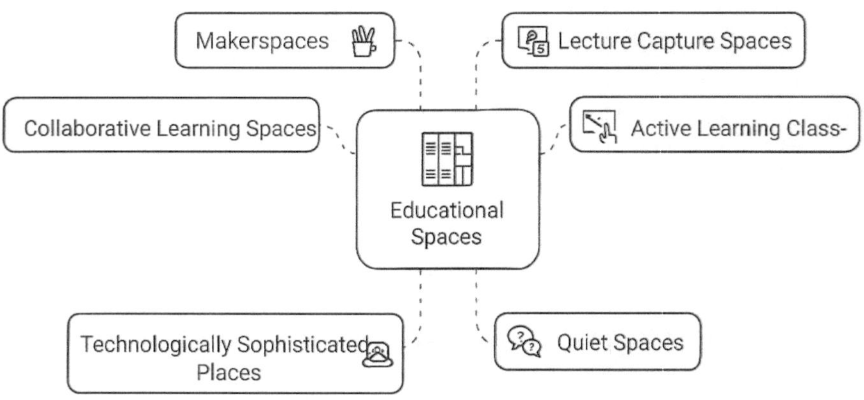

Fig. 1.6 Educational spaces and their functions.

4.5 *Virtual Learning Environments*

Online systems called Virtual Learning Environments (VLEs) have transformed digital education and training (Aeni et al., 2024). Their extensive and engaging platforms allow students and teachers to participate in many educational activities. VLEs, often known as LMSs or online learning platforms, are digital ecosystems that create, deliver, and manage educational material and activities in online or blended learning settings. Teachers, students, and administrators may communicate and access information on these platforms regardless of location.

The VLEs are popular because they can adapt to changing education and training needs, making them more flexible, accessible, and scalable. Content management, communication, evaluation, and analytics are among their features and tools to improve learning and results.

4.6 Advanced Security Systems

Advanced security systems defend assets, information, and people from numerous threats and dangers using advanced technology, methods, and measures (Putri & Sain, 2025). These solutions improve physical and digital security. They use cutting-edge technology, including automation and AI, to protect against many security threats.

Modern education relies on advanced security technologies to protect students, staff, and resources. Security is more crucial than ever as schools employ digital technology and online learning platforms. Advanced security technologies protect digital and physical educational environments. By using biometric authentication, state-of-the-art surveillance, physical campus security, data protection, and cyber security, these solutions establish a safe ecosystem that optimizes learning.

5. Discussion

Because they boost efficiency, security, and learning, AI-driven smart buildings are revolutionizing today's educational institutions. Incorporating AI technology into educational institutions has far-reaching effects on both students and teachers, preparing them for a future when AI and other emerging technologies rule the globe. Moving beyond the traditional classroom setting, these technologies provide a more interactive, interesting, and tailored educational experience.

Learning environments can be revolutionized by smart facilities driven by AI. In today's modern, interactive classrooms, students have the opportunity to engage with content in a real-time manner. Smartboards powered by AI can assist teachers to cater to students with varying learning styles by recognizing handwriting, translating texts in real time, and providing concise lesson summaries. With this adaptability, students may get immediate feedback and work at their own pace while still benefiting from individualized instruction. AI-powered instruments monitor how well students are doing and adjust lessons accordingly. Education becomes more accessible to all students when they take a more individualized approach to learning.

Learning has shifted from a focus on textbooks and lectures to one that is more interactive and hands-on, with the help of AI-powered platforms. Students may construct, evaluate, and enhance AI models using tools like OpenAI GPT, IBM Watson, and Google TensorFlow. These platforms allow students to apply their theoretical knowledge of AI to real-world scenarios. The skills necessary for the workforce of the future, such as comprehension, critical thinking, problem-solving, and creativity, are enhanced through this practical method. Students are prepared to tackle challenges in healthcare, business, and the arts through coding exercises and simulations powered by AI.

Furthermore, adaptive learning systems powered by AI are revolutionizing both the delivery and reception of educational content. In order to personalize the curriculum for every student, these technologies examine student data. Platforms like edX, Udacity, and Coursera allow for individualized instruction that takes into account each student's background and areas of weakness. Students with varying abilities can benefit from this personalization as it allows them to learn AI concepts at their own pace. By filling up gaps in knowledge, AI improves education and ensures that no student falls behind. Students may assess their own performance and make adjustments in real time with the aid of these AI technologies.

Artificial intelligence is being included in non–computer science courses as part of interdisciplinary curricula. AI education was once restricted to the realm of computer science and technology. But AI-powered smart buildings have prompted schools to adopt AI in a wide range of subjects. Students of economics may use machine learning to foretell market trends, while students of biology can use AI to study genetic sequences. Students in the medical field may use AI to analyse imaging studies and make predictions about patients' prognosis, while creative types can use it to create music, art, and digital narratives. This interdisciplinary study demonstrates how AI has the potential to shake up and transform long-standing industries.

Smart facilities powered by AI enhance safety for both students and employees. School security concerns and unauthorized entrance are monitored by real-time AI-powered surveillance systems, such as facial recognition and anomaly detection. By limiting entry to authorized users (such as registered students, faculty, and staff), these technologies help keep sensitive information safe. AI-driven (EMS) employee management system have the capability to assess potential dangers such as medical crises and fires, providing first responders with up-to-the-minute information to act swiftly and effectively. The classroom is now safer and more secure, which is reassuring for everyone involved: children, teachers, and parents.

In today's schools, AI-powered technologies enhance security while optimizing the use of school resources and energy. On the basis of occupancy and environmental factors, AI controls the temperature, lighting, and ventilation in smart buildings. AI has the potential to reduce operating costs and enhance sustainability by utilizing real-time data to decrease energy waste. AI is crucial for reducing water waste, finding leaks, and managing school water systems. In addition to saving money that may be used towards better teaching, these technologies also assist the school to achieve worldwide environmental goals.

Facility management in schools is also impacted by AI. Time and effort were required for traditional administration tasks such as scheduling, grading, and resource allocation. Administrators are now free to concentrate on strategy and education thanks to AI-powered systems that have simplified several of these jobs. Maximizing space usage, reducing conflicts, and optimizing resource allocation may be achieved through the automation of scheduling using AI. Using AI-powered grading systems, teachers may save time by having students' work graded automatically and receiving immediate feedback. Teachers are able to spend more quality time with students and provide more individualized lessons, thanks to this technology.

Education is being revolutionized by AI-driven chatbots and virtual assistants that provide students with on-demand assistance. Learning is made more flexible and accessible by AI-powered technology, which can answer enquiries, clarify hard topics, and offer real-time explanations. They are available around the clock to help students with review, coding, and problem-solving even when class is not in session. Underprivileged students or students in far-flung areas benefit greatly from this accessibility since they often lack access to adequate educational resources and qualified instructors. Through the use of AI-driven technology, the educational aid gap may be closed, allowing all students to experience customized learning.

More and more schools are incorporating AI into their curricula. There is little question that education will be revolutionized by AI-powered mentors, hologram teachers, and fully automated classrooms. More tailored learning opportunities will be available to students as a result of these innovations, and mentors driven by AI will be there to guide, inspire, and assist them.

Smart facilities powered by AI are revolutionizing education in today's classrooms. From more efficient and secure administration to more engaging and individualized learning environments, AI is improving education in every way. To better prepare students for the dynamic and ever-changing job market, schools are embracing new technologies to create more student-centred learning environments that foster innovation, creativity, and problem-solving. Schools powered by AI have the potential to equip the next generation to face the challenges and seize the opportunities of the future.

6. Conclusion

The incorporation of smart facilities powered by AI is completely changing the face of education by improving school operations, facilitating individualized learning, and making the school a safer and more efficient place for everyone. This is because these amenities are making such a big difference. The use of AI in modern educational institutions goes well beyond the four walls of a typical classroom; it improves administrative processes, maximizes the use of available resources, and creates unique educational programmes for each student. Technologies such as adaptive learning tools that let students progress at their own pace and AI-powered monitoring systems that boost safety are transforming the way schools function. AI-enabled smart buildings also regulate water, energy, and other resource usage, which leads to more eco-friendly and cost-effective practices. Further development of AI promises to make education more engaging, more widely available, and better prepared to meet the dynamic needs of the modern world. The potential for AI to completely transform education is become more and more apparent as the technology advances. Schools are preparing students for careers in technology and AI by integrating AI into their digital and physical infrastructures. They are also building dynamic learning environments where every student can reach his/her full potential. Education in the future will be more personalized, efficient, and secure, thanks to intelligent facilities powered by AI. Modern educational institutions throughout the globe are looking at this as a benchmark.

References

Adel, A. (2024). The convergence of intelligent tutoring, robotics, and IoT in smart education for the transition from industry 4.0 to 5.0. *Smart Cities*, *7*(1), 325–369.

Adiguzel, T., Kaya, M.H. and Cansu, F.K. (2023). Revolutionizing education with AI: Exploring the transformative potential of ChatGPT. *Contemporary Educational Technology*, *15*(3), 1–14.

Adipat, S. and Chotikapanich, R. (2024). Advancing higher education with the transition to smart universities: A focus on technology. *Shanlax International Journal of Education*, *12*(3), 29–36.

Aeni, N., Khang, A., Al Yakin, A., Yunus, M. and Cardoso, L. (2024). Revolutionized teaching by incorporating artificial intelligence Chatbot for higher education ecosystem. In: Prathamesh Churi, Shubham Joshi, Mohamed Elhoseny, and Amina Omrane. *AI-centric modeling and analytics.* (pp. 43–76). CRC Press.

Aithal, P.S. and Maiya, A.K. (2023). Innovations in higher education industry—Shaping the future. *International Journal of Case Studies in Business, IT, and Education (IJCSBE)*, *7*(4), 283–311.

Akobe, O.D., Adeboye, F.L., Dauda, B.A. and Yacim, H. (2025). Improving library and information science education in Nigeria: Using smart technology and AI-powered tools for better learning. In: Nayef J. Jomaa, *Using AI Tools in Text Analysis, Simplification, Classification, and Synthesis.* (pp. 377–410). IGI Global Scientific Publishing.

Alam, A. (2022). Employing adaptive learning and intelligent tutoring robots for virtual classrooms and smart campuses: Reforming education in the age of artificial intelligence. In: Dr. Ashraf Alam, *Advanced Computing and Intelligent Technologies: Proceedings of ICACIT 2022.* (pp. 395–406). Singapore: Springer Nature Singapore.

Alsbou, M.K.K. and Alsaraireh, R.E.A.I. (2024). Data-driven decision-making in education: Leveraging AI for school improvement. In *2024 International Conference on Knowledge Engineering and Communication Systems (ICKECS)* (Vol. 1, pp. 1–6). IEEE.

Badshah, A., Ghani, A., Daud, A., Jalal, A., Bilal, M. and Crowcroft, J. (2023). Towards smart education through internet of things: A survey. *ACM Computing Surveys*, *56*(2), 1–33.

Cao, W., Wang, Q., Sbeih, A. and Shibly, F.H.A. (2020). Artificial intelligence based efficient smart learning framework for education platform. *Inteligencia Artificial*, *23*(66), 112–123.

Dimitriadou, E. and Lanitis, A. (2023). A critical evaluation, challenges, and future perspectives of using artificial intelligence and emerging technologies in smart classrooms. *Smart Learning Environments*, *10*(1), 12.

Ejjami, R. (2024). Revolutionizing Moroccan education with AI: A path to customized learning. *International Journal for Multidisciplinary Research*, *6*(3), 1–15.

George, B. and Wooden, O. (2023). Managing the strategic transformation of higher education through artificial intelligence. *Administrative Sciences*, *13*(9), 196.

Ioannou-Sougleridi, E., Kopsidas, S., Vavougios, D., Simos, C., Avramopoulos, A. and Kanapitsas, A. (2024). Revolutionizing learning management systems: Architecture of an AI-based LMS with instructor-driven personalized content generation. *International Journal of Advanced Multidisciplinary Research and Studies*. *13*(1): 3456–3470

Kamalov, F., Santandreu Calonge, D. and Gurrib, I. (2023). New era of artificial intelligence in education: Towards a sustainable multifaceted revolution. *Sustainability*, *15*(16), 12451.

Kavitha, R.K., Krupa, C.R. and Kaarthiekheyan, V. (2024). AI-powered digital solutions for smart learning: Revolutionizing education. In: Kavitha, R.K., Krupa, *Cybersecurity and Data Science Innovations for Sustainable Development of HEICC* (pp. 213–227). CRC Press.

Liando, N.V.F. and Tatipang, D.P. (2024). Enlightened minds: Navigating the nexus of artificial intelligence and educational modernization. *Penerbit Tahta Media.*

Makinde, A.I., Adeleye, S.A., Oronti, A.O. and Jimoh, I.T. (2024). Revolutionizing education. *Artificial Intelligence for Wireless Communication Systems: Technology and Applications*, 103–120.

Makinde, A.I., Adeleye, S.A., Oronti, A.O. and Jimoh, I.T. (2024). Revolutionizing education: AI in next-generation mobile management. In: Makinde, *Artificial Intelligence for Wireless Communication Systems* (pp. 103–125). CRC Press.

Mishra, P. (2025). Revolutionizing education with AI and ML. In: Mishra, P, Springher, *Internet of Behavior-Based Computational Intelligence for Smart Education Systems.* Mishra, P, Springher (pp. 53–94).

Palanivel, K. (2020). Emerging technologies to smart education. *International Journal of Computer Trends and Technology*, *68*(2), 5–16.

Putri, D.F. and Sain, Z.H. (2025). From chalkboards to chatbots: Revolutionizing education with AI-driven learning innovations. *Educative: Jurnal Ilmiah Pendidikan*, *3*(1), 1–10.

Qazi, S., Kadri, M.B., Naveed, M., Khawaja, B.A., Khan, S.Z., Alam, M.M. and Su'ud, M.M. (2024). AI-driven learning management systems: Modern developments, challenges and future trends during the age of ChatGPT. *Computers, Materials & Continua*, *80*(2).

Rahiman, H.U. and Kodikal, R. (2024). Revolutionizing education: Artificial intelligence empowered learning in higher education. *Cogent Education*, *11*(1), 2293431.

Raza, M.M. and Alam, M.F. (2025). Transforming education with artificial intelligence: Innovations, challenges, and future directions. Latest Trends and Techniques of ICT in Education (pp.11-31) Publisher: Scholars' Press

Saaida, M.B. (2023). AI-driven transformations in higher education: Opportunities and challenges. *International Journal of Educational Research and Studies*, *5*(1), 29–36.

Salla, S., Pasumarthy, A., Tadikonda, D., Parsha, T. and Mandal, S.K. (2025). Private AI in education: Critical challenges and aspects of enhancement strategies. In: Salla, S., Pasumarthy, A., *Sustainable Development Using Private AI* (pp. 109–131). CRC Press.

Seldon, A. and Abidoye, O. (2018). *The Fourth Education Revolution.* Legend Press Ltd.

Selvakumar, P., Hemalatha, C., Indumathy, I., Gandhimathi, S. and Mujra, P. (2025). Smart education and sustainable learning environments. In: Selvakumar, P., Hemalatha, *Smart Education and Sustainable Learning Environments in Smart Cities* (pp. 381–402). IGI Global Scientific Publishing.

Shafiq, Z. (2025). Revolutionizing education: The transformative power and challenges of artificial intelligence in learning. In: Shafiq, Z., *Integrating Technology in Problem-Solving Educational Practices* (pp. 241–260). IGI Global.

Srinivasa, K.G., Kurni, M. and Saritha, K. (2022). Harnessing the power of AI to education. In: Srinivasa, K.G., Kurni, M., *Learning, Teaching, and Assessment Methods for Contemporary Learners: Pedagogy for the Digital Generation* (pp. 311–342). Singapore: Springer Nature Singapore.

Strielkowski, W., Grebennikova, V., Lisovskiy, A., Rakhimova, G. and Vasileva, T. (2024). AI-driven adaptive learning for sustainable educational transformation. *Sustainable Development.*

Swargiary, K. (2024). *Information and Communication Technology (ICT) in the Indian education system in 2023.* GOOGLE.

Swargiary, K. (2024). *Integrating IoT in Education: Transforming Learning through Smart Technologies.* EdTech Research Association, US.

Tiwari, S.P. and Fahrudin, A. (2024). Evolving School Dynamics and Emerging Technologies in Education: Critical Success Factors. SciFormat Publishing Inc. 2734 17 Avenue Southwest Calgary, Alberta, Canada

Zeeshan, K., Hämäläinen, T. and Neittaanmäki, P. (2022). Internet of Things for sustainable smart education: An overview. *Sustainability*, *14*(7), 4293.

CHAPTER 2

Artificial Intelligence in the Development Of Education Industry 4.0: A Systematic Review

Manjushree Nayak[1*], *Shweta Patel*[2] and *Awotude Joseph Bamidele*[3]

1. Introduction

Artificial intelligence (AI) has become a disruptive force in the field of education technology (EdTech) during the 2005-2025. AI was only slightly present in education in the early 2000s, mostly in the form of rule-based systems and intelligent tutoring frameworks like AutoTutor, which used natural language processing (NLP) to mimic human tutoring (Graesser et al., 2005). The foundation for the global movement towards data-driven and customized education was established by these early systems. More adaptive learning experiences were made possible by developments in big data analytics and cloud computing by the early 2010s. AI-driven content delivery systems that modify learning paths in real time based on student performance and behaviour were invented by EdTech companies (Knewton, 2014).

The entry of AI into the mainstream of education policy discussions brought about a dramatic change between 2016 and 2020. Institutions and governments began to see how AI could change the way that people learn and teach. While national policies, like India's National Education Policy 2020 (Ministry of Education, 2020), stressed incorporating AI and coding into school curricula to prepare students for roles that are ready for the future, the United Nations Educational, Scientific and Cultural Organization (UNESCO) and international research organizations started promoting the ethical use of AI in education. Simultaneously, Luckin et al. (2016) emphasized the potential of AI in developing intelligent learning environments that could improve teacher decision-making and student comprehension. AI's use in

---

1 Associate Professor, Amity School of Engineering & Technology, Amity University Chhattisgarh, India. https://orcid.org/0000-0001-6383-780X

2 Amity Business School, Amity University Chhattisgarh, India. https://orcid.org/0000-0002-5241-9984

3 Faculty of Information and Communication Sciences, Department of Computer Science, University of Ilorin, Nigeria. https://orcid.org/0000-0002-1020-4432

* Corresponding author: manjushreesai44@gmail.com

DOI: 10.1201/9781003639367-2

remote and hybrid learning environments increased in 2021–2023, the year after the COVID-19 pandemic.

The AI tools for accessibility, language translation, and immediate feedback were integrated into EdTech platforms such as Coursera and Khan Academy to support a diverse student body from different regions (Holmes et al., 2022). AI is being used more and more to monitor emotional engagement, identify learning gaps, and even assist students with disabilities, in addition to personalizing content. Growing awareness of AI ethics during this period also led researchers to investigate how machine learning algorithms might exacerbate digital divides or reinforce bias.

Traditional industries, including education, have undergone a revolution because of the global shift to Industry 4.0, which is defined by the integration of cyber-physical systems, the Internet of Things (IoT), automation, and real-time data exchange (Lasi et al., 2014). The Fourth Industrial Revolution is pushing education systems to change from traditional classrooms to digital learning ecosystems that emphasize adaptability, individualized instruction, and technology-aligned skills (Schwab, 2016). To better prepare students for the skills required for the workforce of the future, this shift has forced educators and legislators to re-evaluate curricular structures and pedagogical frameworks. AI, a key component of EdTech innovation, is at the centre of this evolution in education. Personalized education at scale is now possible, thanks to AI-powered platforms that offer intelligent tutoring systems, adaptive learning, automated grading, and real-time feedback (Zawacki-Richter et al., 2019).

India's Union Education Minister Dharmendra Pradhan gave it importance in 2024, integrating AI with education in a way that is in line with Indian values. At the PadhAI Conclave, he called for the development of AI devices in Indian languages (Pradhan, 2024). Pradhan called AI a "force multiplier" and a "catalyst for innovation", which can add synergy and technology. He also underlined several steps taken by the Government of India to promote AI integration in education, such as creating AI centres of excellence and making AI tools available in Indian languages. According to Pradhan, these actions are facilitating a transition "from chalkboards to chipsets", thereby indicating a step towards a learning environment extended by technology. He insisted that including AI into class is now an important part of contemporary education and invited experts to give policy-level suggestions for the responsible implementation of AI in classrooms. Researchers are now focusing on AI's socio-cultural influence and, in addition, its role in creating inclusive and justified educational systems. Government initiatives like the AI Centre of Excellence in Education signalled the start of a new era centred on developing EdTech solutions that are culturally appropriate, create inclusive and equitable educational systems, and enhance domestic AI capabilities.

When considered collectively, AI's development in EdTech spans from rules-based tutors to emotionally intelligent virtual assistants, reflecting the progress and evolving aspirations of educational technology. As we move forward in the next phase of Industry 4.0, the question is whether AI or non-intelligence can be successfully and responsibly involved in education, and if yes, then "how".

Even with the increasing level of research and experimentation on the use of AI in Industry 4.0 education, the body of existing literature remains fragmented, multidisciplinary, and broad. Research often focuses on discrete innovations without a combination of analytics, AI-managed testing, or intelligent tuition systems; these elements do not work together to change institutional processes and educational outcomes. Additionally, a large part of the current literature, ignoring sociological, moral, and relevant factors, tends to focus on algorithm performance or technical feasibility, especially in India and other developing countries (Zavki-wide et al., 2023). It is important to assess the advantages and disadvantages of consolidated research, identify emerging trends, and compare various educational approaches because the application of AI in education is moving from theoretical use to real-world implementation. At this stage, a systematic review is necessary to provide consistency and clarity in the field. This industry can help map AI development in education within the context of Industry 4.0, highlighting key patterns, potential applications, and evidence intervals.

Additionally, it makes it possible to assess objectively the ways in which AI supports learner engagement, scalability, inclusivity in education, and institutional transformation. Few scoping reviews have attempted a focused and methodologically rigorous synthesis of AI's role in shaping education under the structural and philosophical shifts introduced by Industry 4.0, as most have focused on either general EdTech developments or specific AI models (Holmes et al., 2022; Cukurova et al., 2023). This study's use of a systematic review methodology not only offers a scholarly compilation of recent research but also lays the groundwork for future empirical studies, policy development, and ethical frameworks needed to guide responsible AI integration in education.

2. Literature Review

In recent years, the application of AI in the field of education within Industry 4.0 has been a major research issue. AI-based educational platforms are transforming traditional paradigms towards personalized learning, intelligent tutoring systems, predictive analytics, and automated administrative services. Furthermore, the keystones of Education 4.0, based on learner-centric, adaptable, and technology-enhanced learning, are highly backed by AI technologies, which can lead to real-time adjustment and data-based teaching. In a systematic review, Ans-Wang et al., (2023) elaborate on the blossoming stream of interdisciplinary research concerned with the AI-driven adaptive learning and smart feedback mechanism. The report cites the emergence of AI-centric learning management systems (LMSs) that tailor content and speed to real-time student progress. Moroianu et al., (2022) noted that the AI tools are not only enhancing the learning outcomes, but they are also becoming the key factor in changing the pedagogical method by predictive models and sentiment analysis.

Because of advances in technology, there are both opportunities and problems. The world has undergone a significant transformation in the digital age (Patel et al., 2023). Begum (2024) specifically addresses how AI contributes to the goals of Education 4.0 and provides distinct perspectives on the efficacy of AI in fostering student engagement, automating assessment, and creating lifelong learning ecosystems. Mustafa et al. (2024) conducted a meta-review involving findings from previous independent reviews and outlined a future research agenda, which underscores the necessity for more empirical verifications and ethical concerns in AI inclusion. With the development of AI, its ethical aspects and its standardization have become an issue of primary concern. Tong et al. (2024) investigated the requirement of strong industry standards and interoperability protocols, whereas Ogunleye et al. (2024) examined the implications of generative AI technologies like ChatGPT on academic integrity and on the design of assessment items. Meanwhile, Xu et al. scrutinize the evolution of dependence on large language models (LLMs) in education and note their potential to serve as both co-instructors and virtual assistants. It explores the dual pillars of remote experimentation and learning analytics in education.

Porayska-Pomsta et al. (2024) highlight the ethical quandaries of AI in education, including data privacy, algorithmic bias, and the possibility of dehumanized learning experiences. These concerns are like those presented in a related work (Xu et al., 2021), which highlights the need for transparency and accountability in the application of AI. Bandara and Senanayaka (2024) provide a categorization of AI in education literature over the past decade, including learning activities, teaching support, assessment, and educational management. This research shows a clear trend towards automated, student-centred learning environments that are underpinned by real-time analytics and decision-support systems.

Artificial Intelligence in Education (AIED) focuses on adaptive learning, personalized tutoring, intelligent evaluation, and profiling the learners. A systematic review of 2,223 articles along with content analysis of 125 papers by Wang et al. (2024) highlighted key research areas alongside theoretical frameworks and methodologies. The findings showcased the multidisciplinary nature of AIED alongside the numerous underexplored research areas, emphasizing growing potential for future studies in this domain (Wang et al., 2024).

In alignment with the goals set forth by Education 4.0, which aims to integrate industry-driven technologies, AI has transformed the education sector over the past decade (2010–2020). Educational institutions seek a paradigm that promises personal assistance, technology integration, and flexibility that is woven into the entire learning process. The story of AI technology integration showcases innovation along with increasing responsibility in education systems today, beginning with intelligent tutoring systems (ITS). Liu et al. (2025) categorized ITS into three main clusters: software-based ITS, robot tutoring systems (RTS), and multimodal intelligent robotics tutor systems.

However, this progress also brings important challenges. As AI becomes more integrated into education, concerns about morality, data privacy, fairness, and transparency are increasing. A review of 118 studies by Yan et al. (2023) found that many AI solutions still lack a human-centric design approach. This aligns with Riordan et al. (2023) who emphasized the need to include teachers and learners in the design process to create confidence and promote fairness. A growing consensus is that AI should complement, not replace, the role of teachers, enabling them to focus more on creative teaching and advice.

At the same time, adaptive learning platforms driven by AI are showing promising results. A 2023 synthesis found that more than 86% of AI-based educational systems led to improved learning outcomes, particularly for students who are were risk of falling behind (Begum, 2025). Features such as individual learning dashboard and

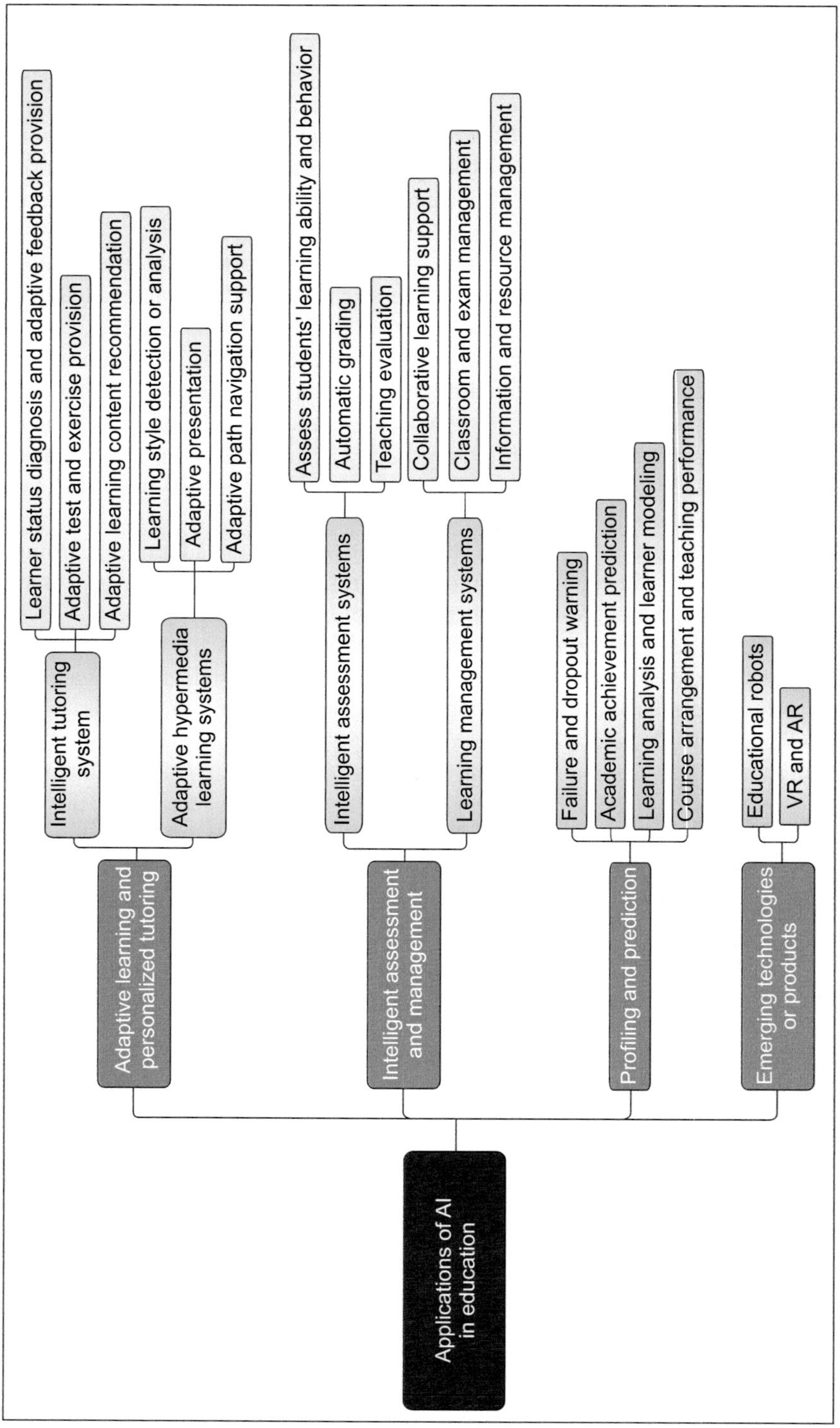

Fig. 2.1 Application of AI in education.
Source: Wang et al. (2024)

predictive analytics are helping teachers to more effectively identify and support students who struggle. The learning analytics framework by Greller and Drachsler (2012) also provides a structured approach to using students' data to improve course design, evaluation, and overall educational effects. In short, the current research portrays the picture of an education system that is being converted by AI—by becoming more personal, data-informed, and inclusive. However, it also emphasizes the need for responsible implementation. As we move forward, it is important to maintain the right balance between technological innovation and moral responsibility.

Future studies should focus on how AI can further support emotional learning, scale educational access, and foster trust among all stakeholders—teachers, students, and institutions.

3. Defining Industry 4.0 and its Emergence

For centuries, every industrial revolution has carried forward humanity, the way we live, and work. Today, we are in the middle of the Fourth Industrial Revolution, which is known as Industry 4.0. This new era is about bringing the digital world together with physical and even organic, which creates smart ways to produce goods, distribute services, and connect worldwide. The term 'Industry 4.0' first appeared in the Hanover Fair in Germany in 2011 (Kagermann et al., 2013)." refers to the original report introducing the term, which is the first reference (the DIN.de report). Industry 4.0 was initially introduced as part of Germany's national strategy to promote digital transformation in manufacturing. Since then, the concept has expanded beyond Germany and evolved into a global movement

The industry is all about a combination of advanced digital technologies with traditional ones. Think of such factories where machines talk to each other, fix problems on their own, and what is happening in real time is favourable. It is possible due to devices like cyber-physical systems, IoT, AI, Robotics, Cloud Computing, and Big Data Analytics (Liao et al., 2017). One of the most important ideas in Industry 4.0 is the interrelation not only among machines but also among entire departments, businesses, and supply chains. Everything is connected through data, from sensors on the factory floor to enterprise planning systems (Kagermann et al., 2013b). Technologies like digital twins—virtual models of physical objects—make it easier to simulate and monitor systems in real time (Tao et al., 2019). Add to that edge computing and 5G, and you get lightning-fast communication between devices, which is essential for today's high-speed manufacturing (Lu, 2017). Altogether, these innovations allow companies to respond quickly to market changes, reduce waste, improve efficiency, and even shorten the time it takes to develop new products.

4. Emergence and Evolution

Industry 4.0 did not just appear out of nowhere—it builds on a long journey of innovation. The First Industrial Revolution, back in the late 1700s, introduced steam engines and mechanized production. The Second, in the late 1800s, brought electricity and assembly lines, making mass production possible. Then came the Third, in the late 1900s, when computers and automation started reshaping factories. Now, in the Fourth

Industrial Revolution, we are going a step further by connecting machines, data, and intelligence. The key phases of the emergence of Industry 4.0 are summarized in Table 2.1.

Table 2.1 Key phases of emergence of Industry 4.0

Time Period 2011–2020	*Technological Milestones*	*Academic Reference*
2000–2005	RFID, programmable logic controllers (PLCs), embedded systems	Lee (2003)
2011	The term "Industry 4.0" was introduced; focus on CPS-Cyber-Physical Systems and IoT	Kagermann et al., (2013b)
2014–2016	Rise of digital manufacturing, AI integration	Lasi et al. (2014); Qin et al. (2016)
2017–2019	Emphasis on digital twins, machine learning, cloud systems	Xu et al. (2018); Lu (2017)
2020–2023	Acceleration due to Covid-19: remote monitoring, automation	Xu et al. (2021)
2024–2025	Green manufacturing, AI governance, sustainable Industry 4.0	Mittal et al. (2024)

5. Key Technologies Involved in Industry 4.0

Table 2.2 provides a clear picture of Industry 4.0 and the technologies that work together to create an efficient and reliable industrial system.

Table 2.2 Key technologies involved in Industry 4.0

Technology	*Function/Description*	*Applications in Industry 4.0*	*Key Reference*
Internet of Things (IoT)	Connects machines, sensors, and systems for real-time data exchange	Predictive maintenance, real-time monitoring, asset tracking	Xu et al. (2014)
Artificial Intelligence (AI)	Enables machines to learn, reason, and adapt using algorithms and data	Quality control, demand forecasting, autonomous decision-making	Lu (2017)
Big Data Analytics	Analyses large volumes of structured and unstructured data for insights	Process optimization, customer behaviour analysis, production forecasting	Wamba et al. (2015)
Robotics (including Cobots)	Automates physical tasks with precision, often using sensors and AI	Assembly lines, material handling, collaborative tasks with humans	Rüßmann et al. (2015)
Cloud Computing	Provides scalable and remote access to computing, storage, and applications	Digital twin hosting, enterprise system integration, SaaS platforms	Zhong et al. (2017)

Source: Created by Author

The centre of this change is the IoT. Imagine machines, sensors, and even products that can "talk" to each other in real time. This continuous stream of information helps factories to monitor their performance, identify problems, and make fast decisions. Then comes AI, which adds "brain" to this system. AI can analyse data on a large scale, detect patterns, and even predict future issues. This leads not only to faster production but also to better quality and lower delays. Big data analytics play an important supporting role. Every other factory makes mountains of data, from machine temperature to customers' reactions. Big data tools help to understand all this information, converting raw numbers into insights that can guide everything from product design to maintenance programmes. Robotics takes automation to a whole new level. Unlike traditional machines, modern industrial robots are flexible, adaptive, and can work safely with humans. They can handle repetitive or dangerous tasks at the same time, increasing efficiency and workplace safety. Finally, the cloud computing industry serves as the backbone of the digital ecosystem of 4.0. This allows all this data and technology to be stored, shared, and accessed from anywhere, which supports cooperation and real-time decision making in places and teams.

6. Transformation of Educational Systems Under Industry 4.0 and Shift from Traditional to Intelligent Learning Environments

The rise of Industry 4.0 is doing more than bringing revolution in factories and industries—it is also telling us how we teach and learn. AI, IoT, Robotics, Cloud Computing, and Big Data Analytics continue to change the world around us. The education sector is suited to prepare students with the necessary skills in this rapidly changing digital age. These include not only digital literacy, but also creativity, important thinking, and the ability to solve the problems of the real world. There has also been a shift from traditional classroom instruction towards smart learning environments. Not long ago, most classrooms followed a familiar setup: the teacher gave lectures, listened to the students, and everyone went at the same speed, whether it was individual needs or learning styles. But that model is no longer enough. In today's world, education should be more dynamic, more individual, and deeply integrated with technology. This is where an intelligent teaching environment comes in. It adjusts the material because of how students learn; education makes it not only more efficient, but also more attractive.

Take the Indian Institute of Technology (IIT) Madras, for example. They have developed an AI-powered assistant called "IIT PAL", which offers video lessons and quizzes tailored to how students perform, helping aspirants of the Joint Entrance Examination (JEE) study smarter, not just harder. In the same spirit, Indian EdTech company BYJU'S uses AI to personalize learning videos, adjusting the pace and difficulty level for each student so that no one feels left behind. Globally, we' are seeing similar progress. In China, Squirrel AI has built adaptive learning platforms that track how students respond in real time and modify the instruction until each student fully understands the concept. This approach has led to measurable improvements in student outcomes.

These examples point to a major shift: from rote learning to real skill-building, from teacher-led instruction to student-centred autonomy, and from standardized tests

Table 2.3 Traditional to intelligent learning environments

Year	*Educational Transformation*	*Technological Shift*	*Example/Initiative*	*Reference(s)*
2000–2005	Introduction of basic e-learning systems	Learning Management Systems (LMS)	Blackboard, Moodle	Liao et al. (2017)
2006–2010	Digital content adoption in higher education	Multimedia, Flash, and HTML-based courses	NPTEL (India), MIT OCW	Liao et al. (2017)
2011–2015	Rise of Massive Open Online Courses (MOOCs)	Cloud-based learning, video streaming	Coursera, edX, SWAYAM (India)	Panigrahi et al. (2021)
2016–2018	Personalization using AI and data analytics	Adaptive Learning Platforms, AI tutors	BYJU'S (India), Squirrel AI (China)	Zhou et al. (2021); MHRD (2020)
2019–2020	Emergency remote learning due to Covid-19	Video conferencing, virtual classrooms	Google Meet, Zoom, Microsoft Teams	World Economic Forum (2020)
2021–2022	Growth of hybrid and blended learning models	AI, cloud, mobile apps	National Education Policy (NEP) 2020 (India), smart classrooms	Panigrahi et al. (2021)
2023	VR/AR integration into classroom learning	Immersive learning environments	Class VR in the UK, immersive science labs	Lu (2017)
2024–2025	Focus on AI-driven personalized and competency-based learning	AI, analytics, and digital twin for education	IIT PAL, AI assessment tools, National Digital University (India)	MHRD (2020); World Economic Forum (2020); Zhong et al. (2017)

Source: Created by Author

to ongoing, real-time feedback. As the job environment values creativity and digital fluency, schools and universities are rethinking what they teach and how they teach it—aligning academics with the skills demanded by Industry 4.0.

7. Role of Artificial Intelligence

In the 1970s, the ITS began with early experiments, seeking to repeat one-on-one tuition experiences. The role of AI in education has changed dramatically over time. The use of AI technologies has increased in recent years, conventional learning methods have been changed to those that are more dynamic, personal, and data-driven. These days, adaptive teaching platforms tailor course content to fit each student's specific learning preferences and pace, resulting in increased engagement and improved outcomes. AI-based evaluation tools further reduce teachers' workload by automatically assessing student performance and providing real-time feedback on assignments and learning activities. Institutions can achieve significant insights into student behaviour through future-stolen and learning analysis, which help predict performance, identify learning gaps, and make quick interventions. Additionally, AI-run chatbots and virtual assistants provide round-the-clock academic assistance, often answering questions and assisting students as they progress through their educational

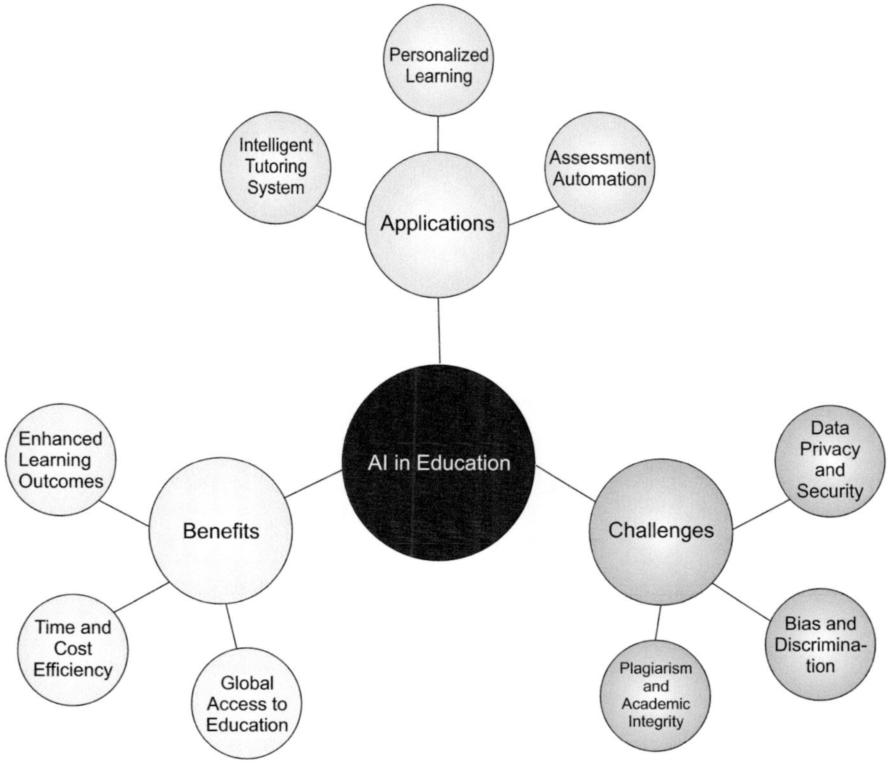

Fig. 2.2 Multifaceted impact of AI in education.

career. The AI also contributes significantly to the course design by assessing students' needs and educational trends to produce current, relevant course materials. AI can improve education for students worldwide, which develops by increasing access, inclusion, and efficacy.

8. Benefits of AI Integration in Education

Education stands to benefit greatly from the transformative effects of AI, which can enhance the quality and effectiveness of instruction. AI enables personalized learning by continuously evaluating each student's performance, making education more individualized and flexible. Instead of teaching everyone in the same way, AI tools now adjust lessons based on each student's learning style, speed, and requirements (Rane et al., 2023; Singh, 2023). With the help of AI, students get the right material at the right time—whether they need more support or are ready for a challenge (Anis, 2023; Khonturaev, 2023).

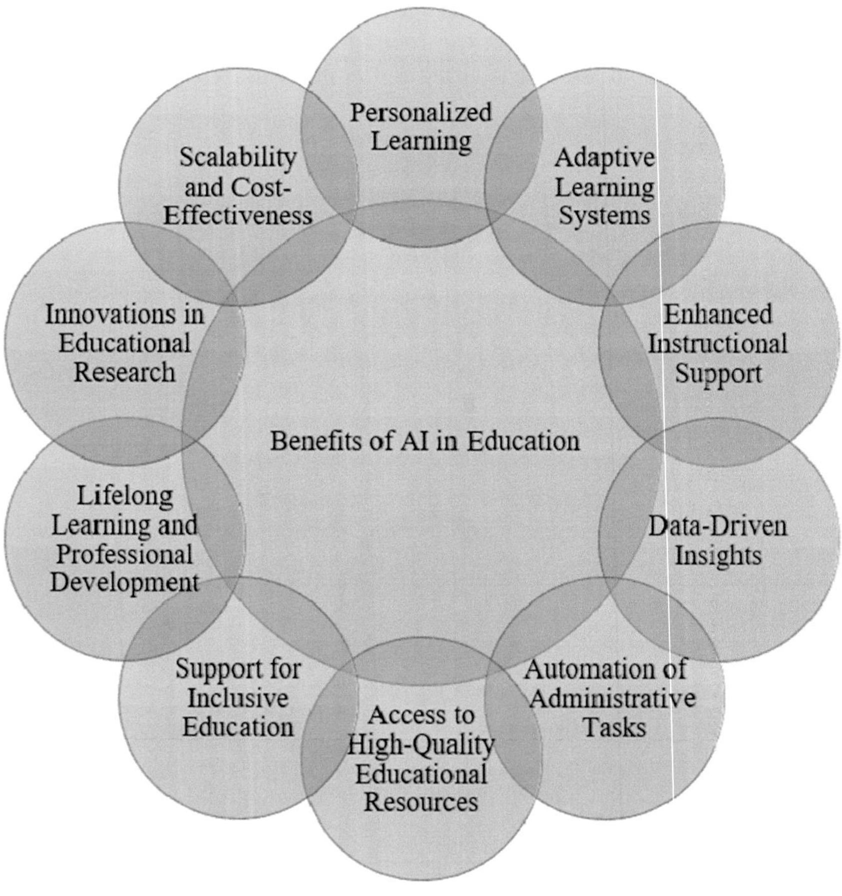

Fig. 2.3 Application of AI in Education.

The ITS function as digital personal tutors, providing feedback, guiding students through lessons, and keeping them motivated (Akyuz, 2020; Conati et al., 2021). These systems also help to make learning more inclusive by supporting various teaching styles—such as visual, listening, or learners on the hands (Kabudi et al., 2021). Outside the classroom, AI is also supporting teachers by creating customized content, tracking student progress, and even grading assignments or managing schedules (Ahmad et al., 2023; Hassoun et al., 2023). Tools such as NLP are being used to develop learning materials tailored to each student's level, while AI-powered assessments offer instant feedback to help students improve (Babitha et al., 2022). By pairing with virtual reality (VR) and interactive tools, AI makes learning more engaging by turning lessons into 3D experiences. This is helping schools and colleges move away from traditional teaching methods towards more student-centred, efficient, and future-focused education (Alam & Mohanty, 2023; Saklani, 2023).

9. Case Studies and Practical Implementations

The following case study and real-world applications help to explain the practical effects of AI in education in different types of references and applications.

- *Carnegie Learning*: Students of middle and high schools can get individual mathematics instructions from Carnegie Learning's ITS using Mathia (USA) AI tool. To meet each student's unique learning needs, it constantly evaluates their performance and modifies the material in real time. According to research, students who use Mathia often show high benefits in mathematics proficiency than those who use more traditional approaches. China's Squirrel AI is a prime example of an adaptive teaching platform that uses cognitive science and machine learning to personalize learning content for each student.

In very large classes, this system analyzes student performance at different points and adjusts the course content dynamically to boost engagement and learning outcomes.

- *Gradescope (USA)*: By automatic testing, quiz and assignment grading, this AI-based evaluation tool help teachers. This grading makes a lot of cuts on time and increases stability by supporting both written and multiple-choice reactions. Universities have adopted a large number of Gradescope to customize evaluation processes.
- *Learning Analytics at the University of Michigan*: The University increased retention rates by using predictive analytics and providing quick intervention.
- Jorgia State University's "Pence" Chatbot is an AI-investigated virtual assistant to interact with students to respond to inquiries about interaction, financial assistance, and entry. The Chatbot showed the value of AI in student aid services by reducing summer melt (students who intend to enrol in college but do not).

10. Future Research Directions

Future research should focus on certain areas to improve AI efficacy, inclusion, and moral application as it continues to change the academic landscape. To demolish AI decision-making in assessment, tuition, and course design, an important avenue is the creation of transparent and clear AI models that can trust teachers and students. To guarantee the consequences of fair learning for people of various socio-economic and cultural backgrounds, research should also pay attention to methods of reducing algorithm bias through the creation of representative and diverse training datasets. Another important area is enhancing emotional intelligence in AI systems, enabling them to recognize and respond to teachers' emotional states, thereby improving connectivity and personalized support. AI-managed multimodal learning environment that integrates text, voice, gesture, and facial expressions has the ability to further change the student interaction and response system. Scalability and adaptation capacity of AI devices for low-resource educational settings, especially in rural or developing areas, should also be focused upon for future research. Research on AI-Abumed Teacher Support System, which can help with different instructions without replacing lessons planning, real-time classroom analytics, and human teachers, is also becoming more and more necessary. To assess the long-term effects of AI, longitudinal research is vital.

11. Policy and Implementation Recommendations

Well-defined policies and implementation strategies are important to ensure moral and successful integration of AI in education. When using the AI tool, policymakers should determine clear rules that put data privacy, accountability, and transparency at the top. To prevent digital division from increasing, it is mandatory to create institutional and national structures that promote proper access to AI technologies, especially in rural areas. To guarantee that teachers and administrators know how AI equipment is used efficiently and responsibly, training and capacity-builiding initiatives should be undertaken. To develop an AI solution to suit educational objectives and moral principles, cooperation is needed among government organizations, educational institutions, technology developers, and civil society.

12. Conclusion

The application of AI in education, through intelligent assessment, real-time analysis, automated assistance systems, and personal instructions has demonstrated its ability to bring revolution in teaching and learning. Analysis of important applications indicates how AI can promote student engagement, improve results, and reduce administrative burden. Examples include AI-operated chatbots, adaptive learning platforms, predictive analytics, and ITS. This important applications examination suggests how AI can reduce administrative burden and enable students to increase connectivity and improve results. There are some examples of ITS, predictive analytics, adaptive learning platforms, and AI-run chatbots. Machine Learning Algorithms (MLAs) can use these techniques to improve the course and to distribute resources more effectively, teachers can use them to better understand and support diverse learners, and technologists can use them to develop inclusive, scalable, and moral solutions.

References

Ahmad, T., Hassan, R., Arshad, M. and Ahmed, A. (2023). AI in curriculum design: A data-driven approach. *Educational Futures*, *15*(1), 11–25.

Akyuz, H.I. (2020). AI-based intelligent tutoring systems in e-learning. *Turkish Online Journal of Distance Education*, *21*(3), 122–138.

Alam, R. and Mohanty, S. (2023). AI applications in EdTech: Opportunities and challenges. *International Journal of Emerging Technologies*, *14*(2), 90–104.

Anis, R. (2023). Personalized learning and adaptive AI: Future of education. *AI & Learning Journal*, *9*(1), 45–57.

Babitha, K., Sushma, K. and Gudivada, V. (2022). Smart assessments using AI: Transforming feedback systems. *Journal of Digital Education*, *6*(3), 88–95.

Bandara, N. and Senanayaka, C. (2024). *Use of artificial intelligence in education: A systematic review. ResearchGate.* [https://www.researchgate.net/publication/381368858_Use_of_Artificial_Intelligence_in_Education_A_Systematic_Review],https://www.researchgate.net/publication/381368858_Use_of_Artificial_Intelligence_in_Education_A_Systematic_Review)

Begum, R. (2025). How AI is revolutionizing Education 4.0: Opportunities and challenges. *International Journal of Educational Technology and AI*, *11*(1), 22–39.

Conati, C., Jaques, N. and Muir, M. (2021). AI in education: Realizing the potential through intelligent systems. *AI & Education Review*, *10*(4), 200–219.

Cukurova, M., Luckin, R. and Kent, C. (2023). Artificial Intelligence in education: Addressing ethics and equity. *AI and Society*, *38*(1), 1–15.

Graesser, A.C., Chipman, P., Haynes, B.C. and Olney, A. (2005). AutoTutor: An intelligent tutoring system with mixed-initiative dialogue. *IEEE Transactions on Education*, *48*(4), 612–618.

Greller, W. and Drachsler, H. (2012). Translating learning into numbers: A generic framework for learning analytics. *Educational Technology & Society*, *15*(3), 42–57.

Hassoun, M., Lu, X. and Deng, L. (2023). Administrative automation in schools: The role of AI. *Journal of Smart Education Management*, *5*(1), 33–46.

Hermann, M., Pentek, T. and Otto, B. (2016). Design principles for Industrie 4.0 scenarios: A literature review. *Procedia CIRP*, 52, 167–172.

Holmes, W., Bialik, M. and Fadel, C. (2022). *Artificial Intelligence in Education: Promises and Implications for Teaching and Learning*. Center for Curriculum Redesign.

Kabudi, T., Pappas, I.O. and Olsen, D.H. (2021). AI in personalized learning: A systematic review. *Computers & Education*, *165*, 104150.

Kagermann, H., Wahlster, W. and Helbig, J. (2013a). Recommendations for implementing the strategic initiative Industrie 4.0. Final Report of the Industrie 4.0 Working Group. https://www.din.de/blob/76902/8c8cac8834185d285bc5e52f3e4420c6/recommendations-for-implementing-industry-4-0-data.pdf

Kagermann, H., Wahlster, W. and Helbig, J. (2013b). Recommendations for implementing the strategic initiative INDUSTRIE 4.0: Securing the future of German manufacturing industry. Final Report of the Industrie 4.0 Working Group.

Khonturaev, R. (2023). Machine learning for adaptive educational content. *Eurasian Journal of Educational Research*, *99*(1), 121–134.

Knewton. (2014). **Adaptive learning technology**. Retrieved September 10, 2025, from https://www.knewton.com

Lasi, H., Fettke, P., Kemper, H.G., Feld, T. and Hoffmann, M. (2014). Industry 4.0. *Business & Information Systems Engineering*, *6*(4), 239–242.

Lee, J. (2003). Smart factory systems and enterprise integration. *IFAC Proceedings Volumes*, *36*(6), 217–220.

Liao, Y., Deschamps, F., Loures, E.d.F.R. and Ramos, L.F.P. (2017). Past, present and future of Industry 4.0—A systematic literature review and research agenda proposal. *International Journal of Production Research*, *55*(12), 3609–3629.

Liu, V., Latif, E. and Zhai, X. (2025). Advancing education through tutoring systems: A systematic literature review. *Educational Technology Research and Development*, *73*(2), 88–104.

Lu, Y. (2017). Industry 4.0: A survey on technologies, applications and open research issues. *Journal of Industrial Information Integration*, *6*, 1–10.

Luckin, R., Holmes, W., Griffiths, M. and Forcier, L.B. (2016). *Intelligence Unleashed: An Argument for AI in Education*. Pearson Education.

Ministry of Education. (2020). National Education Policy 2020. Government of India.

Mittal, S., Khan, M.A., Romero, D. and Wuest, T. (2024). Smart manufacturing and sustainability: Integrating AI and circular economy. *Journal of Cleaner Production*, *421*, 138421.

Mustafa, M., et al. (2024). *A Systematic Review of Literature Reviews on AIED: A Roadmap to a Future Research Agenda*. Available at https://www.researchgate.net/publication/386564845. Sept 10, 2025

Nayak, Manjushree. (2025). Remote experimentation in engineering education: Integrating virtual labs and learning. In *Revolutionizing Education with Remote Experimentation and Learning Analytics*, 397.

Ogunleye, B. et al. (2024). *A Systematic Review of Generative AI for Teaching and Learning Practice*. arXiv.

Patel, S. and Mishra, P. (2023). New skills for sustainability in the socially challenging era of digital transformation: Opportunities and challenges. In R. Pettinger, B.B. Gupta, and A. Roja (Eds.), *Handbook of research on digitalization solutions for social and economic needs** (pp. 305–312). IGI Global. Editors: The editors of the handbook are Richard Pettinger, Brij B. Gupta, and Alexandru Roja. The publisher is IGI Global. February 27, 2023, pp. 305-312.

Porayska-Pomsta, K., Holmes, W. and Nemorin, S. (2024). *The Ethics of AI in Education*. arXiv.

Pradhan, D. (2024). AI in education must align with Indian values, says Dharmendra Pradhan at PadhAI Conclave. *Times of India*, 29 May. Available at https://timesofindia.indiatimes.com/education/ai-in-education-must-align-with-indian-values-says-dharmendra-pradhan-at-padhai-conclave/articleshow/121487505.cms

Qin, J., Liu, Y. and Grosvenor, R. (2016). A categorical framework of manufacturing for Industry 4.0 and beyond. *Procedia CIRP*, *52*, 173–178.

Rane, R., Choudhary, P. and Rane, K. (2023). The impact of AI on student-centric learning models. *Education & AI Research*, *11*(1), 15–29.

Riordan, A., Tang, Y. and Zhou, Y. (2023). Human-centred learning analytics and artificial intelligence in education: A review of stakeholder-focused research. *British Journal of Educational Technology*, *54*(1), 45–62.

Rüßmann, M., Lorenz, M., Gerbert, P., Waldner, M., Justus, J., Engel, P. and Harnisch, M. (2015). *Industry 4.0: The Future of Productivity and Growth in Manufacturing Industries*. Boston Consulting Group.

Saklani, A. (2023). AI in virtual and immersive learning spaces. *Journal of Advanced Learning Systems*, *10*(2), 74–89.

Schwab, K. (2016). *The Fourth Industrial Revolution*. World Economic Forum.

Singh, S. (2023). Artificial intelligence and learner empowerment. *AI for Education Review*, *12*(1), 33–48.

Tao, F., Qi, Q., Liu, A. and Kusiak, A. (2019). Data-driven smart manufacturing. *Journal of Manufacturing Systems*, *48*, 157–169.

Tong, X. et al. (2024). *Developing and Deploying Industry Standards for AI in Education*. arXiv.

Wamba, S.F., Akter, S., Edwards, A., Chopin, G. and Gnanzou, D. (2015). How "big data" can make big impact: Findings from a systematic review and a longitudinal case study. *International Journal of Production Economics*, *165*, 234–246.

Wang, S., Wang, F., Zhu, Z., Wang, J., Tran, T. and Du, Z. (2024). Artificial intelligence in education: A systematic literature review. *Expert Systems with Applications*, *252*, 124167.

Xu, L.D., He, W. and Li, S. (2014). Internet of Things in industries: A survey. *IEEE Transactions on Industrial Informatics*, *10*(4), 2233–2243.

Xu, X., Lu, Y., Vogel-Heuser, B. and Wang, L. (2021). Industrial big data analytics and edge computing in smart manufacturing: A survey. *IEEE Transactions on Industrial Informatics*, *17*(4), 2464–2475.

Xu, X., Xu, C. and Li, L. (2018). Industry 4.0 and digital twins: A new paradigm for intelligent manufacturing. *Procedia CIRP*, *72*, 380–385.

Yan, L., Xu, M. and Ho, C.M. (2023). Practical and ethical challenges of large language models in education: A scoping review. *Computers & Education: Artificial Intelligence*, *4*, 100102.

Zawacki-Richter, O., Marín, V.I., Bond, M. and Gouverneur, F. (2019). Systematic review of research on artificial intelligence applications in higher education—Where are the educators? *International Journal of Educational Technology in Higher Education*, *16*(1), 39.

Zhong, R.Y., Xu, X., Klotz, E. and Newman, S.T. (2017). Intelligent manufacturing in the context of Industry 4.0: A review. *Engineering*, *3*(5), 616–630.

Chapter 3

A New Era of Machine Intelligence in Education Industry: Towards a Sustainable Multifaceted Revolution

Prerna Mishra[1*]

1. Introduction: Education at the Crossroads of Transformation

Education, as a cornerstone of societal development, has continuously adapted to the needs of different historical epochs. In the current era, marked by rapid digitalization, global interconnectivity, and complex socio-economic dynamics, traditional educational structures are facing unprecedented scrutiny. The conventional model of mass education characterized by standardized curricula, fixed pacing, and uniform assessment was historically suited to an industrial logic that prized conformity and predictability. However, such rigidity is increasingly incompatible with the demands of contemporary learners and labour markets (Anderson & Dron, 2011). Educators, policymakers, and learners alike are calling for a paradigm shift towards more responsive, personalized, and context-sensitive approaches to teaching and learning.

At the heart of this transformation is the emergence of machine intelligence (MI) which is a convergence of advanced computational methods, including machine learning, natural language processing, and cognitive systems, that go beyond traditional artificial intelligence (AI) frameworks. Unlike earlier forms of educational automation, MI enables systems to adapt dynamically to learner behaviour, preferences, and environments, thus fostering genuinely personalized educational experiences. The fusion of such intelligent technologies is not merely a matter of augmenting existing pedagogical practices but represents a fundamental rethinking of how knowledge is generated, shared, and internalized within digital ecosystems.

[1] Department of CSE: Computer Science and Engineering, Indian Institute of Information Technology, IIIT Nagpur, India. https://orcid.org/0000-0003-2895-6146

* Corresponding author: pmishra@iiitn.ac.in

DOI: 10.1201/9781003639367-3

1.1 Historical Shifts in Educational Paradigms

The design and delivery of education have historically mirrored dominant societal models. During the Industrial Age, formal schooling systems were engineered to produce standardized human capital. Learners were grouped accordingly by chronological age. They were then exposed to identical content, and evaluated through uniform metrics. Instruction prioritized more on rote memorization, obedience, and skills that were primarily essential for factory-based economies (Robinson, 2011). Such systems were teacher-directed and left minimal or no room for individual learning differences.

With the onset of the Information Age, fuelled by computing and communication technologies, cracks in this paradigm became more apparent. With time, new educational philosophies emerged, advocating for learner autonomy, decentralized knowledge creation, and continuous feedback loops. Onset of digital tools enabled flexible access to information, collaborative learning, and rich instruction which utilized multimedia. Theories such as constructivism and connectivism emphasized that learning occurs most effectively in active, socially networked, and technologically mediated environments (Siemens, 2005). Despite these advancements, the structural remnants of industrial-era schooling still prevail, often hindering the full adoption of intelligent, adaptive technologies in mainstream education.

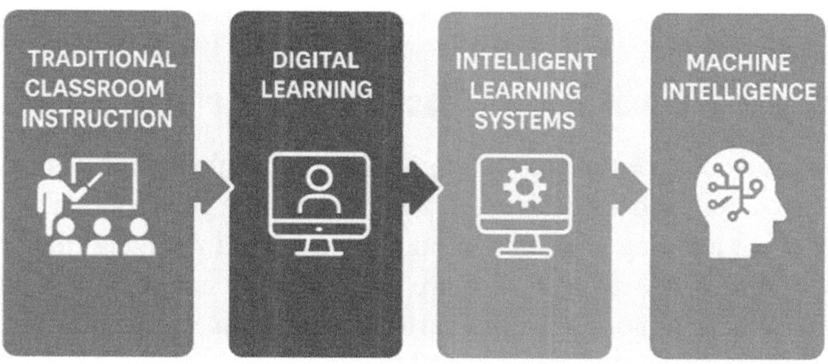

Fig. 3.1 Evolution of educational paradigms towards machine intelligence.

1.2 Emergence of Machine Intelligence: Beyond Traditional AI in Education

The recent convergence of machine learning (ML), big data analytics (BDA), natural language processing (NLP), and generative models has given rise to "machine intelligence" (MI), a more evolved form of AI characterized by autonomy, context-sensitivity, and emergent learning. While traditional AI in education has focused on rule-based tutoring systems, predefined feedback loops, and scripted decision trees, the MI-based systems demonstrate remarkable capabilities such as:

- continuous learning from diverse learner behaviours and modalities;
- real-time personalization of content and assessments;
- generative capabilities for content creation (auto-generating quizzes, summarizing texts, simulating lab experiments); and
- emotion and sentiment detection to tailor socio-emotional support.

In educational contexts, MI has begun to support adaptive learning environments, predictive analytics for student outcomes, automated curriculum mapping, and AI-driven formative assessment tools (Luckin et al., 2016). Importantly, MI is not just a technological upgrade; it also embodies a new cognitive relationship between the learner and the digital ecosystem (Figure 3.1). MI frameworks increasingly incorporate multi-agent systems, probabilistic reasoning, and generative AI to move beyond prescriptive models of learning into co-constructive knowledge environments where learners, instructors, and algorithms form an active triad. This paradigm shift requires not only technological adaptation but also epistemological and ethical reflection, especially concerning data privacy, learner autonomy, and algorithmic bias in education.

This chapter delves into how MI is shaping the future of the education industry not merely through technological intervention but as a force that reconfigures pedagogical, curricular, and institutional logics. By anchoring the discussion in historical, theoretical, and socio-technical contexts, the chapter aims to explain fundamental difference between MI and traditional AI in educational applications, ways the MI systems can serve as an enabler of sustainable, inclusive, and adaptive learning systems. The chapter also discusses the intersection of ML, generative AI, and learning analytics in curriculum and instructional design, while analysing the implications of MI-enabled personalization in assessment, content delivery, and learner engagement.

2. Conceptual Foundations of Machine Intelligence in Education

Machine intelligence represents a paradigm shift in the application of computational technologies to educational systems. Till now, AI has long played a role in automating certain aspects of learning and instruction ranging from intelligent tutoring systems to adaptive quizzes; MI signifies a more integrated, dynamic, and human-centric evolution of these technologies. Unlike earlier computational tools that relied on static models and pre-programmed responses, MI systems continuously adapt, interpret, and co-evolve with learners in real time. This section delves into the foundational understanding of MI, distinguishing it from traditional AI and ML approaches, and highlights its key characteristics that make it particularly relevant for the transformation of education.

2.1 What is Machine Intelligence?

Machine intelligence can be defined as the convergence of advanced technologies such as AI, ML, cognitive computing, NLP, and real-time analytics into a unified, adaptive, and context-aware ecosystem. It is characterized by its ability to simulate elements of human cognition, including reasoning, decision-making, and even creativity. In the educational context, MI systems are capable of perceiving learner behaviour, interpreting diverse multimodal inputs (e.g., text, speech, biometrics), and adjusting instructional strategies dynamically to optimize learning outcomes (Luckin et al., 2016; Mitchell, 2019).

Unlike conventional systems that operate on static rules or require extensive pre-labelling of data, MI leverages self-learning algorithms that evolve through interaction. These systems can generate new content (e.g., exercises, explanations), predict learner needs before they manifest, and provide feedback based not just on correctness but also on a deep understanding of learner cognition and emotion. MI thus acts not merely as a computational assistant, but also as an intelligent collaborator in the learning journey (Holmes et al., 2019).

2.2 Distinction from Traditional AI and ML Paradigms

The distinction between MI and traditional AI/ML lies in both the scale of integration and the depth of adaptivity. Classical AI in education primarily focused on rule-based systems, expert systems, and decision trees method to deliver programmed instructions or diagnose student errors. These systems lacked the flexibility to deal with unstructured data or novel learning situations. ML improved upon this by introducing pattern recognition and data-driven predictions, but often required large labelled datasets and lacked interpretability and context-sensitivity (Fischer et al., 2020).

Machine intelligence integrates adaptive learning algorithms with cognitive models, enabling real-time personalization. It incorporates ongoing learner feedback, environmental data (such as location, device usage, and time of day), and even affective states to adjust pedagogical strategies. MI systems also benefit from transfer learning, semi-supervised learning, and reinforcement learning techniques that allow them to learn effectively in low-data or evolving contexts. This holistic intelligence makes MI systems more resilient, scalable, and contextually aware than their AI/ML predecessors (Anderson & Dron, 2011). Table 3.1 highlights the applicability of MI across different education levels.

Table 3.1 Applications of machine intelligence across educational levels.

Educational Level	*Application Example*	*Learning Objective*	*AI/ML Technique Used*
Primary School	AI chatbots for literacy support	Foundational reading and comprehension	NLP, Speech Recognition
Secondary School	Adaptive math platforms	Skill mastery in algebra	Reinforcement Learning
Higher Education	AI teaching assistants in MOOCs	Concept reinforcement, doubt resolution	Knowledge Graphs, LLMs
Vocational Training	AI simulations for skill practice	Hands-on learning in controlled settings	Computer Vision, AR: Augmented reality
Workplace Learning	Personalized upskilling pathways	Employee performance enhancement	Recommendation Engines, Predictive Analytics
Special Education	Emotion-aware learning tools	Behaviour monitoring and social interaction	Affective Computing, Eye Tracking

2.3 Characteristics of MI-driven Education

Educational systems when powered by MI bring forth a transformative learning environment by embodying key characteristics that enhance both learner agency and systemic flexibility. Among the most notable features are personalization, adaptivity, context awareness, and sustainability. Each of these features is discussed here.

1. *Personalization* in MI-driven education entails the continuous tailoring of instructional pathways based on an individual's cognitive profile, learning history, and engagement patterns. Rather than providing a static curriculum, intelligent systems dynamically adjust content, its difficulty level, format, and sequence to align as per the learner's pace and style. Such personalized trajectories improve motivation, facilitate deeper understanding, and support inclusion by catering to diverse educational needs (Holmes et al., 2019).
2. *Adaptivity* refers to the system's ability to recalibrate itself in response to learner interaction. MI systems monitor indicators such as accuracy, hesitation, or time taken on a task completion to detect cognitive fatigue or conceptual difficulty. These insights drive real-time modifications by altering feedback style, switching learning modalities, or introducing supportive scaffolds which promote continuity in learning.
3. *Context-awareness* enhances the adaptivity by embedding environmental and situational variables into the decision-making loop. The factors in consideration are learner's location, device usage, time, or even emotional signals derived from sentiment or expression analysis, which are at last integrated to fine-tune instructional delivery. This responsiveness ensures that the learning experience remains relevant to the learner's immediate reality.
4. *Sustainability* is another feature of MI-driven education. Automating the repetitive instructional and administrative process eventually reduces the educator workload while at the same time ensuring instructional scalability. Additionally, intelligent content generation and adaptive scheduling support lifelong learning models that accommodate fluid career transitions and evolving learner goals (Anderson & Dron, 2011).

2.4 Positioning MI in Educational Theory

The integration of MI within educational paradigms indicates the shift from static content delivery to dynamic, learner-centric ecosystems. Theoretically, MI aligns and extends the post-constructivist view, particularly within frameworks like connectivism, which conceptualize learning as a decentralized, network-mediated, and context-driven activity (Siemens, 2005).

Unlike traditional pedagogies that rely on hierarchical knowledge transfer, MI supports a distributed learning process where interaction with digital systems plays a critical role in knowledge construction. These systems not only provide content but also function as metacognitive tools which track the progress, highlight cognitive gaps, and suggest relevant learning strategies based on real-time analytics.

The MI technologies empower learners by offering reflective insights into their own behaviours and learning patterns. Learner dashboards enable them to visualize their strengths, areas of improvement, and motivational shifts, enabling self-regulated learning. This transparency transforms the learner from a passive recipient to an active architect of their educational journey.

For educators, MI acts as a cognitive partner rather than a replacement. Through predictive diagnostics, real-time performance data, and instructional recommendations, MI enables teachers to respond proactively to students' performance and their needs. It repositions the teacher's role from direct instruction to facilitation and strategic mentorship, upheld by intelligent support tools.

In essence, MI is not merely a technical augmentation of existing educational methods but it also embodies a philosophical evolution. It introduces an environment (as shown in Figure 3.2) where human and machine intelligences interact symbiotically to produce adaptive, equitable, and scalable learning conditions.

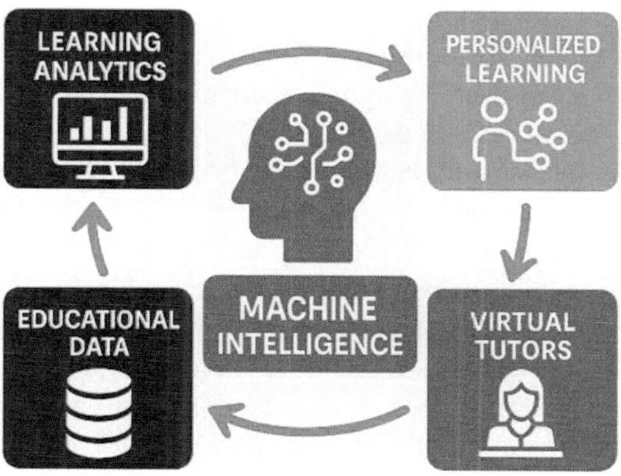

Fig. 3.2 Machine intelligence ecosystem in education.

3. Dimensions of Machine Intelligence Impacting Education

From curriculum architecture to learner engagement, from assessment to administration, MI is enabling a significant shift towards intelligent systems that are adaptive, personalized, and analytically robust. This section explores six major domains where MI is exerting transformative influence.

3.1 Curriculum Development and Personalization

Modern curriculum development is increasingly being informed by intelligent systems that respond to diverse learner profiles and evolving skill requirements. Various MI tools contribute to syllabus personalization by identifying knowledge gaps and cognitive patterns across learner cohorts, enabling the dynamic organization of content to better match individual trajectories (Roschelle & Pea, 2013).

Adaptive course structures are generated using data from learners' past performances, preferences, and behavioural signals. The topic sequencing and difficulty level of these course structures are adjusted based on real-time cognitive diagnostics, helping students progress at optimal rates. Furthermore, content recommendation engines which are powered by large-scale repositories and intelligent filtering techniques provide contextually appropriate resources including videos, simulations, and practice modules that align well with each learner's zone of proximal development (Zou et al., 2019).

A key enabler of such personalization is "learner modelling", where each student is represented as a continuously altering data profile. These profiles are built using algorithms that interpret patterns of interaction, engagement, and knowledge depth, offering the solid foundation for real-time instructional adaptation at different scale (Grawemeyer, 2017).

3.2 Pedagogical Shifts

Machine intelligence is driving significant pedagogical evolution from static content delivery towards responsive, learner-aware environments. A central example of this is the development of intelligent tutoring systems (ITS), which replicate the rich feedback environment typically provided as one-on-one human tutoring. These systems often adjust instruction based on student misconceptions, learning pace, and engagement levels (Koedinger et al., 1997).

Incorporating models from reinforcement learning and sequential decision-making, these adaptive learning systems can revise content flow and interaction strategies in real time, based on learners' accuracy, attention span, and hesitation frequency. Such adaptability enhances learner agency, helping maintain motivation while fostering conceptual mastery, especially in self-paced or hybrid learning environments (Heffernan & Heffernan, 2014]. These innovations signal a pedagogical turn which is away from linear, mass-standardized instruction but towards individualized, data-driven facilitation that evolves with the learner's performance patterns.

3.3 Assessment and Feedback

Machine intelligence systems are also contributing in how student assessment is designed, delivered, and interpreted. One of the most immediate transformations is observed in the area of automated grading, where NLP models are applied to assess essays, programming code, and open-ended responses with a high degree of reliability and speed (Burrows & Gurevych, 2015). Formative assessments are increasingly being supplemented with real-time analytics that provide students with immediate, targeted feedback. This approach fosters an iterative learning process where misconceptions can be corrected immediately rather than after the learning activity (Shute, 2008). Beyond assessing performance, MI systems also support predictive evaluation. Using historical academic records, behavioural patterns, and engagement data, predictive models are trained to flag potential academic risks such as underperformance or likelihood of course withdrawal. Generally, such forecasts allow instructors and institutions to intervene proactively, offering tailored support strategies to at-risk learners (Campbell & Oblinger, 2007).

3.4 Cognitive and Emotional Intelligence

One of the most groundbreaking applications of MI in education is its ability to perceive and respond to the emotional and cognitive states of learners. Emotion detection systems leverage computer vision, speech analysis, and physiological sensors to infer emotional cues such as confusion, frustration, or enthusiasm

(D'Mello & Graesser, 2014). These systems are used to drive affect-aware tutoring environments that adapt not only to what a student knows, but also to how they feel about the learning experience (Porayska-Pomsta et al., 2013). Empathy modelling, which simulates human-like emotional understanding, plays a critical role in building trust and relational engagement in digital learning environments. MI systems capable of expressing empathy through personalized feedback or supportive gestures can enhance motivation and reduce anxiety, particularly among vulnerable learner groups.

3.5 Teacher Augmentation

Decision support systems for educators analyse classroom-level and individual learner data to recommend instructional strategies, resource allocations, and intervention plans (Holmes et al., 2019). These systems help teachers personalize instruction efficiently, even in large and diverse classrooms (as shown in Figure 3.3). AI-based teaching assistants, such as conversational agents and intelligent help desks, can handle routine queries, deliver supplementary explanations, and manage course logistics. This delegation of administrative and repetitive tasks frees educators to focus on higher-order pedagogical functions and mentorship roles (Winkler and Soares, 2020).

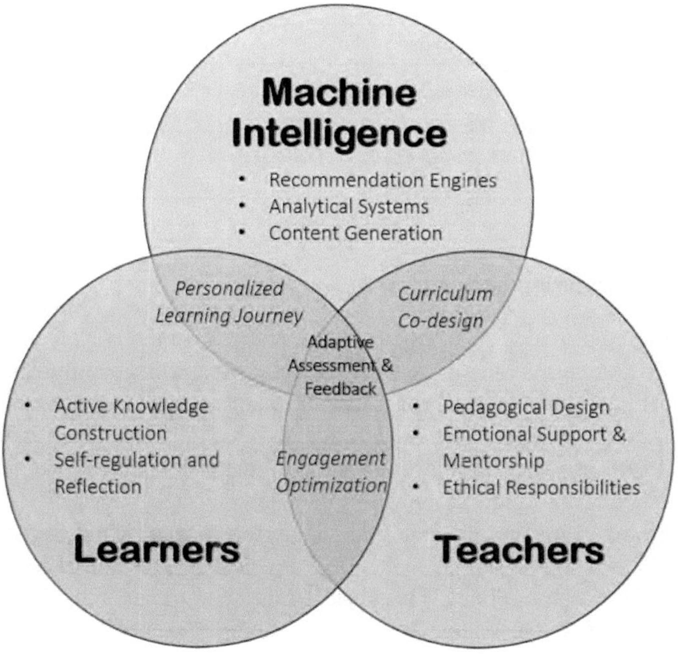

Fig. 3.3 Human-machine collaboration in intelligent learning environments.

3.6 Administrative and Institutional Intelligence

The MI technologies are also streamlining administrative functions within educational institutions. From AI-driven scheduling and automated admissions screening to fraud detection and resource optimization, MI supports data-informed decision-making at the institutional level (Chen & Lin, 2018]. Intelligent systems can forecast enrolment trends, optimize faculty allocations, and monitor compliance metrics, fostering operational efficiency and governance transparency. These administrative applications exemplify how MI extends beyond the classroom to shape policy-making, institutional strategy, and educational equity at scale.

4. Technological Enablers of Machine Intelligence in Education

The realization of MI in educational contexts is predicated on the convergence of multiple technologies that collectively enable adaptive, responsive, and scalable learning experiences. These enablers form the digital infrastructure that supports intelligent systems capable of learning from, responding to, and augmenting human educational processes. Table 3.2 highlights the various technologies used in MI system.

Table 3.2 Key technologies powering machine intelligence in education

Technologies	*Description*	*Educational Use Case*
Natural Language Processing	Understanding and generating human-like text	Automated essay scoring, chatbot tutors
Computer Vision	Visual data interpretation	Gesture recognition, virtual labs
Reinforcement Learning	Learning through feedback/ rewards	Personalized content pacing in gamified apps
Federated Learning	Privacy-preserving collaborative learning	Training models across institutions
Knowledge Graphs	Semantic data relationships	Intelligent tutoring systems

4.1. Cloud Computing and Edge Intelligence

Cloud computing provides the computational backbone for large-scale deployment of MI-based educational services. By enabling elastic storage and compute power, cloud platforms facilitate the training, deployment, and maintenance of AI-driven applications across institutions. Services like AWS Educate, Google Cloud for Education, and Microsoft Azure provide tailored cloud environments for educational institutions. However, reliance on cloud alone introduces latency and data privacy concerns, particularly in real-time learning analytics. Thus, the concept of edge computing was used where all the computations are performed on local devices like tablets, interactive whiteboards, or classroom hubs. This emerged as a complementary paradigm. Edge intelligence allows for low-latency decision-making and localized personalization without constant cloud dependency, thus enhancing both responsiveness and privacy (Nguyen & Kumar, 2022).

4.2 Natural Language Processing and Conversational Agents

Natural language processing plays a pivotal role in enabling ITS, question-answering modules, essay grading, and language learning apps. Recent advances such as transformer-based architectures allow MI systems to understand context, semantics, and intent at near-human proficiency. NLP-powered conversational agents (like virtual tutors or AI teaching assistants) engage students in real-time dialog learning, provide scaffolding, and clarify doubts in natural language. These agents also collect nuanced interaction data, which feed back into learner models to refine content delivery and assessment strategies (Hu & Zhang, 2023).

4.3 Learning Analytics and Educational Data Mining

The core intelligence of MI systems stems from their ability to detect patterns and infer insights from large volumes of heterogeneous data. Learning analytics and educational data mining enable the modelling of student engagement, progression, emotional states, and risk factors. Techniques include clustering learners by behaviour, anomaly detection for academic risk, and time-series prediction of dropouts. Advanced analytics frameworks often integrate psychometric data, clickstream logs, and assignment histories to generate rich learner profiles. These profiles are foundational for adaptive content recommendation and intelligent intervention systems (Papamitsiou & Economides, 2022).

4.4 Generative AI and Content Creation Tools

Generative AI, particularly large language models (LLMs) and diffusion models, has revolutionized educational content creation. Teachers and learners can co-create quizzes, summaries, diagrams, and even multimedia learning experiences. Platforms like ChatGPT, Google's NotebookLM, and GitHub Copilot (for programming education) act as creative collaborators. Moreover, generative models support multilingual and inclusive education by instantly translating or simplifying complex texts. In assessment, they are being explored for generating parallel test sets and custom assignments tailored to individual learning gaps (Yang & Sun, 2023].

4.5 Sensor Technologies and Multimodal Learning Interfaces

The integration of multimodal sensing technologies such as eye-tracking, facial recognition, EEG headbands, and haptic feedback devices enables MI systems to capture deeper cognitive and emotional data from learners. These inputs are critical for modelling attention, frustration, boredom, and engagement in real time. Smart learning environments equipped with Internet of Things (IoT) sensors and Augmented Reality/Virtual Reality headsets extend education beyond traditional interfaces, supporting experiential, kinesthetic, and immersive learning. Such multimodal setups align with MI's emphasis on personalized, adaptive, and emotionally responsive instruction (Li & D'Mello, 2023].

4.6 Interoperable Learning Platforms and Open Standards

Technological integration is vital for MI adoption at scale. Interoperable Learning Platforms (ILPs) such as Moodle, Canvas, and Blackboard increasingly support Application programming interface (APIs) and plugins that allow seamless integration with MI services. Standards like Experience API and Learning Tools Interoperability enable consistent tracking and communication between diverse educational technologies. These frameworks are essential for deploying MI across institutional contexts, ensuring that learner data can be securely transferred, interpreted, and acted upon across platforms.

4.7 Security and Infrastructure Resilience

With increasing dependence on cloud-based MI services, ensuring the robustness and security of underlying infrastructure is non-negotiable. Distributed denial-of-service attacks, data breaches, and system outages pose existential risks to digital classrooms. Recent innovations in blockchain for education, zero-trust architectures, and federated learning approaches are gaining traction to safeguard sensitive learner data while maintaining algorithmic performance across decentralized learning systems.

5. Ethical, Policy, and Privacy Considerations

As MI is becoming increasingly embedded within education systems, there are now concerns around ethics, data governance, and policy frameworks gaining critical prominence. While MI offers unprecedented potential in education, its implementation is not without significant ethical challenges. Concerns regarding its fairness, transparency, accountability, and the autonomy of learners have become increasingly prominent (Zelinger & Tan, 2022). These concerns stem, in part, from the opaque nature of many algorithmic systems, which rely on massive datasets encompassing students' behavioural, emotional, and cognitive traits. When these systems functions as a "black boxes", it becomes difficult for educators and learners to understand how decisions are made or to contest outcomes that may seem unjust.

5.1 Bias, Fairness, and Equity in MI Models

Numerous studies have drawn attention to the ways in which MI applications may perpetuate or even exacerbate existing educational disparities. Models trained on non-representative datasets often risk producing biased outcomes, disproportionately impacting students from historically marginalized or underserved groups (Holstein & Aleven, 2023). For instance, dropout prediction algorithms have been found to perform inconsistently across different racial or socio-economic demographics, raising the issue of algorithmic bias embedded within data or design assumptions.

There is need for more holistic view of fairness which goes beyond statistical debiasing and algorithmic correction in order to address these disparities. Hence, equity in MI must begin with inclusive design practices that actively involve diverse stakeholders, teachers, students, and community representatives. Such participatory involvement can help ensure that the data used is representative and that interventions are grounded in the social realities of learners (Haider & Sharma, 2023).

5.2 Data Privacy and Surveillance Concerns

Machine intelligence applications depend heavily on the continuous collection and processing of learner data including biometric data from wearables, keystroke dynamics, voice interactions, and emotional cues. While this enables powerful learning insights, it also blurs the boundaries between education and surveillance (Truong & Park, 2022). The increasing deployment of emotion AI and gaze-tracking in virtual classrooms has sparked intense debate about the intrusion of such systems into students' cognitive privacy. Key privacy challenges include informed consent, data minimization, anonymization, and data ownership. It is essential that MI systems adhere to established data protection standards like General Data Protection Regulation and Family Educational Rights and Privacy Act while ensuring that privacy safeguards are auditable and student-centred.

5.3 Accountability, Transparency, and Explainability

A significant ethical barrier in MI deployment in education is the lack of transparency in how decisions of content recommendation, adaptive difficulty levels, or behavioural interventions are made. Many current models are black-box in nature, making it difficult for educators and learners to trust or contest outcomes (Berendt, 2022). To address this, there is increasing interest in Explainable AI (XAI) frameworks tailored for education. These frameworks not only clarify the reasoning behind machine outputs but also allow educators to intervene meaningfully. Furthermore, building trust requires clearly defined accountability chains, i.e., responsibility when an MI-based intervention causes harm or fails.

5.4 Policy Frameworks and Regulatory Readiness

Despite rapid technological advancement, regulatory frameworks have not kept pace with the deployment of MI in educational settings. Recent international efforts like the 2022 guidelines on AI in education by the United Nations Educational, Scientific and Cultural Organization (UNESCO) and the European Commission's AI Act draft (2023) seek to regulate risk and promote ethical AI adoption in schools and universities (UNESCO, 2022; European Commission, 2023). However, national education systems still face challenges in localizing these policies to their institutional and cultural contexts. A multi-stakeholder approach involving policymakers, technologists, ethicists, and educators is required to co-create governance models that are adaptable, enforceable, and aligned with pedagogical goals.

5.5 Towards Responsible and Sustainable MI in Education

A responsible approach to MI in education involves not just ethical AI development but also the embedding of digital literacy and ethical reasoning within curricula. Students should be made aware of how MI systems influence their learning and be equipped with tools to question, critique, and co-design such technologies. Moreover, sustainable MI deployment should prioritize long-term human development goals over short-term efficiency metrics. This includes designing systems that are inclusive, privacy-preserving, culturally responsive, and pedagogically sound.

6. Policy, Governance, and Standardization

There is a growing need for robust policy frameworks and institutional governance mechanisms. Educational policymakers must address challenges related to data privacy, algorithmic transparency, and ethical deployment of these intelligent systems in classrooms. It must be ensured to provide equitable access to MI-driven learning technologies to prevent further marginalization of underserved learners and communities. Governance models should explicitly define accountability for data use, system errors (if any made), and outcomes generated by autonomous educational tools. Global institutions such as UNESCO, Organisation for Economic Co-operation and Development (OECD), and Institute of Electrical and Electronics Engineers (IEEE) are actively developing international guidelines and ethical standards for AI in education. These frameworks emphasize on responsible design, bias mitigation, and interoperability of AI systems across borders. National governments and accreditation bodies must collaborate with Educational Technology (EdTech) providers to certify these intelligent platforms, enforce regulatory compliance, and build public trust in MI adoption. Moving forward, a participatory governance approach that shall involve educators, technologists, and civil society will be crucial in shaping a safe, inclusive, and sustainable digital education future. Table 3.3 discusses the limitations, risks, and governance of AI in education.

6.1 Need for New Educational Policies and Accreditation Systems

Existing educational policies and accreditation mechanisms were designed for analogue or early digital systems, which do not account for the dynamic, autonomous, and personalized nature of MI-driven tools. Policies must now address novel issues such as algorithmic accountability, AI explainability in assessment decisions, and equitable access to intelligent learning platforms (Knox, 2022). Accreditation systems also need to evolve beyond traditional credit-hour models to accommodate competency-based and adaptive learning pathways enabled by MI. Micro-credentialing and digital badges often validated by MI algorithms require a formal recognition by national and international educational authorities. This calls for the redefinition of learning outcomes and quality benchmarks that are inclusive of AI-mediated instructional design (OECD, 2021).

Table 3.3 Ethical and operational challenges in MI-driven education systems

Challenge	*Description*	*Suggested Mitigation Strategy*
Algorithmic Bias	Discrimination based on demographic features	Diverse datasets and bias audits
Data Privacy	Sensitive student information misuse	Federated learning, General Data Protection Regulation (GDPR) compliance
Teacher Deskilling	Overreliance on automation	Professional development and hybrid models
Tech Infrastructure Gap	Lack of access in rural/low-income areas	Low-cost devices, government partnerships
Model Transparency	Black-box nature of deep models	Explainable AI, transparent decision trees

6.2 Role of Governments and Educational Technology Bodies

Governments play a foundational role in creating policy blueprints that integrate MI into national education strategies. For instance, India's National Education Policy (NEP 2020) outlines provisions for AI in teaching and learning, while the European Commission's "Digital Education Action Plan" emphasizes the safe and ethical adoption of emerging technologies in education (European Commission, 2022). To operationalize these visions, national education technology agencies and quality assurance bodies must (i) define legal frameworks for student data usage and algorithmic transparency; (ii) set procurement standards for EdTech solutions embedding MI; (iii) provide funding incentives for pilot programmes that demonstrate equitable MI-based learning. Such coordinated interventions prevent the rise of fragmented and unregulated AI-driven educational markets that may exacerbate inequality or propagate opaque systems.

6.3 Global Institutional Governance: UNESCO, OECD, and Others

International organizations such as UNESCO, OECD, and the World Economic Forum are actively shaping the discourse around ethical and inclusive AI in education. UNESCO's 2021 publication on "AI and Education: Guidance for Policymakers" advocates for human-centric and rights-based AI integration in curriculum, while OECD's framework for AI in education promotes transparency, interoperability, and lifelong learning models (UNESCO, 2021; OECD, 2022). These institutions serve multiple roles such as,

- harmonizing global standards for AI tools used in education;
- providing capacity-building programmes for low- and middle-income countries; and
- facilitating intergovernmental cooperation and data-sharing protocols for AI ethics and governance in education.

Furthermore, international collaborations can help address cross-border challenges such as digital sovereignty, intellectual property of generative content, and ethical deployment of learning analytics systems.

6.4 Need for Ethical and Algorithmic Standardization

Given the opaque nature of many ML models, there is a growing consensus that algorithmic standardization is essential for trust and fairness in MI-based education systems. This includes a few points such as:

- establishing standards for bias detection, fairness metrics, and interpretability of MI algorithms;
- creating regulatory sandboxes where innovative educational AI systems can be tested under controlled environments before wide-scale rollout;
- promoting open-source and audit-friendly AI tools in publicly funded institutions.

Organizations such as IEEE and International Organization for Standardization The Artificial Intelligence (AI) for Kindergarten to 12th Grade initiative are already working on technical standards specific to educational contexts, including curriculum-level integration of algorithmic thinking and ethical reasoning. Policy, governance, and standardization form the scaffolding necessary to responsibly integrate MI into the educational mainstream. By aligning local educational reforms with international frameworks, institutions can create equitable and future-ready learning ecosystems. Without these safeguards, even the most advanced MI technologies risk reinforcing existing disparities or introducing new forms of algorithmic injustice in education.

7. Integrating Machine Intelligence into Curriculum and Adaptive Learning Systems

As the educational paradigm evolves from static content delivery to personalized and adaptive learning ecosystems, MI is now a transformative enabler. This section explores how MI technologies are being harnessed to redesign curriculum frameworks and implement responsive, learner-centred pedagogical systems.

7.1 Integrating Machine Learning, Data Analytics, and Generative AI into Curriculum and Learning Design

Traditional curriculum models were designed for standardization and mass delivery, often failing to adapt to individual learner needs or interests. In contrast, MI-based approaches offer a paradigm shift by enabling dynamic, data-informed, and learner-specific curriculum pathways (Holmes & Tuomi, 2022). ML models trained on historical learning data can uncover patterns in learner behaviour, preferred modalities, and knowledge gaps. These insights can then inform AI-assisted content curation, aligning learning materials with students' cognitive profiles and learning trajectories. Generative AI, such as LLMs, further enhances content adaptability by producing on-demand explanations, examples, and formative quizzes tailored to curriculum goals (Lu and Wang, 2023).

Moreover, curriculum mapping tools embedded with NLP capabilities can scan syllabus and instructional objectives to ensure alignment with learning standards and competencies. For instance, AI can suggest course-level revisions when policy standards (e.g., Bloom's Taxonomy or Common European Framework of Reference for Languages [CEFR] levels) change or when students consistently underperform in a topic cluster. An essential component is data analytics integration which could include for instructors which allows real-time monitoring of learner progress, enabling micro-adjustments to instructional pacing, depth, and modality. Collectively, these innovations signal a move from "one-size-fits-all" syllabus to intelligent, evolving learning frameworks.

7.2 Machine Intelligence–enabled Adaptive Learning Systems in Education

Unlike traditional systems that follow predefined instructional flows, MI-enabled platforms adjust learning content, pacing, and strategy based on real-time insights into the learner's cognitive and behavioural profiles (Zhai et al., 2021). Central to these systems is the dynamic learner model as a representation that reflects not only academic performance but also user engagement, emotional signals, learning preferences, and contextual factors such as time of access or device usage.

These platforms personalize learning trajectories using various algorithmic techniques. For example, a student facing difficulty in mathematical reasoning might receive content modified in both complexity and representation, such as shifting from textual explanations to visual aids or any gamified exercises. On the other hand, a student performing well may be presented with special enrichment modules, exploratory assignments, or adaptive challenges that promote his deeper learning without repetition. Several educational platforms exemplify these approaches. ITS like Carnegie Learning's MATHia, and large-scale adaptive tools such as Knewton and Squirrel AI, incorporate methods like reinforcement learning, Bayesian knowledge tracing, and advanced learner analytics to deliver targeted instruction (Chen et al., 2023). Beyond customization, such adaptive systems promote educational "inclusiveness". By responding to the varied needs of learners if stemming from language barriers, learning differences, or inconsistent access, they help bridge equity gaps. Educators are now thus empowered not merely as content providers but also as facilitators of individualized learning journeys, supported by intelligent systems.

Ultimately, adaptive learning and curriculum innovation together constitute two prime pillars of education reform through MI. While curriculum reforms focus on evolving content and learning outcomes, adaptivity ensures that the delivery of this content is responsive, equitable, and engaging for all learners.

8. Future Outlook and Research Gaps

The trajectories of MI are set to redefine not only how learning is delivered but also how it is conceptualized, governed, and continuously improved. However, this transition is neither uniform nor without challenges. This section outlines the forward-looking trends that are likely to shape the educational landscape and highlights critical gaps that require urgent scholarly and institutional attention.

8.1 Future Trajectories of Machine Intelligence in Education

The immediate future of MI in education is poised for several paradigm-defining developments. First, the rise of multimodal learning environments which integrates audio, video, haptics, and text is expected to gain traction, allowing learners to engage across sensory modalities and learning styles (Li and Zhao, 2023). Second, context-aware learning systems, leveraging data from IoT devices, geo-location, and real-time emotional cues, are likely to personalize learning experiences beyond the screen, integrating seamlessly into daily activities (Wang and Lin, 2023). Another promising trajectory involves federated learning and privacy-preserving AI in education, which will allow training of adaptive models across institutions without sharing sensitive student data (Chandra & Bhatia, 2022). These advancements will be crucial in addressing privacy concerns while still benefiting from collaborative AI systems.

In parallel, AI-based formative assessment is likely to become more nuanced and able to assess soft skills such as creativity, collaboration, and ethical reasoning through simulations, game-based environments, and natural language analysis (Raj & Ghosh, 2022). This will represent a move towards whole-person education, essential for twenty first–century learning outcomes.

8.2 Key Research Gaps and Challenges

Despite the rapid proliferation of MI tools and platforms, there are significant research gaps too, such as:

- *Theory–Practice Disconnect*: Much of the current discourse is technologically driven, with limited grounding in educational psychology or learning sciences. There is a need for interdisciplinary research that integrates pedagogical theory with AI/ML capabilities.
- *Algorithmic Bias and Equity*: Models trained on homogeneous or biased datasets can perpetuate inequalities in learning outcomes, particularly for marginalized or linguistically diverse learners (Holmes and Zhang, 2023).
- *Longitudinal Impact Studies*: Most studies evaluate AI interventions over short durations. There is limited evidence about long-term impacts on learning efficacy, cognitive development, or socio-emotional growth.
- *Educator Readiness and Professional Development*: Teachers remain underprepared to interpret AI-generated insights or to work collaboratively with AI systems. Thus, there is an urgent need for new models for professional training.
- *Ethical and Governance Frameworks*: There is an absence of globally agreed-upon ethical standards for AI use in education. Issues around consent, data ownership, surveillance, and algorithmic transparency remain inadequately addressed or un-addressed (Selwyn & Jandri´c, 2022).

8.3 Towards a Research and Innovation Agenda

To meaningfully advance the field, a coordinated research and innovation agenda must be adopted. Such an agenda should focus on:

1. co-design of AI systems with educators, students, and policymakers to ensure contextual fit and pedagogical alignment;
2. development of open, diverse, and representative datasets for educational AI training and validation;
3. investment in explainable and transparent AI systems that can justify recommendations and facilitate trust;
4. building global coalitions (e.g., UNESCO AI4EDU) to foster equitable and sustainable AI adoption across educational contexts.

Certainly, MI is not merely a tool for education but it is also reshaping the fundamental grammar of learning. Its future depends on our ability to navigate its technical potential with critical pedagogy, robust governance, and inclusive innovation. By addressing the gaps outlined above, researchers and practitioners can ensure that the next generation of educational systems is not only intelligent but also just, humane, and contextually responsive.

9. Summary

The convergence of MI and education marks a transformative phase in the history of learning systems. As demonstrated throughout this chapter, the evolution from industrial-age pedagogy to information-age and intelligence-driven education is not merely about digitization but also about redefining the entire architecture of learning. MI which comprises adaptive systems, cognitive computing, NLP, emotional intelligence, and data-driven personalization offers unprecedented opportunities to reimagine curriculum design, pedagogical strategies, learner engagement, and institutional governance. While early applications of AI in education were centred on automating specific tasks, the emergence of MI has marked a significant paradigm shift. MI enables the development of holistic, responsive systems that are contextually aware and emotionally attuned. However, this transformation does have its own challenges. The deployment of MI in education structure raises critical concerns, including algorithmic bias, disparities in digital access, insufficient educator preparedness, and, most importantly, fragmented regulatory frameworks. This chapter has underscored the importance of addressing both technological capabilities and the broader socio-technical and governance issues that shape the successful and ethical integration of MI in education systems.

To realize the full potential of MI in education, a collective and multidisciplinary effort is required. A multidisciplinary approach from technologists, educators, learners, and policymakers must develop open standards, foster inclusive innovation, and advance research grounded in real-world educational needs. The evolution brought about by MI is not intended to replace educators but to support and enhance their roles through intelligent systems that are empathetic, adaptive, and grounded in human values. At its heart, the rise of MI indicates a shift not only in educational technology but also in how we conceive and design future-ready, resilient, and learner-centred educational cultures.

References

Anderson, T. and Dron, J. (2011). The three generations of distance education pedagogy. *International Review of Research in Open and Distributed Learning*, *12*(3): 80–97.

Berendt, B. (2022). AI ethics and explainability in educational technology. *AI & Society*, *37*(3): 1231–1246.

Burrows, S. and Gurevych, I. (2015). Automated essay scoring using NLP. *Computational Linguistics*, *41*(4): 459–478.

Campbell, J.P. and Oblinger, D.G. (2007). Predicting student success in higher education. *EDUCAUSE Quarterly*. Available at https://er.educause.edu/articles/2007/1/predictingstudent-success-in-higher-education (accessed on 06 June 2025).

Chandra, S. and Bhatia, R. (2022). Federated learning in education: Privacy-preserving AI for student modeling. *Journal of Educational Data Mining*, *14*(3): 1–20.

Chen, J. and Lin, G. (2018). AI in education: An expanding role for administration. *Educational Management Administration & Leadership*, *46*(4): 589–602.

Chen, L., Wang, Y. and Shao, Y. (2023). Adaptive learning systems using AI: A systematic review and future research agenda. *Educational Technology Research and Development*, *71*(2): 411–437.

D'Mello, S. and Graesser, A. (2014). Affective tutoring systems: Modeling learner emotion. *International Journal of Learning Technologies*, 4(1): 57–72.

European Commission (2022). European Commission: Directorate-General for Education, Youth, Sport and Culture, Digital education action plan 2021-2027 – Improving the provision of digital skills in education and training, Publications Office of the European Union, 2023, https://data.europa.eu/doi/10.2766/149764

European Commission. (2023). AI act proposal: Regulation on artificial intelligence. Available at https://artificial-intelligence-act.eu (accessed on 11 June 2025).

Fischer, G., Redmiles, D. and Sumner, T. (2020). Machine learning for education: Challenges and opportunities. *Journal of Educational Data Mining*, *12*(3): 1–12.

Grawemeyer, B. (2017). Learner modeling for adaptive instructional systems. *IEEE Transactions on Learning Technologies*, *10*(2): 240–252.

Haider, S. and Sharma, P. (2023). Towards inclusive AI for education: Participatory design and data justice. *Journal of Learning Analytics*, *10*(2): 90–108.

Heffernan, N.T. and Heffernan, C.L. (2014). The future of adaptive learning with assistments. *User Modeling and User-Adapted Interaction*, *24*(5): 361–384.

Holmes, W., Bialik, M. and Fadel, C. (2019). *Artificial intelligence in education: Promise and implications for teaching and learning*. Center for Curriculum Redesign.

Holmes, W. and Tuomi, I. (2022). AI and curriculum: Redesigning learning in the age of machine intelligence. *British Journal of Educational Technology*, *53*(5): 1057–1072.

Holmes, W. and Zhang, H. (2023). AI and educational equity: Challenges and opportunities in algorithmic learning environments. *AI & Society*, *38*(1): 55–70.

Holstein, K. and Aleven, V. (2023). Fairness and accountability in AI-powered learning systems. *AI in Education Journal*, *1*(1): 45–67.

Hu, L. and Zhang, Q. (2023). Conversational AI in education: Advances in NLP for intelligent tutoring. *Computers & Education: Artificial Intelligence*, *4*: 100089.

Knox, J. (2022). AI and policy in education: Emerging directions and tensions. *Learning, Media and Technology*, *47*(1): 1–15.

Koedinger, K.R., Anderson, J.R., Hadley, W. and Mark, M.A. (1997). Intelligent tutoring goes to school in the big city. *International Journal of Artificial Intelligence in Education*, *8*(1): 30–43.

Li, J. and Zhao, M. (2023). Multimodal learning in education: An empirical review and future directions. *Computers & Education*, *190*: 104707.

Li, Y. and D'Mello, S. (2023). Multimodal learning analytics: Capturing engagement, cognition, and emotion in smart classrooms. *IEEE Transactions on Affective Computing*. Early Access.

Lu, S. and Wang, J. (2023). Generative AI in education: Opportunities and challenges of GPT-based tools. *Computers and Education: Artificial Intelligence*, *4*: 100136.

Luckin, R., Holmes, W., Griffiths, M. and Forcier, L.B. (2016). Intelligence unleashed: An argument for AI in education.

Mitchell, M. (2019). *Artificial Intelligence: A Guide for Thinking Humans*. Farrar, Straus and Giroux.

Nguyen, H. and Kumar, P. (2022). Edge intelligence in smart classrooms: Enhancing privacy and responsiveness in educational applications. *IEEE Transactions on Learning Technologies*, *15*(2): 109–120.

OECD. (2021). Artificial intelligence in education: Promises and implications for teaching and learning. OECD Education Working Papers, 239.

OECD. (2022). OECD framework for classifying AI use in education. OECD Education Policy Perspectives, *48*.

Papamitsiou, Z. and Economides, A. (2022). Educational data mining and learning analytics: Trends and opportunities in the post-pandemic world. *British Journal of Educational Technology*, *53*(4): 776–795.

Porayska-Pomsta, K., Mavrikis, M. and Pain, H. (2013). Affectsensitive intelligent tutoring systems. In *Proceedings of the International Conference on Artificial Intelligence in Education* (pp. 23–31). Springer.

Raj, A. and Ghosh, R. (2022). Assessing soft skills with AI: Toward automated evaluation of creativity and collaboration. *British Journal of Educational Technology*, *53*(6): 1341– 1360.

Robinson, K. (2011). *Out of Our Minds: Learning to Be Creative*. Capstone.

Roschelle, J. and Pea, R. (2013). Personalized learning: A framework for inquiry and implementation. *Journal of Learning Sciences*, *22*(3): 428–477.

Selwyn, N. and Jandrić, P. (2022). The ethics of artificial intelligence in education: Critical perspectives and emerging frameworks. *Learning, Media and Technology*, *47*(1): 1–13.

Shute, V.J. (2008). Focus on formative feedback. *Review of Educational Research*, *78*(1): 153–189.

Siemens, G. (2005). Connectivism: A learning theory for the digital age. *International Journal of Instructional Technology and Distance Learning*, *2*(1): 3–10.

Sümer, Ö., Goldberg, P., D'Mello, S., Gerjets, P., Trautwein, U. and Kasneci, E. (2021). Multimodal engagement analysis from facial videos in the classroom. IEEE Transactions on Affective Computing, 14(2), 1012–1027.

Truong, K. and Park, A. (2022). Student surveillance or support? Ethical implications of sensor-based learning systems. *British Journal of Educational Technology*, *53*(5): 1182–1198.

UNESCO. (2021). *AI and Education: Guidance for Policymakers*. United Nations Educational, Scientific and Cultural Organization.

UNESCO. (2022). Guidance for the ethical use of artificial intelligence in education. Available at https://unesdoc.unesco.org/ark:/48223/pf0000380803 (accessed on 11 June 2025).

Wang, P. and Lin, Z. (2023). Context-aware AI for education: Bridging physical and virtual learning spaces. *IEEE Transactions on Learning Technologies*, *16*(2): 198–210.

Winkler, R. and Soares, W. (2020). Artificial intelligence for teaching assistants: Challenges and opportunities. *Journal of Educational Technology Systems*, *49*(3): 474– 490.

Yang, J. and Sun, X. (2023). Generative AI in education: Potentials, pitfalls, and practical guidelines. *Educational Technology Research and Development*, *71*: 1145–1170.

Zelinger, B. and Tan, Y. (2022). Ethical issues in AI-powered education: What should educators and designers consider? *Computers & Education: Artificial Intelligence*, *3*: 100047.

Zhai, X., Chu, S.T. and Wang, M. (2021). Artificial intelligence in adaptive learning: A review of empirical studies. *Journal of Educational Computing Research*, *59*(6): 1135–1163.

Zou, X., Ma, W. and Baker, R.S. (2019). Helping teachers select optimal content for students: A data-driven approach. In: Isotani, S., Millán, E., Ogan, A., Hastings, P., McLaren, B., Luckin, R. *Proceedings of the International Conference on Artificial Intelligence in Education (AIED)* (pp. 133–142). Springer.

Chapter 4

Artificial Intelligence in Education Technologies: New Developments and Innovative Practices

Kanimuthu A. R. M[1], *Navaneethakrishnan M. V.*[1], *Nishath J. P.*[1], *Devi Priya R.*[1*] and *Keshav Dahal*[2]

1. Introduction to AI in Education Technologies

In today's age of speedy education revolution, transformation of teaching and learning processes is being increasingly influenced by the developments in artificial intelligence (AI). Instead of remaining limited to exploratory domains or expensive research institutes, AI has now been incorporated in the mainstream learning sphere, wherein it is helping in meeting complex and diverse scholarly demands (Correia et al., 2024). With the conventional forms of education increasingly put to test by the imperatives of personalization, skilfulness, and universal access, scalable AI-based solutions are being embraced to improve efficiency and student focus.

The presence of AI is visible across the length and breadth of the realm of education. From the use of machine learning (ML) algorithms for processing student performance data to the implementation of natural language processing (NLP) in conversational tutoring systems and language-learning applications, AI is being leveraged to provide a range of instructional support. Visual recognition functionalities driven by computer vision are being used to enable real-time feedback, automated grading, and accessibility enhancements. At the same time, reinforcement learning has been used to enable adaptive learning environments and gamified learning experiences that ensure learner engagement. With these cornerstone technologies, interactions among students, teachers, and learning content are being redefined at a foundational level.

This next section discusses how the core AI technologies—ML, NLP, computer vision, and reinforcement learning—are shaping pedagogical change. The third

[1] Department of Computer Science and Engineering, KPR Institute of Engineering and Technology, Tamil Nadu, India.

[2] School of Computing, Engineering and Physical Sciences, University of the West of Scotland, UK. ORCID ID of Devi Priya R is 0000-0002-6216-2166[1].

* Corresponding author: scrpriya@gmail.com

DOI: 10.1201/9781003639367-4

section gives an overview of to the application of AI in the classroom, including intelligent tutoring systems, adaptive learning systems, and AI-facilitated personalized instruction. The recent breakthroughs like generative AI and immersive technologies like augmented reality and virtual reality are also discussed in an attempt to demonstrate how recent breakthroughs continue to expand opportunities for learning. Aside from technology use, focus is laid heavily on core subjects like data privacy, algorithmic discrimination, and fair access. Addressing these concerns is crucial to ensure that AI applications in schools are ethically compliant and inclusive (Cukurova, 2024). In the latter part of the chapter, visionary discussion is presented, including the development of new fields of research, incorporation of pedagogical policies, and strategic initiatives at the global level to enable responsible and scaled-up use of AI technologies. Current studies focus on the need for creating ethical guidelines that maintain AI integration into education as transparent, accountable, and in line with instructional best practices (Porayska-Pomsta et al., 2024).

Overall, the chapter aims to offer a balanced and nuanced description of AI's impact on today's education. By placing the tools and technologies in context, as well as their implications, benefits, and other factors, the discussion aims to prepare the educators, technology developers, and policymakers with a level-headed and informed understanding of the changing landscape of AI's influence on learning environments.

2. Foundational AI Technologies Driving Modern Education

The introduction of AI to the learning ecosystem has transformed traditional models of learning and instruction, offering unprecedented potential for adaptive learning, automated grading, and personalized learning experiences. As learning organizations globally embark on digital transformation, educators, administrators, and policymakers need to understand the basic AI technologies that underpin these innovations. This section addresses four core AI technologies that have been found to be significant in learning institutions: ML, NLP, computer vision, and reinforcement learning (see Figure 4.1).

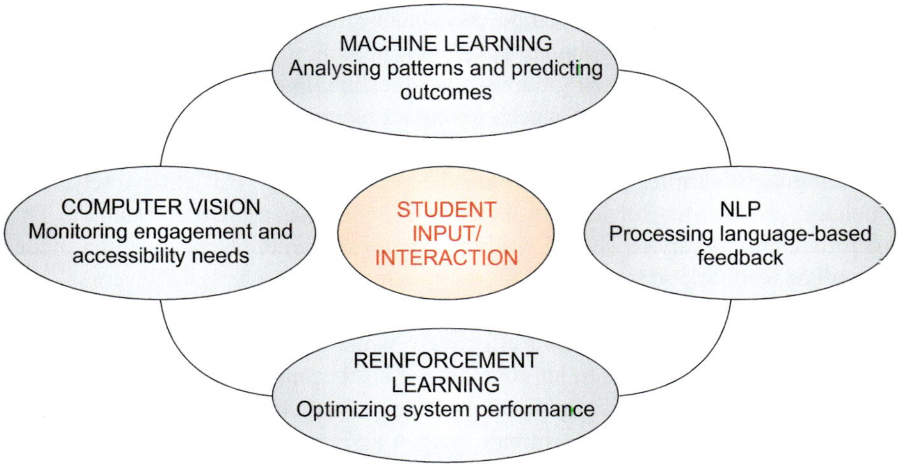

Fig. 4.1 Interconnection between AI models.

2.1 Learning Patterns Through Algorithms: Machine Learning's Educational Role

Machine learning (ML) represents the foundation of technology that enables computers to learn patterns from the given data without explicitly programming for every scenario into it. In educational contexts, the ML algorithms analyse large amounts of student data which are collected through interaction with the students to identify learning patterns, predict academic outcomes, and customize educational experiences based on individual outcomes.

The most important application of ML in the education field is through adaptive learning platforms that can adjust content difficulty and presentation style based on individual student performance. Carnegie Learning's MATHia platform exemplifies this approach by continuously analysing student responses to mathematical problems, identifying knowledge gaps, and providing targeted interventions. The system employs intellectual modelling techniques to understand how every student learns mathematical concepts and adapts to the instruction accordingly. Predictive analytics represents another significant ML application, where algorithms analyse historical student data to identify at-risk learners before they fail. In recent times, applications of ML algorithms and methods at the level of higher education show that the predictive power of the technology has been enhanced compared to the previous times. For instance, Advanced eAdvisor systems have been created to examine students' course-taking patterns, grades, and demographic data, which allows the institutions to make predictions about students' success rates and prescribe the right interventions for the students (Essa and Ayad, 2023). The advantages of ML in learning include improved personalization, the early detection of learning challenges, and data-based decision support for curriculum development. Despite this, there are several issues that need to be addressed such as data privacy breaches, algorithmic bias that could potentially lead to a disadvantage for certain student groups, and the necessity for large investments in this technology infrastructure to properly implement the ML systems.

2.2 Natural Language Processing for Educational Enhancement

Technology such as NLP helps computers to enable the capabilities of understanding, interpreting, and generating human languages, making it a valuable technology for language learning, automated grading, and other intelligent tutoring systems. This technology processes text and speech to extract meaning, sentiment, and linguistic patterns that inform educational interventions.

Language learning platforms have been embracing NLP extensively, with Duolingo (a tool used for learning languages) serving as a premier example. This platform employs various NLP algorithms to analyse learner responses and provides immediate feedback on pronunciation and grammar; it also adapts lesson complexity based on individual progress. Speech recognition capabilities of this tool allow the students to practise pronunciation effectively while also receiving some real-time corrections and suggestions to improve; thus it has become one of the popular tools that uses NLP. Automated essay scoring systems represent another significant NLP application, where algorithms evaluate written assignments based on criteria such as coherence, vocabulary usage, and argument structure. These NLP-based engines illustrate how the technology can offer prompt, consistent feedback on student writing and pinpoint specific areas where improvement is needed (Hartman, 2024).

Intelligent tutoring systems capitalize on NLP to teach students with natural language conversations, responding to questions and offering explanations through human-like dialogue. Such systems are capable of examining student inquiries in terms of conceptual challenges and offering proper scaffolding assistance. The benefits of NLP in teaching are scalable language models, instant feedback supply, and lower grading effort for teachers. However, its limitations are resistance to processing colloquial speech, linguistic and cultural bias in language models, and weak handling of contextual and subtle meaning in students' responses.

2.3 Computer Vision Applications in Educational Settings

The computer vision technology gives power to the computers to read and interpret visual information from user input like images and videos, thereby opening up new possibilities for educational assessment. Visual information is interpreted to recognize objects, faces, gestures, and activity, providing valuable insights into student learning and engagement behaviour. By using automated proctoring systems using computer vision, we can observe remote tests and look for possible cheating techniques like straying away from the screen, unauthorized materials, or multiple people taking the test. As controversial as these tools are, Proctorio and Responds Monitor have insinuated themselves amid times of remote instruction, but they raise legitimate concerns about privacy and equity.

Computer vision has helped with mathematics education through the applications of recognition of handwritten equations and step-by-step solving. Microsoft Math Solver and other such apps can solve problems from handwritten mathematical expressions, offer step-by-step solutions, and describe solution steps, thus promoting student learning outside of the classroom. Machines are able to translate visual information to audio descriptions for visually impaired students, detect sign language for deaf students, and provide real-time text recognition from printed materials. The benefits of computer vision are increased access for disabled students, objective observation of behaviour for research, and innovative evaluation methods that could be more effective than the design of conventional testing. The drawbacks include privacy invasion concerns, technical problems in recognizing intricate handwriting patterns, and inherent prejudice in face recognition systems that can bias the rights of certain ethnic groups.

2.4 Reinforcement Learning for Adaptive Educational Systems

Reinforcement learning (RL) is an AI paradigm in which systems learn optimal action through error-and-trial interactions with their environment and receive rewards or penalties as performance outcomes. In learning environments, RL allows systems to learn successful teaching methods by constantly modifying model instruction based on learning results from students.

Adaptive testing systems make use of RL to streamline question selection in tests, selecting the items which yield greatest information regarding ability levels of students without taking much time for testing. The systems learn from patterns of responses of students to recommend better questions in the future and give precise ability estimates. Personalized curriculum sequencing is another application of RL where systems can learn the best content sequencing for individual students. Through observation of which content sequences can result in improved learning, RL algorithms can suggest personalized learning pathways that maximize the educational impact for each learner's individual set of characteristics and preferences. Recently, empirical research has shown that RL-based tutoring systems can bring about substantial improvements in students' learning outcomes. Research shows that adaptive systems produced learning gains of around 23% over more conventional fixed-sequence instructional methods, showing how the technology can increase educational efficiency (Nguyen, 2024). The benefits of RL are systems that are really adaptive and get better over time, optimal allocation of resources for maximum learning leverage, and data-driven discovery of effective teaching. The challenges include the need for long periods of data collection before systems are effective, challenges in specifying good reward functions for complicated learning goals, and systems optimizing for easily quantifiable goals while overlooking key but harder-to-measure education goals.

2.5 Implementation Evidence and Future Directions

Empirical research has shown that measurable learning results can be improved with the use of AI technology. A comparative analysis of core AI technologies is given in Table 4.1.

With reference to Chen et al. (2020), the integration of artificial intelligence in education has significantly enhanced administrative efficiency, customized learning and instructional quality. In the same way, comparative research of NLP-based language learning software has indicated 34% higher vocabulary learning rates of users compared to conventional instruction (Vesselinov & Grego, 2023). These foundational AI technologies are also evolving rapidly, with new applications to virtual reality instruction, emotion recognition for assessment of engagement, and multimodal learning analytics integrating multiple AI paradigms. As these technologies evolve, their integration into learning architecture should be simpler and more streamlined, but they must be approached with caution and sensitivity to issues of ethics, equity, and pedagogical integrity to ensure effective implementation.

3. Innovative AI Applications in Education

This section discusses how the reshaping of the educational system is done with the help of AI technologies. There used to be a situation in the past when the students would depend on how technology would reshape educational delivery and outcomes across diverse learning contexts.

3.1 Intelligent Tutoring Systems (ITS)

Intelligent tutoring systems are one of the most sophisticated applications of AI within the education system, acting as sophisticated electronic mentors providing customized help and guidance. These systems utilize cognitive science principles and methods and AI techniques to construct learning environments that respond dynamically to

Table 4.1 Comparative analysis of core AI technologies

Technology	*Primary Purpose*	*Educational Use Cases*	*Key Benefits*	*Major Limitations*
Machine Learning	Pattern recognition and prediction	Adaptive learning platforms, predictive analytics, personalized recommendations	Customized instruction, early intervention identification, data-driven insights	Data privacy concerns, algorithmic bias, infrastructure requirements
Natural Language Processing	Language understanding and generation	Language learning apps, automated grading, intelligent tutoring	Scalable language instruction, immediate feedback, natural interaction	Cultural bias, context limitations, colloquial language challenges
Computer Vision	Visual information processing	Automated proctoring, handwriting recognition, accessibility tools	Enhanced accessibility, objective analysis, innovative assessment	Privacy invasion, recognition accuracy issues, potential demographic bias
Reinforcement Learning	Optimal action discovery	Adaptive testing, curriculum sequencing, strategy optimization	Truly adaptive systems, optimized learning paths, continuous improvement	Data collection requirements, reward function complexity, narrow optimization focus

learners' needs in real time, much like human teachers, but with the added advantage of round-the-clock availability and consistency. ITS architecture typically consists of four dependent components: a subject matter expert-based domain model, a student model that tracks individual learning, a pedagogical model that defines pedagogical strategy, and an interface model that controls student interaction. Integrated systems allow systems to diagnose the learning issue, select appropriate instructional material, and provide individualized feedback based on student behaviour.

Cognitive Tutor created by Carnegie Learning is perhaps the best example of the potential of ITS in mathematics learning. The software applies cognitive task analysis to learn the methods that students use in answering mathematics problems, recognizing common mistakes made by students and offering immediate feedback. More recent studies with cognitive tutoring systems have shown they are valuable tools in mathematics education. The research shows that students who use these AI-based tools have strong performance gains on standardized assessments, with effect sizes between 0.2 and 1.0 standard deviations compared to more traditional forms of instruction (Ritter et al., 2023). ITS applications today extend beyond mathematics to physics, computer programming, and language arts. These programs help in continued learning and show significant improvement in their understanding of individual learners through ML algorithms that monitor patterns of interaction, response times, and error rates in an attempt to provide better instruction.

3.2 Individualized Learning Platforms

The platforms that use AI to develop tailored educational experiences based on students' learning style and learning preferences come under individualized learning platforms. Platforms evaluate each student's performance information and make adjustments in content delivery, complexity level, and teaching methods in real time by using advanced algorithms which makes them distinguishable from the conventional one-size-fits-all strategy. The base level of tailored learning comes under complete learner analysis, where the AI systems collect and process multiple data points like academic grades, engagement levels, learning styles, and behavioural trends. This data is fed into recommendation engines that propose best learning pathways and activities for every student.

Khan Academy's personalized learning dashboard demonstrates this approach effectively by keeping track of student progress across multiple subjects and identifying knowledge gaps. The platform's AI algorithms suggest specific practice exercises, instructional videos, and remediation activities based on an individual's performance patterns. Recent implementation studies on individualized learning platforms have shown important results. This analysis indicates that students utilizing adaptive learning features have 1.5 times greater learning gains compared to students following standard curricula approaches (Roschelle et al., 2024). DreamBox Learning is another successful implementation in elementary mathematics education. This platform adapts in real time to student responses, adjusting problem complexity, providing hints, and selecting appropriate visual representations based on individual learning patterns. Longitudinal studies of adaptive mathematics platforms have revealed significant benefits for consistent users. Research has proven that students learning with help from these systems for at least 60 minutes per week demonstrate prominently higher achievement gains on assessments (Huang et al., 2023).

3.3 AI in Language Learning

Artificial intelligence has re-evolved language learning by providing excessive and interactive experiences that simulate natural language acquisition processes. Language learning applications combine speech recognition, NLP, and adaptive algorithms with AI to create personalized instruction that responds to individual pronunciation and grammar development needs.

Similar language learning platforms use sophisticated speech recognition technologies to evaluate pronunciation accuracy and provide immediate feedback. These systems analyse acoustic features of student speech, comparing their speech with native speaker models and identifying specific areas for improvement. NLP algorithms test the written responses for grammar accuracy, vocabulary usage, and semantic coherence. Duolingo's AI-driven approach exemplifies innovative language learning methodologies through its use of spaced repetition algorithms and adaptive difficulty adjustment. The platform's AI system analyses individual forgetting curves for vocabulary items and optimizes review schedules to maximize retention. Conversational AI tutors represent an emerging frontier in language learning, where students engage in dialogue with AI-powered virtual conversation partners. These systems use advanced natural language understanding to maintain context-based conversations while providing few corrections and language modelling. Companies like Speeko and Elsa Speak have demonstrated the effectiveness of AI-powered pronunciation coaching, with users showing measurable improvement in their accent reduction and fluency development.

3.4 AI for STEM Education

Science, technology, engineering, and mathematics (STEM) education has adopted AI applications tackling the multifaceted, multi-layered nature of STEM learning. Such implementations vary from simulated laboratory environments to smart problem-solving aides that walk students through intricate mathematical and scientific reasoning steps. AI-based STEM education platforms usually embed simulation environments where students perform virtual experiments, control variables, and view outcomes in simulated digital environments. The systems hold physics engines and mathematical modelling to establish realistic experimental conditions and offer AI tutors that assist in hypothesis building, interpretation of data, and conclusion.

The PhET Interactive Simulations programme at the University of Colorado Boulder helps in visualizing how AI can augment physics and chemistry learning through adaptive simulations. The tools adapt levels of complexity by student interaction and offer intelligent suggestions when students get stuck in learning. Implementation studies by Perkins et al. (2024) proved that students working with PhET simulations evinced far greater conceptual understanding than those working with normal laboratory experiences. In mathematics learning, AI tools such as Assessment and Learning in Knowledge Spaces (ALEKS) apply advanced algorithms to plot student knowledge and determine the best order of learning. The system constantly monitors student comprehension and adjusts teaching in response, requiring prerequisite concepts to be mastered prior to covering more advanced topics. Research by Fang et al. (2023) illustrated that students who learned with ALEKS exhibited significant improvements in algebraic reasoning and problem-solving abilities.

3.5 AI-enhanced Gamified Learning

The integration of AI and game-based learning has resulted in compelling learning experiences that encourage learners through interactive challenges and tailored reward systems. These AI programs study the players' pattern of behaviour to fine-tune game mechanics, allowing for proper levels of challenge to keep them interested and encourage them to play again and again without leading them into frustration. AI in gamified learning platforms paves the way for generating dynamic storylines that respond to students' decisions and levels of performance. With the use of RL algorithms, the systems learn what encourages individual students and tune the game's mechanisms that also include story development and reward systems.

DragonBox is a learning game for algebra which illustrates how AI can make abstract mathematical concepts accessible in the form of fun interaction. The game has ML to examine student strategies in problem solving and adjusts the puzzle difficulty to keep it at optimal levels. Students learn algebraic manipulation by playing symbolic games so that they acquire knowledge of mathematical techniques. Classcraft is a gamified classroom management system that uses AI to monitor student engagement and their characteristic patterns.

The role of AI technology in education illustrates the revolutionary movement towards more individualized, effective, and inclusive learning processes. Table 4.2 illustrates the applications, key benefits, and challenges of AI implementation in education. Every application presents distinct advantages from the one-to-one instruction to virtual reality immersion. Ubiquitous challenges are found across implementations, which range from data privacy issues and the requirement of significant technical infrastructure to the need for safeguarding human control over the learning processes. These technologies keep changing because of which their implementation will be considered based upon their teacher training and balanced integration in human life instead of teaching human-oriented academic principles.

Table 4.2 Comparative analysis of AI applications in education

Application	*Primary AI Technologies*	*Key Benefits*	*Main Challenges*
Intelligent Tutoring Systems	Cognitive modelling, expert systems, natural language processing	Personalized instruction, immediate feedback, consistent availability	High development costs, domain expertise requirements, limited subject coverage
Personalized Learning Platforms	Machine learning, recommendation systems, learning analytics	Adaptive content delivery, individual pacing, data-driven insights	Data privacy concerns, algorithm transparency, digital divide issues

Contd.

Table 4.2 *Contd.*

Application	*Primary AI Technologies*	*Key Benefits*	*Main Challenges*
AI in Language Learning	Speech recognition, natural language processing, conversational AI	Pronunciation feedback, immersive practice, cultural context integration	Accent bias, limited conversational depth, cultural sensitivity challenges
AI for STEM Education	Simulation engines, knowledge representation, intelligent tutoring	Virtual experimentation, conceptual modelling, safe learning environments	Technical complexity, hardware requirements, authentic assessment difficulties
AI-enhanced Gamified Learning	Reinforcement learning, behavioural analytics systems	Increased engagement, intrinsic motivation, social learning features	Potential distraction, assessment validity, addictive behaviour risks

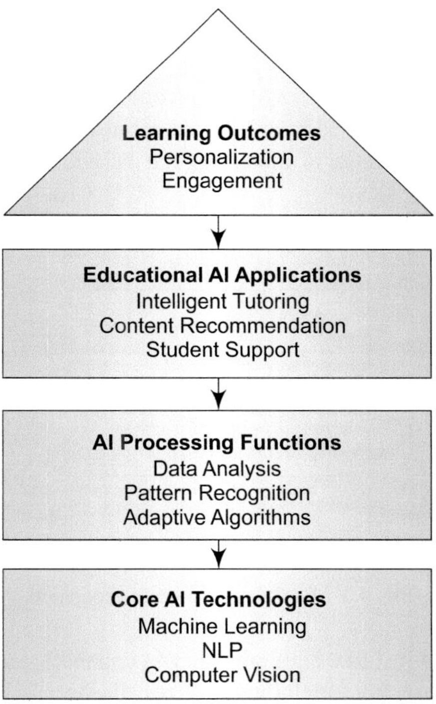

Fig. 4.2 AI integration levels in educational applications.

The framework given in Figure 4.2 depicts how fundamental AI technologies are being utilized to advance educational outcomes. Starting from the foundation level, ML, NLP, and computer vision are used to enable AI processing capabilities like data analysis, pattern detection, and adaptive algorithms. These are utilized through learning AI applications, such as intelligent tutoring systems, content recommendation systems, and learning support tools. In turn, improved learning outcomes like personalization and engagement are obtained.

4. AI in Education Content Creation and Management

In this section, the operation of the AI system will be analysed within the fields of educational content development and administration. The role of AI in changing traditional textbooks to smart textbooks will be highlighted along with how the automation of processes in different fields is enabled.

4.1 Automated Content Generation

Artificial intelligence is dominating the field of content generation with the help of technologies like generative adversarial networks (GANs) and large language models (LLMs) such as GPT-4. The automated creation of educational materials is facilitated with these models, including videos, graphics, quiz questions, and flashcards. In addition, AI can provide customized explanations and generate problems for students based on their needs, as depicted in Figure 4.3.

Key Benefits

- AI reduces the manual effort involved in content creation
- Content can be personalized based on student needs
- A wide range of educational materials can be generated, which may help in creating free textbooks

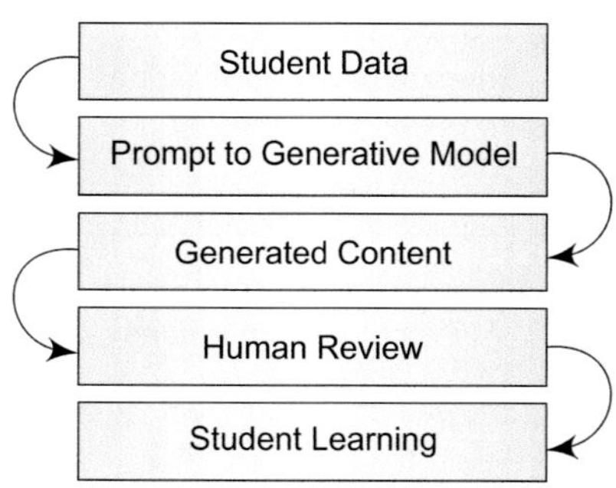

Fig. 4.3 Workflow of content generation.

A Digital Learning Institute highlights that AI enhances value by creating content for digital learning programs, including quizzes, flashcards, simulations, and case studies, thereby accelerating the entire process.

4.2 Smart Textbooks

One of the greatest advancements in traditional education materials is the introduction of smart textbooks along with AI features. This helps the students to incorporate their learning styles and interactions. Digital tools use embedded analytics to cope with student engagement, adjusting content complexity and presentation.

One such innovation is Pearsons's MyLab platform where adaptive algorithms used respond to student performance and identify the knowledge gaps.

As shown in Table 4.3, smart textbooks get instant simulations and feedback from students to create a more effective learning space for students than traditional textbooks. Various subjects that are mainly focused upon in schools have adapted to smart textbooks. Reasonable improvements are made in student performance due to the implementation of AI-enhanced technologies in education. Recent studies have reported that there is an increase in course completion rates of up to 15% when these customizable learning technologies are integrated into curricula (Li & Zhao, 2024).

Table 4.3 Comparison between traditional textbook and smart textbook

Feature	*Traditional Textbook*	*Smart Textbook*
Feedback	End-of-chapter only	Instant adaptive feedback
Multimedia	Static illustrations	Interactive visuals/AR: Augmented Reality
Adaptivity	None	Based on learner progress

4.3 Curriculum Design with AI

A curriculum plan integrated with AI is a learning plan model that relies on ML and data science to optimize sequences of learning, content selection, and skill learning (see Figure 4.4). From this data analysis of student performance and curriculum needs, the systems will provide optimal instruction pathways. Knewton is a company that uses predictive modelling which charts relations between learning concepts that guarantee tailored curriculum suggestions depending on variables like concept complexity and learner preparedness. Colleges that have adopted AI-based curriculum design systems report that there has been an improvement in student results. Recent case studies prove that such technologies are capable of realizing up to a 25% decrease in course completion time while maintaining education quality standards (Johnson et al., 2024).

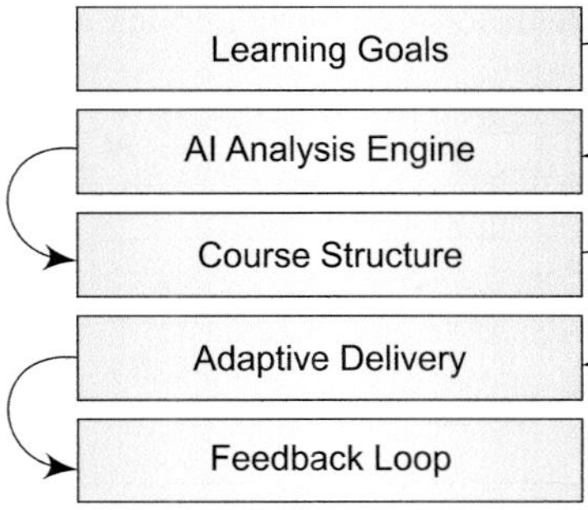

Fig. 4.4 Workflow of curriculum design with AI.

Challenges

- *Content Accuracy and Bias*: It is difficult to ensure accuracy and cultural awareness. The pedagogical suitability of AI content and human monitoring must be used to counteract bias.

- *Digital Equity and Privacy*: Smart textbooks are difficult in regards to digital equity and privacy issues.
- *Technological Hindrances*: The cost for the implementation of these technologies in schools would be huge.
- *Alignment with Local Standards*: Implementation of these technologies in local fields requires human intervention.

5. Recent Trends and New Developments

5.1 AI with AR/VR in Education

The use of AI in education combined with AR and VR helps create an immersive experience for all ages, including children, making them engage without their minds getting distracted. The use of appropriate animations and cartoons helps them learn and grasp content better. It also helps adults in a way that every concept gets embedded into the mind when the subject taught is more practical, real-world experience. AR and VR enable this. AI can be used to manage the user's eye engagement to determine the kind of content they like; this helps to find and deliver the content that the user would prefer. It adjusts the difficulty level and offers content suggestions based on the detection. Companies like VR education and Zspace are at the forefront of this innovation where students can visualize and get immersed in the engaging concepts that once felt hard for them, such as virtual labs, exploring historical events, and visualizing complex concepts.

They are proven to have a significant impact on the learning journey and develop practical skills, particularly in science and technology (Zhang, 2024). A medical student gets a chance to practice surgery with immersive technologies; by repeating this, AI can make sure that he can readily operate with a good confidence level. History is a difficult subject because of its vast content, dates, and names of people. It is often seen just as text, without feeling empathy or a real connection. This lack can be addressed through AR/VR, as represented in Figure 4.5. It makes learning more engageable and it prepares them for real-world applications.

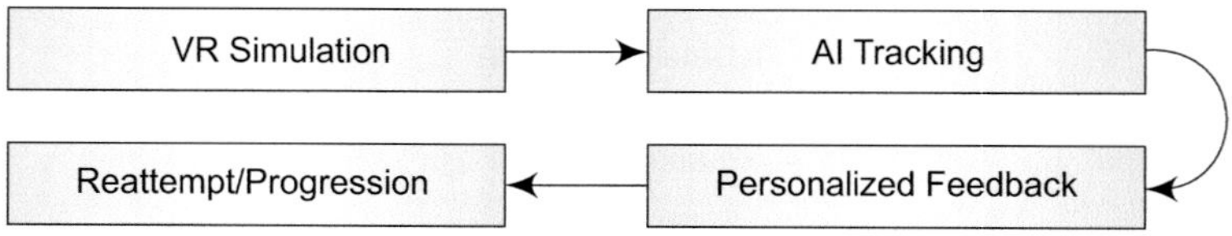

Fig. 4.5 VR-AI feedback mechanism for skill-based learning.

5.2 Use of Generative AI

Generative AI, like Generative Pre-trained Transformer (GPT), is the main player in the AI industry because of the deep knowledge gained through years of training in large corpora. It helps provide tailored responses and feedback to the user in the form of a chat interface, chatbots, voice assistants, and many more. AI-integrated applications like Duolingo make the complex task of learning a language easier and achievable for absolute beginners; Khan Academy's Khanmigo offers support in various subjects. Also, one of the huge benefits is the accessibility of AI solutions 24/7. This allows the student to learn at a faster pace and make decisions frequently whereas in a school education set up students need to wait to get a response from their teacher (see Table 4.4). Also, many researchers have proven that students using AI technologies for their learning tend to have enhanced skills in problem-solving and engagement with learning materials (Chen et al., 2024).

Table 4.4 Application, benefits, and risks of the usage of generative AI

Application	*Benefits*	*Risks*
Content Creation	Time-saving, scalable	May hallucinate facts
Feedback	Immediate, personalized	Context misinterpretation
AI Tutoring	On-demand support	Over-dependency

Even with a large number of benefits, there are some areas where AI lacks such as factual errors, misinterpretation, and over-dependency. It should be used carefully to make the best of it.

5.3 AI in Microlearning and Lifelong Learning

For the current generation, there are plenty of resources available, and everyone is competitive, improving their knowledge day by day unlike the olden days when books were the main source of knowledge. AI helps us to learn the required content and information quickly without wasting a lot of time surfing to find better resources.

Microlearning is the technique wherein, instead of spending a lot of time from our busy schedule, we can ask AI to clear the concepts along with some examples in a shorter way. This makes the content fruitful and helps to improve our knowledge retention capabilities. *Lifelong learning* is the ongoing process of one's pure interest in learning and increasing knowledge and skill. Platforms like Axonify and Grovo are known for delivering useful courses of a shorter time span with quality content, and they also update their content and course duration based on the user interaction. Recent investigations show that students who follow microlearning with the help of AI can enhance their knowledge retention by up to 50% compared to the traditional methods (Kumar & Singh, 2024).

Challenges

- The advanced technological products and solutions are high cost which makes real-world implementation challenging.
- Spending plenty of time motionless and straining of eyes can cause health issues for a user that need to be considered.
- Students rely on AI, making it harder for individuals to improve their own skills.

6. Real World Implementations and Case Studies

6.1 Global Initiatives

Implementation of the AI solution globally, making it a feasible and also the best solution, was a huge task. The Digital Education Action Plan of the European Union

(EU) and the AI and Education initiative by the United Nations Educational, Scientific and Cultural Organization (UNESCO) leverage AI for educational improvement (see Figure 4.6). Many learners today are eagerly looking for an AI-implemented learning experience, as it has a huge role in the global education market. Most of the large-scale implementations done in the field of education have stated that education integrated with AI has a significant improvement in their system. They also documented the performance, stating that performance improvements were up to 30% compared to the traditional way of teaching (Li et al., 2024). The information as mentioned above, are the ones that give significant data on how crucial it is to use AI in education. Meanwhile, Finland's action on this matter gets significant attention. They focused on the ethical consideration of AI implementation and stated the impact that real teachers can have on the students. They demonstrated that AI cannot replace teachers as they help to build creativity in education University of Eastern Finland. (2025, September 3).

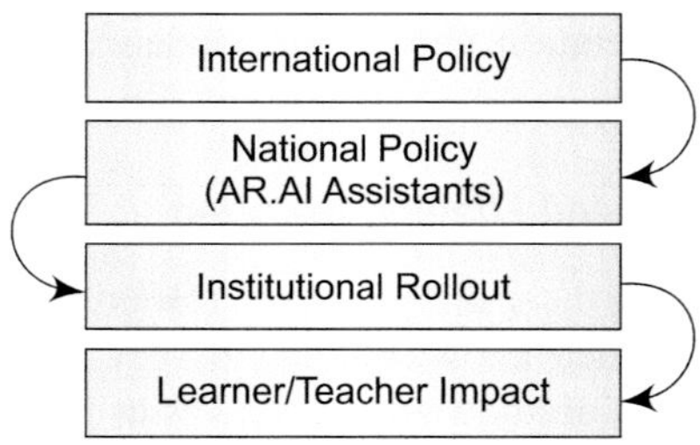

Fig. 4.6 Global AI education strategy.

6.2 National Project (DIKSHA)

India's great initiative named the DIKSHA platform is truly a game-changer in the educational sector. It helps more than 35 million users by delivering educational content seamlessly. This platform also integrates AI to recommend the learning material, understand how the students learn, provide feedback on where they can improve, and provide tailored personalization to each student. One of the main goals of DIKSHA is to bridge the gap in education for every person in a fair way. It involves bringing quality content to every student who is in need of that, regardless of their background. This initiative has brought a great improvement in the lives of many students in the rural areas, making a real impact. As shown in Figure 4.7, the main vision of DIKSHA states that education can be provided to students in rural, hilly regions by means of AI-integrated virtual applications, making the resource available to anyone at any time. Rather, the traditional learning method is not possible in that region. This will make sure that every student in India gets democratized quality access to educational content (Singh & Patel, 2025).

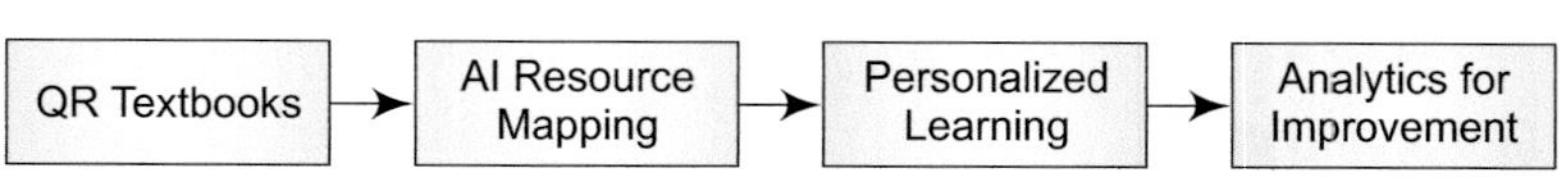

Fig. 4.7 Diksha AI integration.

6.3 EdTech Startups Using AI

The emergence of startups like EdTech based on AI technologies has resulted in a rich and varied ecosystem for educational solutions. A large number of players such as Century Tech, Cognii, and Third Space Learning have created specialized applications focused on particular issues in education. AI-driven learning platforms such as Century Tech combine cognitive science and ML to build customized pathways for students. Their application in the UK schools has seen significant gains, with some schools recording up to 20% improvement in test scores, testifying to the ability of AI to improve learning efficiency by filling learning gaps in real time. AI tutoring system of Cognii's conversational application provides tailored writing instruction and feedback options. As depicted in Figure 4.8, by employing various NLP algorithms, it analyses student responses of each test to offer targeted guidance. This has resulted in improvements in writing skills and overall student satisfaction in higher education. The integration of behavioural analytics in educational platforms has enabled more sophisticated understanding of student engagement patterns, allowing for proactive interventions and personalized support strategies (Chen, 2024). Table 4.5 brings out the point that though generative AI improves content generation, feedback, and mentoring, it also has associated risks such as factual inaccuracies, misinterpretation, and over-reliance. Tactful usage must be ensured so that these advantages and disadvantages are kept in balance.

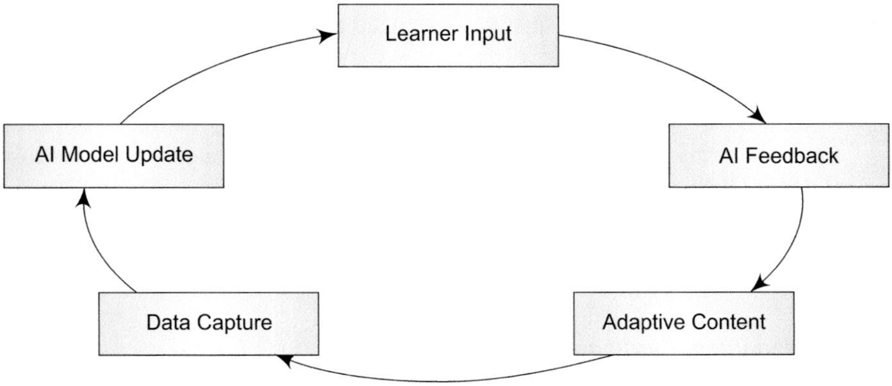

Fig. 4.8 AI EdTech innovation loop.

Table 4.5 Comparison of global initiatives, national projects, and EdTech startups

Category	*Global Initiatives*	*National Projects*	*EdTech Startups*
Scope	Multi-country	Country-wide	Targeted market segments
Stakeholders	NGOs, governments	Ministries, local government	Private sector
Main Benefits	Knowledge sharing	Public access to AI tools	Innovation and customization
Key Challenges	Scale, cultural fit	Infrastructure gaps	Monetization and privacy

Challenges

- Exorbitant development costs and specialized hardware make it difficult for educational applications of new technologies.
- Issues related to motion sickness and digital eye strain must be resolved.
- Academic integrity is the primary concern with potential for students to over-depend on AI.
- Institutions must craft policies that will reconcile the worth of generative AI with the need for critical thinking.
- AI-based microlearning requires extensive resources and surveillance, and these must be coordinated with learning goals.

7. Challenges in Adopting AI in Education

Implementing AI in the education sector offers many benefits but obviously it has a lot of challenges too. The challenges can be summarized as technological, ethical, infrastructural, and pedagogical (see Table 4.6). These should be addressed to apply to AI.

7.1 Technological Challenges

- *Data Quality and Availability*: AI needs a large set of high quality data which may not be practical all the time due to several issues, particularly in villages or yet-to-promote areas. Data quality is the need of the hour as many analyses go wrong if the input is dubious.
- *System Integration*: Combining AI with current education pedagogy, especially with the National Education Policy (NEP 2020), will be very hard to stick to and costly too many times. System integration may lead to disappointment because of current bias.
- *Algorithm Bias*: AI follows some pre-set algorithm which may go wrong sometimes. Though the programming is highly scanned, it still has some drawbacks. The threat is that algorithms can be hacked and lead to bad implementation.

7.2 Pedagogical Challenges

- *Teacher Training*: The teachers who are teaching from kindergarten to college level might not be ready to learn a new technology, especially AI, due to the perceived threat that it will replace them in the near future.
- *Over-reliance on Technology*: Over-dependence on AI will definitely reduce human interaction, which is highly important. The students will not be controlled as a result. Technology may fail sometimes and this should be accepted.
- *Curriculum Adaptation*: The existing traditional curriculum definitely matches with an AI-driven system which requires a much dedicated team to implement to pass many hurdles. Simply describing, it is like pulling a mountain with hands, which is impossible.

7.3 Infrastructural Challenges

- *Digital Divide*: It is certainly not possible that all educational institutions will have the access to implement the required hardware, internet connectivity, and other digital parts or software. Many government institutions are yet to have a digital base.

- *Cost of Implementation*: Though AI is good for the educational sector, the initial investment to implement is not affordable for all institutions, especially government and government-aided schools. The cost is a big factor for many.
- *Scalability Issues*: Customized AI packages developed for some educational sectors will not scale or match with other AI solutions as they are mainly tailor-made AI solutions.

7.4 Ethical and Privacy Concerns

- *Data Privacy*: Accumulating and collecting the students' personal and non-personal data, stored in AI databases, will lead to a lot of security risk as data can be hacked by high-level experts, and will give an opportunity for misuse (Sidiropoulos & Anagnostopoulos, 2024).
- *Transparency*: The usage of AI will raise questions on transparency because many a times the bottom-level workers (here, the teachers) cannot understand the AI-based decisions and happenings because of their limited knowledge.
- *Equity and Inclusion*: AI will definitely widen the existing gaps between the slow learners and fast learners. The slow learners will find it very difficult to adopt AI because the teachers may not have that much patience to explain it fully.

7.5 Resistance to Change

- *Stakeholder Scepticism*: All human beings like teachers, parents, and even students will have the fear that AI will replace the job opportunities in the near future. They also feel that AI will reduce the learning ability.
- *Policy and Regulation Gaps*: Implementing AI in the educational sector lacks clear guidelines which will impact the sector in the wrong way. The teachers, students, and even the management will get confused on which regulations should be followed.

Table 4.6 Summary of challenges in adopting AI

Category	*Key Challenges*
Technological	Data related problems, system package integration and bias in program
Pedagogical	Less adequate training, too much reliance on or trust in AI, syllabus does not match
Infrastructural	Digital platform, too much cost, measurability and scale
Ethical	People privacy risk, less openness, not following rules

8. Future Prospects and Research Directions

This section summarizes the future prospects and research directions while adopting AI in education, especially with respect to India's NEP 2020. The future holds great prospects for implementing AI, especially in the educational sector.

8.1 Future Prospects of AI in Education

Table 4.7 clearly depicts the key research directions of implementing AI in education, where the focus needs to be on its effective usage.

Table 4.7 Key research directions

Area	*Research Topics*
Ethical AI in Education	Data protection, program fairness, legal guidelines
AI for Skill Development	The way AI supports seasonal education, career learning
Cross Linguistic AI Tools	NLP-driven AI across a variety of Indian languages
Learning Analytics	Big data for AI analysis, outcome-based through AI
Human AI Collaboration	Learning the importance of AI like teacher or co-teacher
AI in Assessment	Evaluation automatically in project work, case studies
AI and Equity	Include AI-based solutions for every socio-economic group

Personalized Learning at Scale

- Implementing a personalized learning base will offer tailor-made solutions which align with the students' pace, their interest, talent, and expertise. As a result, their learning will be better.
- AI tutorials and chatbots will provide 24/7 academic assistance in many languages. Chatbots like ChatGPT will benefit everyone positively.

Intelligent Assessment Systems

- With the help of NLP and image recognition, self-directed grading and feedback can be done. Image creation and image recognition both will lead the future, but in terms of education, image recognition will lead the future.
- AI will measure academic knowledge with critical thinking, emotional intelligence and creativity. Emotional intelligence is most important.

AI for Inclusive Education

- There are a few tools for challenged learners, such as voiceover recognition for students needing hearing support, text-to–speech for visually impaired students.
- With the help of real-time translation tools, AI will help to overcome language barriers.
- AI is bound to help those who are suffering from language barriers.

Teacher Support Systems

- It assists teachers in syllabus planning, monitoring and evaluating the students, and also for generating reports. Teachers can gain a lot of benefits by using AI and chatbots.
- Through predictive analytics, it is possible to forecast the learning ability of student. Predictive analytics is a good long-term forecasting system in the business field too.

Gamification and Immersive Learning

- AR and VR play a very significant role in gaming, which is an entirely new experience.
- Virtual laboratory and simulation for low grade to college-level subjects will make the students undergo experience-based learning. The learning will be fun through simulation.

The research directions highlighted in Table 4.7 reflect the growing importance of AI in transforming education. Ethical use of AI must be ensured through proper data protection and fairness guidelines. Skill development with AI will need to support flexible and lifelong learning needs. Language diversity can be addressed using cross-linguistic NLP tools. AI-driven analytics are expected to enhance student performance tracking and feedback. Collaboration between humans and AI must be fostered for better teaching outcomes. Overall, AI research should aim to promote equity and inclusion across all learning communities.

9. Conclusion

The learning environment has been revolutionized at a major level through the integration of AI. The change signifies not just a technological development but also a big move towards customized, adaptive learning spaces fulfilling every individual student's needs. Throughout this exploration, it has been shown how adaptive platforms understand individual learning patterns using ML. Also, communication barriers are being broken by NLP and intuitive interactions are being enabled. New accessibility possibilities are being opened up by computer vision, and educational pathways are being optimized in unprecedented ways through RL. Real-world relevance is already being demonstrated. Millions of students are being reached through India's DIKSHA platform, quantifiable learning improvement is being evidenced by intelligent tutoring systems, and how the access to quality education can be democratized is being illustrated by international EdTech projects, underserved populations being especially favoured. But this heavily burdens the ride with this revolution. Questions of data privacy, algorithm bias, infrastructure requirements, and educator professional development concerns require careful ethical consideration. The digital divide remains on top of the mind—all learners must be benefited by AI, not just those who have been privileged with access to this technology.

Promising futures are being offered by emerging trends like generative AI, AR/VR experiences, and microlearning environments. Yet tomorrow's success depends not on the technology itself but on responsible, ethical implementation aimed at improving learning outcomes for all learners. Educators, technologists, policymakers, and students need to come together to create the future. In this people-centric vision that is being maintained while embracing innovation, the potential of AI can be harvested to impart education that is effective, efficient, engaging, inclusive, and inspiring for students worldwide.

References

Chen, L., Chen, P. and Lin, Z. (2020). Artificial Intelligence in Education: A Review. *IEEE Access*, 8, 75264–75278.

Chen, W. (2024). Classcraft and behavioral analytics: How AI personalizes classroom management. *Learning Media and Technology*, *49*(1), 91–109.

Chen, Y., Liu, Z. and Zhang, H. (2024). Exploring the impact of generative AI in education: A study on GPT-based tutoring systems. *Computers & Education*, *199*, 104781.

Correia, A.-P., Hickey, M.S. and Xu, F. (2024). Beyond the virtual classroom: Integrating artificial intelligence in online learning. *Distance Education*, *45*(1), 65–83.

Cukurova, M. (2024). The interplay of learning, analytics, and artificial intelligence in education: A vision for hybrid intelligence. *British Journal of Educational Technology*, *56*(2), 469–488.

Essa, A. and Ayad, H. (2023). eAdvisor: Using machine learning to predict student success. *Journal of Learning Analytics*, *10*(3), 55–72.

Fang, H., Huang, Y. and Cheng, C. (2023). Using ALEKS to enhance algebra learning: A comparative study. *Journal of STEM Education Research*, *6*(1), 48–62.

Hartman, D.K. (2024). Can artificial intelligence identify reading fluency and level? A comparison of machine and human performance. *Reading & Writing Quarterly*, *40*(1), 19–36.

Huang, Y., Roschelle, J. and Kaput, J. (2023). DreamBox learning impact on elementary mathematics achievement. *Journal of Learning Analytics*, *10*(1), 21–38.

Johnson, L., Becker, S. and Cummins, M. (2024). The future of curriculum design with AI. *Educational Management Administration & Leadership*, *52*(1), 15–34.

Kumar, M. and Singh, R. (2024). Microlearning and AI: Transforming lifelong education through adaptive content. *Educational Technology Research and Development*, *72*(2), 145–162.

Li, Y. and Zhao, J. (2024). Smart textbooks and adaptive learning: A review of Pearson's MyLab platform. *Interactive Learning Environments*, *32*(1), 58–74.

Nguyen, A. (2024). Gamified learning and AI: Using reinforcement learning to optimize engagement. *International Journal of Artificial Intelligence in Education*, *34*(1), 15–33.

Perkins, K., Loeblein, P. and Moore, E. (2024). Impact of PhET interactive simulations on student learning in physics and chemistry. *Journal of Science Education and Technology*, *33*(2), 89–106.

Porayska-Pomsta, K., Holmes, W. and Nemorin, S. (2024). The ethics of AI in education. Available at https://arxiv.org/abs/2406.11842v1 (accessed on 12 July, 2025).

Ritter, S., Anderson, J.R., Koedinger, K.R. and Corbett, A. (2023). Cognitive tutor: Evidence-based effectiveness in math instruction. *Journal of Educational Psychology*, *115*(4), 789–807.

Roschelle, J., Feng, M., Murphy, R.F. and Mason, C.A. (2024). Personalized learning and student gains: Evidence from Khan Academy. *Educational Evaluation and Policy Analysis*, *46*(2), 217–238.

Singh, M. and Patel, K. (2025). DIKSHA and personalized education in rural India: An AI-powered case study. *Education and Information Technologies*, *30*(1), 25–42.

Vesselinov, R. and Grego, J. (2023). Duolingo effectiveness study: Equivalence to college learning outcomes. *Language Learning & Technology*, *27*(3), 23–41.

Zhang, L. (2024). AI and AR/VR integration in STEM classrooms: A review of zSpace technologies. *Educational Technology Research and Development*, *72*(1), 115–134.

.

Chapter 5

Future of AI-based Applications in Education Sector: Challenges and Scope

Aparna Nair M. K.[1*] and *Vineeth V. K.*[2]

1. Introduction

Artificial intelligence (AI) is a branch of computer science that enables machines to mimic human intelligence (Russell & Norvig, 1995). AI is being utilized in various applications ranging from object detection to cyber security, showing its excellence and bringing a huge technological transformation. In the context of education, AI refers to systems capable of performing tasks traditionally requiring human intelligence, such as problem-solving, understanding natural language, and learning from experience. The goal in integrating AI into education is to enhance the outcomes of education. The education sector exploits the possibilities of AI in classrooms, as well as for administrative purposes. The benefits of AI-based education include personalized learning, intelligent tutoring systems (ITS) (Psotka & Mutter, 1988), and administrative automation. Besides these, AI-based education enhances the enthusiasm of both educators and learners, making the teaching and learning process more engaging. Although AI benefits education in many ways, it brings many challenges, like ethical issues, privacy breaches, and the gap between those with access to technology and those without it (Ali et al., 2024).

The use of AI in education began many years ago. Early AI research focused on symbolic reasoning and rule-based systems, and started in the 1950s and 1960s. Researchers came up with expert systems in the 1970s, which could simulate the decision-making ability of human tutors. ITS emerged in the 1980s and 1990s and has brought more progress in adaptive learning. With the introduction of machine learning models in the 2000s, personalized learning platforms (Jian, 2023) were

[1] Department of Electronics and Communication Engineering, Government College of Engineering, Kannur, Kerala; and APJ Abdul Kalam Technological University, Kerala. https://orcid.org/0000-0002-0575-0850

[2] Department of Mechanical and Industrial Engineering, Texas A&M University, USA. https://orcid.org/0000-0003-1378-0619

* Corresponding author: aparnanairmullachery@gmail.com

DOI: 10.1201/9781003639367-5

introduced. Further advancements in machine learning witnessed a revolution in AI-based education with the introduction of chatbots, automated grading, and sentiment analysis using natural language processing (NLP) and deep learning in the 2010s. Today, AI brings a novel method for delivering and consuming knowledge with the introduction of generative AI and multimodal learning interfaces. Figure 5.1 depicts the timeline of AI milestones in education.

Studying the future scope of AI in education is essential for several reasons. First, understanding emerging trends helps policymakers and educators foresee changes and accordingly design curricula and training programmes. Second, while AI offers many benefits, there are potential risks in integrating it into education. Data privacy, algorithmic bias, and making learning less human are significant concerns that need to be addressed through ethics and rules. Third, looking at both the challenges and opportunities helps everyone imagine a fairer education system where technology supports teachers instead of taking their place. This chapter focuses on how AI shapes the future of education. It focuses on current uses, emerging technologies, future opportunities, and the ethical and technical challenges that must be addressed for the effective use of AI in education. The chapter aims to help educators, policymakers, researchers, and tech developers understand how AI can transform education while maintaining the human elements that make learning meaningful and transformative.

> "AI is making education more inclusive by bringing diverse learning tools to millions globally."
>
> —*Satya Nadella, CEO, Microsoft*

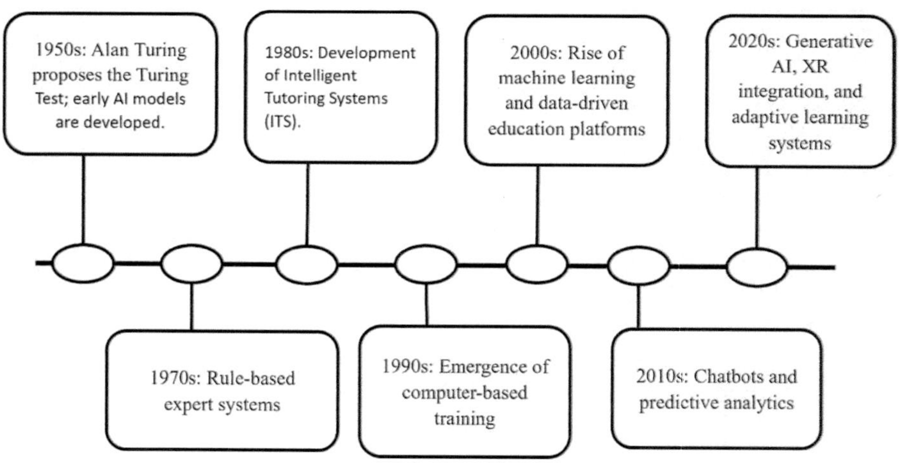

Fig. 5.1 Timeline of AI milestones in education (1950s to Present).

2. AI in Education Today: Where are we Now?

Educational institutes are promoting AI-based teaching–learning process to enhance the quality of both the processes. Learning platforms are introduced to analyse the performance of the students in real time and adapt to the students' capability by adjusting the level of difficulty, making it easier for slow learners to learn. These systems can identify when a student struggles with specific concepts and provide additional resources or alternative explanations specific to their learning preferences. ITS has revolutionized education by providing 24-hour support to learners through conversational AI. Students can raise questions, which will be answered, and a step-by-step explanation will also be provided for complex problems. ITS comprises of four different models: domain, student, tutoring, and user interface (Freedman et al., 2000). The domain model houses the subject matter expertise and curriculum structure that defines what students need to learn. This knowledge base feeds into the student model, which continuously assesses and updates each learner's current understanding, tracking their progress and identifying gaps. The tutoring model then constructs the most apt teaching strategies based on the inputs from the other two models. The user interface model serves as the bridge between human and machine, creating an interactive environment where students can engage with content, receive guidance, and demonstrate their understanding.

The ITS enables educators to provide immediate feedback and adopt necessary changes while monitoring each individual. The learners get a personalized learning experience. However, ITS comes with certain limitations like high cost of development, lack of emotional attachment between learner and educator, and over-dependence on the system, thereby causing a decline in learner–educator interaction. Language learning has become easier with AI-powered applications that provide pronunciation feedback, grammar correction, and conversation practice with virtual native speakers. The AI tools for language learning make the learning easy and interesting, specifically to those who might not have access to human language partners or traditional classroom instruction.

The AI-based tools or methodologies not only make the teaching–learning process efficient, but they are also very advantageous at the administration level as well. Instead of implementing traditional methods for grading the students, AI-based implementations can be done, which enable automatic grading. Such a grading system can help the instructors to assess the students early in the academic year, and also help them to provide timely instructions to the students to avoid any academic failures.

3. Current Applications of AI in Education

Artificial intelligence is playing a very powerful role in transforming the traditional educational system. Some of the applications of AI currently used in the education sector are detailed in this section. Figure 5.2 shows a pictorial representation of different components of AI and their applications in education sector.

3.1 Intelligent Tutoring Systems (ITS)

Intelligent tutoring systems are a one-to-one tutoring platform powered by AI that provides learners with real-time feedback and hints, evaluates their progress, and provides adaptive pathways. ITS analyses various factors of the learners, such as motivation, emotional state, mental processes, existing knowledge, individual capabilities, and learning preferences, and adapts dynamically to the ability of each learner. ITS continuously monitors the performance and growth of the learner and

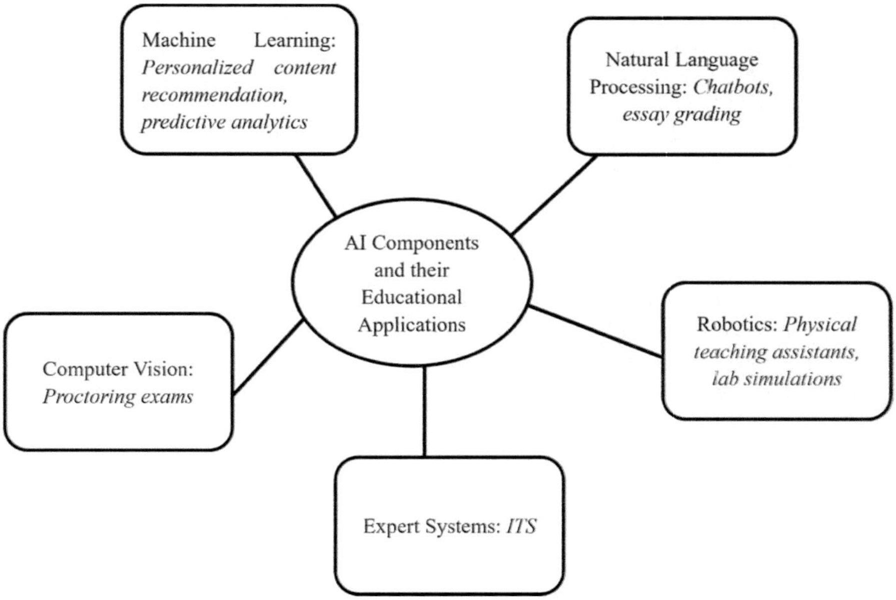

Fig. 5.2 AI components and their applications in the education sector.

offers personalized support through timely feedback and guided assistance. ITS also modifies the learning experience by selecting exercises and tasks that match the current level and learning ability of the student. ITS can diagnose misconceptions, personalize instruction, and offer support more like human tutoring, which is not possible with static e-learning modules. Examples of ITS include MATHia by Carnegie Learning and AutoTutor. The ITS tool MATHia offers mathematics coaching to students on a one-to-one basis, focusing on middle school and high school students. The tool AutoTutor generates courses on demand and provides customized course content to AI Tutor. ITS analyses learner responses and determines what will be the best instructional method for students by utilizing machine learning, Bayesian knowledge tracing, and domain-specific rules. These systems are very useful in large classrooms where individualized attention may not be possible.

3.2 Technology-enhanced Learning (TEL)

Efficient teaching and learning have been made possible with technology-enhanced learning (TEL) (Deng & Benckendorff, 2022). TEL involves applying technology to create more effective learning environments rather than only using digital tools. TEL makes use of interactive web platforms and social media to create experiences like augmented reality (AR) and virtual reality (VR), and also includes game-based learning. This approach is very much welcomed in today's education, where remote and hybrid learning models are becoming common. With TEL, learners gain greater control over their pace and style of learning, leading to more engaged learning and improved outcomes. However, it is essential to ensure that the technology supports learning sustainably.

3.3 Personalized Learning Platforms

Coursera, Duolingo, Squirrel AI, etc., are some of the AI-based personalized learning platforms that customize educational content to the learning pace, style, and preferences of each learner. These platforms use algorithms to analyse user data such as time spent on tasks, accuracy, and learning preferences, and customize the content delivered to each user (Jian, 2023). Squirrel AI in China has access to the world's largest student database and uses the most fine-grained knowledge points and adaptive learning paths to teach students with the accuracy as humans. Duolingo, a language learning app, uses reinforcement learning to suggest exercises that ensure more student involvement and help them to remember what they learn. Coursera, an American massive open online course provider, works with universities and other organizations to offer online courses, certifications, and degrees in a variety of subjects. These platforms have proven effective in increasing motivation and learning efficiency.

3.4 Automated Grading and Assessment Tools

Grading is one of the most time-consuming tasks for educators. AI systems are capable of automating the grading of multiple-choice quizzes, fill-in-the-blank questions, and open-ended responses (Murugan et al., 2025; Gnanaprakasam & Lourdusamy, 2024). Tools such as Gradescope, ExamSoft, and Feedback Studio utilize machine learning and NLP to evaluate student submissions and provide feedback. Google Forms are being widely used in academics for automatic grading. Gradescope provides students with timely and in-depth updates about their performance. Educators utilize these systems to analyse both individual questions and overall assignment trends. Gradescope also offers digital submissions, ensuring that the original work of a student is stored and can be easily accessed from any location, making the entire evaluation process more convenient for everyone involved. ExamSoft provides a secure offline digital assessment platform, while Feedback Studio supports learners in cultivating essential writing, critical thinking, and problem-solving skills. AI can be used for formative assessments, making students aware of their performance and helping teachers adjust instruction accordingly. This reduces workload and also provides a high-quality academic evaluation.

3.5 Chatbots for Student Support

The AI-powered chatbots are virtual assistants that provide support for students anytime. They can answer queries about course content, deadlines, enrolment, and other administrative tasks. IBM Watson (Ferrucci, 2011) is used in several educational institutions to enable smoother communication between students and administrative departments. These chatbots use NLP to understand and respond to student inquiries in real time, improving the efficiency and responsiveness of student support services. This is especially beneficial in large institutions or online learning environments.

3.6 Predictive Analytics for Student Performance

Predictive analytics involves using AI to forecast student outcomes based on historical and real-time data. Institutions use such tools to identify at-risk students, predict dropout rates, and personalize interventions. Civitas Learning and Brightspace Insights offer dashboards that visualize student performance and provide timely suggestions to support them. These systems use data from attendance, assignment submissions, quiz results, and even behavioural patterns on learning platforms. As a result, educators can take necessary measures to support students and prevent their academic failures (Benablo et al., 2018).

4. Emerging Trends and Innovations in AI for Education

Applications of AI in education are rapidly increasing, and several new trends and technologies are shaping the future of teaching and learning. AI-based technologies make learning more personal, interactive, and easier to access. With the introduction of AI-based technologies, students are more engaged in learning and find it more interesting.

4.1 Generative AI and Large Language Models (LLMs)

Generative AI is an AI technology that creates new content, such as text, images, audio, or even video, based on the data it has learned from. Generative AI is capable of generating content that is identical to or closely resembles human-created content. Large language model (LLM) (Brown et al., 2020) is one of the most well-known forms of generative AI. Generative Pre-trained Transformer (GPT), an LLM model, learns patterns, structures, and other important details of human language by getting trained from a huge amount of text from books, articles, websites, and more. GPT produces appropriate text that matches the context by learning from the data. LLM models are capable of creating essays, answering questions, or summarizing content. They are also used to translate languages and simulate conversations. Generative AI and LLMs are benefiting in education sector by assisting students with writing and research, summaries, and providing learning materials. They can support teachers in lesson planning and content creation, enabling personalized learning through interactive chatbots and tutoring systems. Claude, DeepSeek, and ChatGPT are examples of generative AI built on LLMs.

4.2 AI-enhanced Virtual Reality (VR) and Augmented Reality (AR)

Augmented reality and VR in combination with AI make learning fun-filled and easier to understand (Murugan et al., 2025). Students are able to experience lessons in a more interesting way with AI-powered AR and VR rather than just reading from books. VR can take students to a different world where they can visualize lessons and learn better. The learners can also explore places or ideas that would be hard to see in real life. For example, students can:

- look inside the human body to see how organs work;
- visit historical landmarks like the pyramids in Egypt; and
- practise science experiments safely in a virtual lab.

When AI is added, the experience becomes smarter by adjusting the lesson based on what the student needs help with, or offering tips during the activity.

Augmented reality adds digital things to the real world using phones, tablets, or AR glasses. For example:

- In a geography class, students can point their phone at a map and see mountains pop up in 3D.
- In chemistry, they can watch how molecules move and react.
- In language learning, they can see objects labelled in different languages just by scanning them.

Artificial intelligence helps AR tools understand what the student is doing and guide them with helpful feedback or questions to think about. AI-enhanced VR and AR make learning more fun, personal, and memorable, helping students better understand and connect with the material.

4.3 Adaptive Learning Ecosystems

Artificial intelligence is helping create smarter learning environments known as adaptive learning ecosystems, systems that learn about the learner (Derkach et al., 2024). These platforms track how students are doing in real time and adjust the learning experience based on their progress, needs, and understanding. To cite an example, the learning system can identify if the student is facing difficulties with a mathematics concept, and accordingly can slow down. The system can then provide suggestions for extra practice and can also provide explanations in a simpler way with the help of some additional videos. The system is also capable of skipping certain topics or presenting difficult problems if it identifies that a student is moving quickly through the lessons. The adaptive learning ecosystems use tools like learning analytics, knowledge graphs, and AI algorithms to identify the learner's progress. Knewton and Smart Sparrow are examples of adaptive learning ecosystems, where students receive personalized suggestions rather than a one-size-fits-all lesson plan. These learning ecosystems provide students with their progress and encourage them to keep them motivated and complete their learning tasks. With the advent of adaptive learning ecosystems, education has become more flexible, effective, and helps the growth of each individual.

4.4 Emotion AI and Sentiment Analysis

Emotion AI (Kapoor & Verma, 2024), also known as affective computing, is a branch of AI that can detect and respond to human emotions. Emotion AI analyses human emotions from facial expressions, voice tone, body language, word choices, and tries to understand what he/she is feeling. Education sector can utilize this technology to create an emotionally aware learning environment (Reindl, 2022). Such an emotionally aware education system, where an AI-powered tutoring system identifies when a student looks frustrated or bored based on their facial cues or how they are interacting with the content, can help in making learning more engaging. AI tools, based on their analysis, can respond by offering encouragement, easier material, or a short break. Sentiment analysis is capable of determining if the tone is positive, negative, or neutral based on written or spoken language (Sultana et al., 2018). WordNet Affect, SenticNet, and SentiWordNet are examples of tools based on sentiment analysis. Teachers can utilize Emotion AI in virtual classrooms and can accordingly modify the teaching method by getting a real-time understanding of how students are feeling. These technologies are still in the developmental phase, but can certainly offer new ways to make education more human, supportive, and emotionally intelligent.

4.5 AI for Inclusive Education

Artificial intelligence is making education open to all by helping students of all abilities, backgrounds, and learning styles learn in ways that work best for them (Melo-López et al., 2025). Several years ago, learning for students with disabilities was impractical. But now, with the aid of AI tools, learning is no more a dream for them. Immersive Reader by Microsoft supports students with dyslexia or visual impairments by reading text aloud, breaking down syllables, and translating content into different languages. Tools are available that can convert speech to text and thus help students who struggle with writing or have physical disabilities express their ideas more easily. Google Translate or AI-enabled subtitles enable real-time translation and help non-native speakers understand lessons in their language. AI technology can be utilized to develop personalized learning materials based on the cognitive needs of a student, such that every individual will benefit. AI assistant can be made available in virtual classrooms and monitor students to suggest ways to support those who may need extra help.

5. Challenges in Implementing AI in Education

Artificial intelligence in education has both positive effects and negative impacts. It comes with a lot of challenges. These challenges must be identified and addressed for an efficient and secure AI-driven educational system.

5.1 Ethical and Privacy Concerns

Integrating AI into education makes the teaching–learning process efficient, but it also raises important ethical and privacy concerns that must not be ignored (Tahir et al., 2024). AI tools collect information from students like their learning habits, test scores, and emotional responses. Certain tools, like online proctoring, collect biometric information to validate the student's identity before the online assessment starts. This information is used for personalizing lessons for the individual learner. However, the security of this data is not guaranteed. The sensitive information can be exposed if not properly secured. Consider a scenario in which a student's information is leaked due to weak data protection. If the student is a weak learner, the learning difficulties will be exposed, leading to discrimination.

Bias in AI algorithms is another ethical concern that must be addressed. AI models trained on data reflecting existing social inequalities might lead to favouring a group of learners over the others. For instance, consider an AI model for a grading system that evaluates students differently based on their cultural background or language proficiency; or an AI-powered admissions tool, if trained on past data representing students from specific schools or certain socio-economic backgrounds, might favour similar applicants only. This can affect the admission of talented students from other communities. This leads to unfair grading, and the trust in the system will be lost. The AI system in education should be transparent, and the learners and educators must be aware of how the tool is making decisions. To ensure that the algorithms are transparent, the developers must thoroughly examine the training data for any hidden biases and consider all relevant information. Diverse groups can be involved in the design process, and the AI tools can be tested routinely for fairness. Algorithms used in AI systems must hence be fair, such that AI supports education for learners without any inequality.

General Data Protection Regulation (GDPR) in Europe, Family Educational Rights and Privacy Act (FERPA) in the US, and Digital India Act in India are some of the regulatory frameworks to protect student data and privacy when AI is used in education. GDPR sets strict rules about how personal data must be collected, stored, and processed. GDPR also gives individuals control over their information. FERPA similarly ensures that students and parents have rights over educational records and limits who can access this sensitive data. The Digital India Act provides a framework for addressing data protection, transparency, and accountability while integrating AI in education. These regulations help create trust in using AI tools in classrooms. However, many developing countries lack such comprehensive policies, which creates a challenge. Without clear laws, schools and AI companies may not have guidelines for handling student data, increasing the risk of misuse or breaches. For example, in some countries, student data might be shared with third parties without proper consent, which may leave students vulnerable.

Hence, while using AI in education, it must be ensured that the privacy of the end-users is safeguarded, the AI model is not biased, and humans must be involved to build trust and protect students' rights.

5.2 Lack of Infrastructure and Digital Divide

Another major challenge in implementing AI for education, specifically in rural and developing areas, is the lack of infrastructure and the digital divide (Vesna, 2025). AI tools, virtual tutors, and smart classrooms are very well established and transforming education in urban areas and well-connected regions. However, the learners from remote or low-income communities are still not benefiting. Rural areas where internet connection is not stable cannot make use of the advanced AI technologies. Also, if the learners do not have a personalized smart device, they will not be able to enjoy the benefits of AI-based learning platforms. Most of the developing countries are still lacking basic infrastructure facilities like broadband connectivity or computer labs. The gap between those who have these facilities and those who do not is creating an unequal AI-based educational experience. Not only are the learners affected, but also the educators in under-resourced areas may lack proper training to use AI tools, reducing the effectiveness of these tools. A school in a well-connected city utilizes AI to track the performance and progress of students and provide suggestions and measures for boosting performance, while schools without digital access might still depend on outdated teaching methods. It is not only required to implement AI tools in these developing areas, but measures should also be taken to improve the physical infrastructure, affordable connectivity, and digital literacy, such that every student can benefit equally from AI in education. Governments and other non-governmental organizations could take appropriate measures to ensure affordable internet and hardware in their countries. They can also conduct teacher training programmes on AI and make available open-source, lightweight AI tools suitable for low-resource settings.

5.3 Pedagogical Misalignment

Pedagogical misalignment is a term that refers to the mismatch between AI-based education and traditional teaching methods. Educators and learners may prefer a different way of teaching–learning, but the AI tools may be providing a different method of grading or content delivery, or study plan. For example, while assessing a student's performance, the AI platform may issue a very difficult test, while the teacher may prefer creative thinking, collaboration, or social-emotional learning. In such cases, the focus of AI tool would just be measurable outcomes like test scores, whereas the much-needed outcomes like critical thinking or curiosity remain unevaluated. This may also narrow down the learning experience of the students. He/she may be trained to perform only certain tasks and will not understand what the subject is truly about. Above all these, if AI tools are implemented without any input from educators, they can disrupt classroom dynamics, confuse students, or even weaken the teacher's role. To avoid pedagogical misalignment, AI must be integrated in education as only a supportive tool and not as a replacement for the traditional teaching–learning process.

5.4 Teacher and Student Resistance

Although AI-based education shows a transformation in education, its impact will not be realized unless the learners and educators are ready to utilize it. AI-based education is threatening many teachers since they fear that they will be replaced someday or their authority in the classroom will be reduced. AI-based teachers have become a common scenario in most educational institutions now. Also, if the AI-based judgement differs from the teacher's judgement, it can create stress. Teachers might also resist adopting AI since they are not trained well on these tools, or they are not being consulted while implementing them.

Students might also feel uncomfortable using AI tools if learning is more data-driven. AI-based online proctoring has become common in educational institutions, but students may find it difficult to concentrate on the exams with the unnecessary warnings raised by the software. Some students may struggle to engage with AI tools when they find that these tools do not adapt well to their learning style or cultural context. Whether it be with students or teachers, resistance to using AI tools in education arises from a lack of trust, awareness, or involvement in the design and implementation of these technologies. To overcome this, schools and policymakers must involve teachers and students from the beginning of its implementation. Teachers should be offered proper training. Also, timely feedback should be taken, and it should be ensured that AI serves as a supportive tool only. Building this trust is essential to create an educational environment where AI can truly enhance learning.

5.5 Cost and Sustainability

Artificial intelligence–based education requires a huge investment in terms of purchasing software, deploying hardware, training teachers, maintaining infrastructure, regular updating and maintenance of these systems, etc. To set up an AI-driven personalized learning platform, it may require high-performance servers, reliable internet connectivity, and continuous technical support. Many of the schools, particularly in rural or developing regions, might not have sufficient funds for these requirements. Many AI tools operate on subscription models, and hence, long-term access to these tools will depend on the financial capability of the institutes.

Sustainability involves the question of whether these technologies can develop with educational needs over time (Lin et al., 2023). When the curricula get updated or the teaching practices are changed, AI system may become outdated or incompatible. If these tools are not regularly updated or maintained, they may degrade in performance or fail to reflect the latest pedagogical standards. To ensure both cost-effectiveness and sustainability, education systems must adopt adaptable AI solutions and prioritize partnerships that support long-term goals.

Challenges in implementing AI in education and the strategies for overcoming them are tabulated in Table 5.1.

6. Future Scope and Opportunities of AI in Education

With the developments in AI technologies, the educational sector is exploiting them to improve the teaching–learning process. AI is utilized in educational institutions to automate administrative tasks, personalize learning, as well as to create intelligent systems capable of supporting lifelong learning for an individual. This section presents a brief discussion on the different opportunities that AI can provide for learners, educators, and institutions in the coming years.

Table 5.1 Matrix of challenges in implementing AI in education

Category	*Example Challenge*	*Stakeholders Affected*	*Mitigation Strategy*
Ethics and Privacy	Biometric data misuse	Students, parents	GDPR-compliant data policies
Bias and Fairness	Discriminatory predictions	Marginalized student groups	Algorithm audits and diverse datasets
Infrastructure	Poor internet connectivity	Rural schools	Government digital infrastructure plans
Pedagogy	AI misalignment with the curriculum	Teachers, students	Co-design frameworks
Stakeholder Resistance	Job insecurity among teachers	Educators	Training and reassurance
Cost	High licensing fees for AI platforms	School administrations	Open-source solutions. Public-Private Partnerships (PPPs)

6.1 Personalized Lifelong Learning Pathways

Personalized lifelong learning pathways mean the different ways AI can be used to support education throughout the life of an individual, not only in traditional classrooms. AI tools analyse the learner in terms of their strengths, weaknesses, interests, and pace, and create a customized learning experience. For example, an AI tool may be designed in such a way that it is capable of identifying that a student excels in mathematics but is struggling with reading comprehension. The tool can then recommend specific resources to the learner, or it can adjust the difficulty level so that the learner can learn at their own pace. The system can adapt to the ability of the learner, and as the learner grows, it can recommend learning based on their career goals, industry trends, or personal interests. Such an adaptive platform is very much essential in this fast-changing world where individuals can upgrade themselves to remain relevant in their fields. With the help of personalized learning paths, education becomes more flexible, engaging, and open to all. It should be ensured that these AI systems are accessible to all, regardless of socio-economic background, as this is important to prevent educational inequality. Personalized lifelong learning can make the individual empowered at every stage of life, but only with the right infrastructure and proper ethical designs. It can transform education into a continuous and self-directed journey for each individual. Developments such as AI-driven coaches and modular education systems are future directions in terms of personalized life learning. AI-driven Learning Coaches should be able to track the development of a learner from school to career, thereby timely recommending courses, experiences, and micro-credentials. Modular education systems should be developed where AI helps learners build their own curricula by selecting learning units across institutions, platforms, and formats.

6.2 Intelligent and Autonomous Educational Systems

Intelligent and autonomous educational systems refer to AI-driven platforms that can independently manage, adapt, and deliver learning experiences with minimal human involvement. An ITS can monitor a student's performance, identify areas of difficulty, and instantly adjust the content or teaching strategy, much like a human tutor would. Autonomous systems are used in administrative processes such as scheduling, grading, and tracking student engagement. The educators can thus focus more on mentorship and creativity with the help of such autonomous systems. AI chatbots are used in virtual classrooms that can answer student queries, provide immediate support, and reduce wait times. There are systems that utilize AR and VR to simulate classroom environments, thereby providing hands-on learning experiences. These technologies promise efficiency and personalization, but at the same time pose challenges in terms of data privacy, human interaction, and over-dependence on machines. To ensure that these technologies just aid educators and not replace them, human-guided AI decision-making is essential. AI systems should be designed such that they follow ethical principles and the system is transparent, unbiased, and includes all classes of learners. AI tools can change from simple assistive tools to autonomous agents. These agents are capable of managing learning environments without any human supervision. AI systems capable of managing timetables, resource allocation, and making student groupings based on real-time data can be designed. Such autonomous systems can effectively improve the quality of education.

6.3 AI for Global Education Equity

Artificial intelligence for global education equity reduces the learning gaps between different regions, income groups, and learning abilities (Varsik & Vosberg, 2024). Students from many parts of the world, especially in rural areas, face challenges such as a lack of trained teachers, outdated materials, and limited access to quality education. AI tools can be utilized in these regions to overcome these gaps. AI tools can deliver personalized lessons in multiple languages, allowing children in remote villages to learn on their own, even without constant teacher supervision. To support learners with disabilities, tools like speech-to-text and text-to-speech conversions are introduced. AI tools are also introduced to analyse the performance of students, identify the ones at risk of dropping out, and provide timely support to them. These tools can also provide educators with an idea of what they should do to support these learners. With proper infrastructure, AI can promote education equity and support inclusive education systems.

Multilingual learning platforms using real-time translation and speech synthesis, AI tutors for low-resource regions, using offline models that adapt to local curricula and languages, accessibility innovations like AI sign language interpreters or emotion-aware tutors for neurodiverse learners are some of the opportunities that ensure AI for global education. When combined with open educational resources and low-cost devices, AI could help achieve UNESCO's goal of "education for all".

6.4 Integration with Human-centred Design

When AI tools are developed and deployed in education systems, it must be ensured that they are human-centred to serve the needs of learners and educators (Yang et al., 2021). Intervention of humans in AI system development focuses on empathy, inclusion, and real-world usability, prioritizing people over processes. AI systems developed with the input from teachers, students, and parents are most likely to be effective and accepted in the classrooms. For example, instead of designing a one-size-fits-all learning app, developers can work closely with educators to create interfaces that accommodate different learning styles, cultural contexts, and age groups. A human-centred approach ensures that AI supports, not replaces, the teacher's role and also adapts to human needs rather than forcing humans to adapt to the tool.

Potential Innovations:

- *Co-teaching systems*, where AI partners with teachers to co-plan lessons, monitor class engagement, and suggest interventions.
- *Emotional intelligence in AI*, enabling it to detect and respond to student emotions, offering encouragement, or signalling when teacher attention is needed.

6.5 Cross-disciplinary and Intercultural Learning Environments

Artificial intelligence can significantly enhance cross-disciplinary and intercultural learning environments. Intercultural learning environments develop critical thinking among students, improve their creativity, and bring global awareness by combining knowledge from different academic fields and cultural perspectives (Kirk-Lawlor & Allred, 2017). AI supports these learning environments by connecting learners with diverse resources and mentors from around the world. Virtual classrooms powered by AI can facilitate real-time collaboration among students in different countries, bringing about intercultural understanding and breaking down barriers of geography and language. AI can also help design curricula that integrate multiple disciplinary and cultural viewpoints, such as sustainability or social justice. However, AI systems must be designed with sensitivity to cultural context to support cross-disciplinary and intercultural learning. With proper designs, AI can provide diverse learning experiences that prepare students to grow in a complex, multicultural world.

6.6 Real-time Educational Research and Curriculum Evolution

The AI tools speed up the traditional curriculum development process by adapting to the changing needs. Information regarding student performance, current technologies, skills needed for available jobs in markets, etc., will be collected, and AI tools will be modelled based on this data. The tools can then develop curricula in a faster period that meets the demands of the current world. Researchers also exploit AI tools in conducting large-scale studies more efficiently.

6.7 AI in Vocational and Skill-based Education

Vocational courses such as carpentry, electrical work, and healthcare can integrate AI to give the learners a more real-world experience. Instead of investing in expensive instruments, AI-based emulations can be done to gain more understanding of the topics learnt. For example, a student learning about automotive repair can make use of virtual simulations to understand diagnostics and repair. AI tools help workers upgrade themselves by keeping them updated with the skills needed in the current job market. AI-based mobile apps and virtual labs are made available, enabling learning options for learners in remote areas.

The above-mentioned opportunities are promising, but they cannot be made possible without:

- strong ethical frameworks governing data use and algorithmic decisions;
- investment in infrastructure and teacher capacity-building;
- inclusive design involving all stakeholders, including students, parents, and marginalized groups;
- global cooperation to share AI innovations and best practices, especially with developing countries.

7. Policy Recommendations and Strategic Roadmap

Artificial intelligence should be integrated with education only with the support of robust policies and strategic initiatives. This section introduces policy recommendations and proposes a guide to governments, educational institutions, technology developers, and other stakeholders in building ethical, inclusive, and effective AI-driven education systems.

7.1 Establish Ethical Guidelines for AI-based Education

The major needs in AI–education integration are transparency, accountability, and fair practices (Memarian & Doleck, 2023). Hence, the educational policies must ensure:

- *Transparency*: Clear documentation of how AI algorithms make decisions.
- *Explainability*: Tools must be able to justify their outputs to educators, students, and parents.
- *Bias Mitigation*: Systems must be regularly audited to detect and correct biases related to race, gender, or socio-economic background.
- *Data Protection*: Strict adherence to global and local data privacy laws (e.g., GDPR, FERPA) for protecting student information.

Policy Action: National ethics committees and regulatory frameworks must be established by the governments specifically for education, considering existing models such as This document UNESCO's Recommendation on the Ethics of AI can be accessed from UNESCO's digital library (UNESDOC).

7.2 Foster Public-Private Partnerships (PPPs)

Developing and implementing AI technologies on educational systems requires collaboration among governments, educational institutions, private sector, and startups. Governments should introduce regulations to safeguard the rights of learners and should also provide adequate funds for making AI-based education possible. Educational institutions must implement an adaptive curriculum, and the support from private sector and startups is essential for acquiring adequate infrastructure and introducing innovations. PPP models can facilitate:

- co-development of AI tools tailored to local needs;
- shared datasets and benchmarking initiatives; and
- joint teacher training programmes.

Policy Action: Introduce incentives for educational technology companies to collaborate with public schools and rural institutions, ensuring equitable access and context-specific solutions.

7.3 Develop National AI-readiness Frameworks for Education

Countries should adopt a comprehensive AI-readiness assessment model tailored for the education sector. This includes evaluating the availability of digital infrastructure, like the internet or other devices, adaptability to the curriculum, digital literacy of the educator, and availability of data.

Policy Action: Ministries of education should create national action plans for AI in education aligned with broader digital transformation goals, supported by regular progress monitoring.

7.4 Invest in Teacher Training and Professional Development

Teachers are key to successful AI integration. However, many lack the training to effectively use or critique AI tools. Policymakers should provide directions for preservice and in-service training on the fundamentals of AI and its classroom applications and the ethics governing the AI-based education. Professional development courses incorporating AI, such as blended learning and flipped classrooms, must be provided to the educators. The educators must also be given access to peer learning communities and AI education toolkits.

Policy Action: Allocate funding for teacher fellowships, innovation hubs, and collaboration with universities to support research–practice integration.

7.5 Encourage Inclusive and Accessible AI Design

The AI solutions must serve all learners, including those with disabilities, from diverse linguistic and cultural backgrounds, and under-resourced communities.

Policy Action:

- Mandate universal design principles in all AI-based educational tools
- Provide subsidies or incentives for tools developed in local languages or for special education needs
- Include students, parents, and marginalized groups in co-design workshops during AI system development

7.6 Promote Open Standards and Interoperability

To avoid vendor lock-in and to keep data systems connected, AI in education must follow open standards and interoperable frameworks. The benefits of such a framework will be:

- seamless integration of AI tools across platforms Learning Management System (LMS), Content Management System (CMS), and Student Information System (SIS).
- easier scaling and replication of successful solutions
- enhanced data sharing for research and policy decisions

Policy Action: Encourage open-source development and require interoperability compliance in procurement guidelines for educational technologies.

7.7 Create AI Sandboxes and Innovation Zones

To safely experiment with cutting-edge AI applications, governments and institutions should establish "AI sandboxes" (Moraes, 2024), regulated environments for testing new tools without long-term policy implications. Some of the examples include the implementation of pilot projects for AI-based formative assessments in a district or exploiting Emotion AI in hybrid learning environments.

Policy Action: Provide temporary regulatory exemptions for approved innovation zones, along with continuous monitoring and ethical review boards.

7.8 Implement Continuous Evaluation and Feedback Mechanisms

Artificial intelligence in education must grow with stakeholder feedback and based on real-world evidence. Hence, it is needed to establish:

- impact evaluation frameworks tracking learning outcomes, user satisfaction, and equity implications;
- AI education observatories to collect, analyse, and disseminate best practices globally;
- early warning systems for unintended consequences, such as learning gaps or disengagement.

Policy Action: Establish regular feedback among educators, learners, technologists, and policymakers.

8. Conclusion

Artificial intelligence has evolved over the years and is finding applications in almost every field, whether it is entertainment or medical, or education. AI, when integrated with education, is capable of enhancing learning outcomes, personalizing instruction, and bringing about a transformative effect. Various applications of AI in education include ITS to emotion-aware learning platforms. AR and VR can be integrated with AI to provide a more interesting and real-world-like experience to learning.

But AI-based education poses various complex challenges as well. Data privacy, algorithmic bias, and transparency are the concerns that must be addressed for an efficient AI-based education. Robust policies and ethical design principles must be developed to overcome these challenges. Also, AI-based education must be open to all, irrespective of whether the learner or educator is from a well-connected or under-resourced region, to ensure equity in education. Lack of infrastructure, improper training of teachers, and lack of standard practices also affect the efficient integration of AI into education.

Artificial intelligence is a powerful technology, and it can enable sustainable and equitable education systems with the right policies and standards. All the stakeholders must cooperatively work towards ensuring that AI tools are not only innovative but also human-centric and accountable. Technology alone cannot shape the future of education, it greatly depends on how that technology is being applied. AI is capable of making knowledge accessible to all, empowering educators, and inspiring a new generation of lifelong learners. The choices we make today will help build the education for tomorrow.

References

Ali, O., Murray, P.A., Momin, M., Dwivedi, Y.K. and Malik, T. (2024). The effects of artificial intelligence applications in educational settings: Challenges and strategies. *Technological Forecasting and Social Change*, *199*, 123076.

Benablo, C.I.P., Sarte, E.T., Dormido, J.M.D. and Palaoag, T. (2018), February. Higher Education Student's Academic Performance Analysis through Predictive Analytics. In *Proceedings of the 2018 7th International Conference on Software and Computer Applications* (pp. 238–242).

Brown, T., Mann, B., Ryder, N., Subbiah, M., Kaplan, J.D., Dhariwal, P., Neelakantan, A., Shyam, P., Sastry, G., Askell, A. and Agarwal, S. (2020). *Language models are few-shot learners. Advances in neural information processing systems*, 33, pp.1877–1901.

Deng, R. and Benckendorff, P. (2020). Technology-enabled learning. In *Handbook of e-Tourism* (pp. 1–27).

Derkach, O., Bazyka, S. and Yamnyuk, B. (2024). Adaptive educational ecosystems using artificial intelligence for forming students' educational trajectories. *HUMANITAS Zarządzanie*, *3*, 97–106.

Farooqi, M.T.K., Amanat, I. and Awan, S.M. (2024). Ethical considerations and challenges in the integration of artificial intelligence in education: A systematic review. *Journal of Excellence in Management Sciences*, *3*(4), 35–50.

Ferrucci, D.A. (2011). IBM's Watson/DeepQA. *SIGARCH Computer Architecture News*, *39*(3).

Freedman, R., Ali, S.S. and McRoy, S. (2000). Links: What is an intelligent tutoring system? *Intelligence*, *11*(3), 15–16.

Gnanaprakasam, J. and Lourdusamy, R. (2024). The role of AI in automating grading: Enhancing feedback and efficiency. In *Artificial Intelligence and Education-Shaping the Future of Learning*. IntechOpen.

Jian, M.J.K.O. (2023). Personalized learning through AI. *Advances in Engineering Innovation*, *5*, 16–19.

Kapoor, A. and Verma, V. (2024). Emotion AI: Understanding emotions through artificial intelligence. *International Journal of Engineering Sciences & Humanities*, 14, 223–232.

Kirk-Lawlor, N. and Allred, S. (2017). Group development and integration in a cross-disciplinary and intercultural research team. *Environmental Management*, 59.

Lin, C.C., Huang, A.Y. and Lu, O.H. (2023). Artificial intelligence in intelligent tutoring systems toward sustainable education: A systematic review. *Smart Learning Environments*, *10*(1), 41.

Melo-López, V.-A., Basantes-Andrade, A., Gudiño-Mejía, C.-B. and Hernández-Martínez, E. (2025). The impact of artificial intelligence on inclusive education: A systematic review. *Education Sciences*, *15*(5), 539.

Memarian, B. and Doleck, T. (2023). Fairness, accountability, transparency, and ethics (FATE) in artificial intelligence (AI) and higher education: A systematic review. *Computers and Education: Artificial Intelligence*, *5*, 100152.

Moraes, T. (2024), September. Ethical AI Regulatory Sandboxes: Insights from cyberspace regulation and Internet governance. In *Proceedings of the Second International Symposium on Trustworthy Autonomous Systems* (pp. 1-10).

Murugan, T., Periasamy, K. and Abirami, A.M. (eds). (2025). *Adopting Artificial Intelligence Tools in Higher Education: Teaching and Learning* (1st edn). CRC Press.

Psotka, J. and Mutter, S.A. (1988). *Intelligent Tutoring Systems: Lessons Learned.* Lawrence Erlbaum Associates.

Reindl, S. (2022). Emotion AI in education: A literature review. *International Journal of Learning Technology*, *16*, 288–302.

Russell, S.J. and Norvig, P. (eds). (2021). *Prentice Hall Series in Artificial Intelligence*. Englewood Cliffs, NJ: Prentice Hall.

Staff Reporter (2025) Inspiring Quotes About AI in Education: Revolutionizing Learning and Teaching, The Edinburgh Reporter, 3 January. Available at: https://theedinburghreporter.co.uk/2025/01/inspiring-quotes-about-ai-in-education-revolutionizing-learning-and-teaching/?print=print&utm_source=chatgpt.com (Accessed: [9th Sept, 2025]).

Sultana, J., Sultana, N., Yadav, K. and AlFayez, F. (2018), April. Prediction of sentiment analysis on educational data based on deep learning approach. In *2018 21st Saudi computer society national computer conference* (NCC) (pp. 1-5). IEEE.

UNESCO (2022) Recommendation on the ethics of artificial intelligence. Paris: UNESCO. 43 pp. Available at: https://unesdoc.unesco.org/ark:/48223/pf0000381137 (Accessed: [09th September 2025]).

Varsik, S. and Vosberg, L. (2024). *The Potential Impact of Artificial Intelligence on Equity and Inclusion in Education* (No. 23). OECD Publishing.

Vesna, L. (2025). Digital divide in AI-powered education: Challenges and solutions for equitable learning. *Journal of Information Systems Engineering and Management*, *10*, 300–308.

Yang, S.J.H., Ogata, H., Matsui, T. and Chen, N.-S. (2021). Human-centered artificial intelligence in education: Seeing the invisible through the visible. *Computers and Education: Artificial Intelligence*, *2*, 100008.

Chapter 6

Artificial Intelligence Techniques for Supporting Face-to-Face and Online Collaborative Learning

R. Felista Sugirtha Lizy[1*], *R. Suya Padhra Haridha*[2], *S.N. Sithi Shamila*[3] and *Ir. Bambang Sugiyono Agus Purwono*[4]

1. Introduction to AI in Collaborative Learning

Students engaged in collaborative groups exchanging ideas, solving problems, and building knowledge together has long been the crux of collaborative learning. Today, thanks to a little something called artificial intelligence (AI), this process is even more individualized and advanced than ever before. AI in collaborative learning acts as an unseen mediator that nurtures diverse groups by recognizing learning differences between visual, auditory, and kinesthetic learners, as well as differences in cognitive abilities, motivation levels, and prior subject knowledge. AI can also assess engagement levels, recommend appropriate materials, and even notify instructors when an individual is potentially disengaged or not actively participating. By examining group work patterns, AI can assist learners and educators in identifying what is effective and what requires adjustment. So while AI cannot replace the human-to-human connection at the heart of collaborative learning, it can supplement it by harnessing technology to create greater connection and equilibrium, transforming collaboration into a more equitable and productive experience for all.

[1] Department of Computer Science, A.P.C. Mahalaxmi College for Women, Tamil Nadu, India. https://orcid.org/0000-0003-3648-0205

[2] Department of Physics, V. O. Chidambaram College, Thoothukudi- Tamil Nadu, India, 628008. https://orcid.org/0000-0002-9988-0946

[3] Department of Information Technology, Wavoo Wajeeha Women's College of Arts and Science, Tamil Nadu, India. https://orcid.org/0000-0003-0439-2534

[4] Mechanical Engineering Department, State Polytechnic of Malang, Indonesia. https://orcid.org/0000-0002-5982-8983

* Corresponding author: 21felistaa@gmail.com

DOI: 10.1201/9781003639367-6

1.1 What is AI in Education

The integration of machine learning, natural language processing, and other AI-driven approaches in education is made possible through intelligent tools that personalize learning, automate administrative tasks, provide real-time feedback, and deliver extensive support to students and teachers alike. AI is making education more efficient, engaging, and accessible. One of the best examples of this is with advanced, personalized learning systems where students are guided to the content that is best for them, their strengths, and learning style—all powered by AI. Also, automation of administrative tasks like grading, attendance, and scheduling frees up educators' time to teach better. Advanced AI-powered smart tutoring systems offer self-paced complex concept instruction along with immediate feedback and support. AI technologies also foster enhanced engagement in learning through the use of chatbots and gamification which transform passive students into active learners. Furthermore, the incorporation of speech to text, real-time translation, and other tools are beneficial for students with disabilities, promoting access to all. By providing AI with data, useful insights can be attained regarding the use of adaptive teaching methods to improve student performance by overcoming identified barriers to learning. With the use of virtual classrooms and remote learning, students all over the world can access educational materials, giving them global opportunities to learn (see Figure 6.1).

Fig. 6.1 AI in education.

1.2 Differences between Face-to-Face and Online Collaborative Learning

Both face-to-face (F2F) and online collaborative learning are limited through the constraints of group processes. The modalities differ in terms of methods and modes of communication, peer engagement, flexibility of schedule, and tangible learning results.

1.2.1 Mode of Interaction

In in-person learning, students are deeply connected and interacting with each other through verbal and non-verbal cues like discussion, movement, and even facial expressions for coherence. Compared to in-person education, online education relies on technology like video conferences, discussion boards and chat rooms that lack the same level of interactivity and connection. For many learners, historical discrimination and exclusion have created barriers to attending in-person classes at fixed times and locations.

Virtual collaborative learning addresses this issue by allowing people from different backgrounds to join from anywhere, at any time. Students in class have the opportunity to connect and collaborate with fellow students in one physical space and in real time, and often surprisingly, which leads to enriched group projects and a greater sense of collective happiness. Through online platforms, students can connect and collaborate with peers in real time, often leading to enriched group projects and a stronger sense of collective engagement.

However, online education also presents limitations such as technical difficulties, time zone differences, and lack of immediate feedback, which may reduce focus and clarity. In contrast, in-person discussions provide opportunities to offset these disadvantages, although they come with their own barriers of access and scheduling.

Multiple digital platforms, such as video conferencing, shared documents, and online discussion boards, are widely used in online education. In-person education takes advantage of more concrete tools such as whiteboards, paper handouts, and lectures in class.

1.2.2 Collaboration/Team Work

Students are quicker to build connections, engage in experiential learning, and productive team work in in-person classes.

In online learning, collaboration involves more work in arranging virtual meetings, sharing resources, and communicating actively.

1.2.3 Assessment and Feedback

While learners receive instantaneous response from their instructors and fellow learners during F2F interactions, learners in an online set up receive feedback through written notes, quizzes, or recordings, which is often impersonal.

Table 6.1 The main differences between face-to-face and online collaboration

Feature	*Face-to-Face Collaborative Learning*	*Online Collaborative Learning*
Social Interaction	Rich, in-person interaction; strong non-verbal communication; fosters strong social bonds	Digital interaction; potential lack of non-verbal cues; can be more challenging to build social bonds
Feedback	Immediate, real-time feedback and clarification	Feedback may be delayed; relies on digital communication for clarification
Flexibility	Structured schedule; limited flexibility in time and location	Highly flexible; learners can participate at their own pace and location
Accessibility	Limited by geographical constraints; requires physical presence	Accessible to learners in remote areas or with limitations
Technology	Technology may be used as a supplement; less reliance on digital tools	Heavy integration of online tools and platforms

Contd.

Table 6.1 *Contd.*

Feature	*Face-to-Face Collaborative Learning*	*Online Collaborative Learning*
Structure	More structured environment, set times, and places	Less structured environment, relies heavily on individual motivation
Cost	Typically, higher costs due to venue and travel	Potentially lower costs, dependent on required technology
Community Building	Generally easier, spontaneous development of relationships	Needs deliberate action to create the feeling of community
Communication	Relies on verbal and non-verbal communication	Relies heavily on written and sometimes video communications

Table 6.1 summarizes that both F2F and online collaborative learning have their own lists of benefits and shortcomings. In-person learning fosters experience-based, immersive real-time learning and engagement, while online learning offers the flexibility and accessibility that makes so many learners thrive. In the end, how powerful each one actually is greatly varies based on the learners' needs, their level of access to technology, and the overall learning environment they are in.

1.3 Looking to Get Involved: Here is How AI Improves Engagement, Personalization, and Efficiency in Education

Artificial intelligence is rapidly transforming classrooms, creating more immersive, customized, and meaningful learning experiences. AI tools enable learners and educators to engage in a more interactive, personalized, and efficient educational experience.

1.3.1 Enhancing Engagement

The AI-enabled platforms introduce new levels of interactivity with chatbots, gamification, and virtual simulation tools bringing a whole new level of engagement to learning. Even more so than traditional educational games, various AI-infused experiences like adaptive quizzes help students become more motivated with their instant gratification and feedback that they often provide. The model powered by AI, virtual reality/augmented reality (VR/AR) provides immersive learning experiences to make complex topics more absorbing and easier to grasp. Also, AI chatbots give a helping hand to students by answering their questions, guiding their way through the lessons, and generally keeping them engaged.

1.3.2 Enabling Personalization

Artificial intelligence can bring a personalized learning experience to every student based on their own needs, strengths, and pace. Adaptive learning tools use AI to track student performance in real time and then suggest resources, practice exercises, or advanced challenges. So if students are struggling with a particular math concept, then AI will take over and give them extra explanations or practice questions based on how hard it is for them. AI tutors can give as much as needed one-on-one attention tailored to each student's learning style, keeping education one-on-one.

1.3.3 Improving Efficiency

In less creative duties such as assignment grading, attendance monitoring, and task scheduling, AI steps in, allowing the teachers to have the spare time to focus more on education. AI-based analytics pattern areas of student learning, allowing teachers to immediately intervene as needed and address students with targeted support. Next, AI handles admissions and other student queries via communication chatbots, reducing strain on school staff. These efficiencies make integration more seamless for both the student and teacher. As shown in Table 6.2, AI enhances educational processes by improving engagement, supporting personalized learning, and increasing efficiency.

Table 6.2 AI enhances engagement, personalization, and efficiency in education

Aspect	*How AI Enhances It*	*Examples*
Engagement	AI-powered tools make learning interactive and motivating through gamification, simulations, and chatbots	• AI-driven educational games with instant rewards • Virtual reality/augmented reality (VR/AR) for immersive learning • AI chatbots guiding students through lessons
Personalization	AI customizes learning experiences based on individual student needs, strengths, and learning pace	• Adaptive learning platforms recommending resources • AI tutors adjusting teaching methods • Tailored exercises based on student performance
Efficiency	AI automates administrative and teaching tasks, improving productivity for educators and students	• AI grading assignments and tracking attendance • AI-driven analytics identifying learning patterns • Automated chatbots handling student queries

Artificial intelligence increases engagement with interactive resources, personalizes learning through tailored content, and increases efficiency by automating administrative tasks. These developments contribute to better, easier, and more fun learning while preparing the students for a future in technology.

2. Literature Survey: Advanced AI in Education

The development and application of AI in the field of education has received much attention from scholars, especially in the principles of sustainable development, the personalization of learning systems, and digitalization of education.

Pedro et al. (2019) have written on the broader social aspects and development of AI in education (and the opportunities and urgent ethical and equity considerations therein). Similarly, Chen et al. (2020) offer a detailed insight into uses of AI and highlight how it has the potential to improve both engagement and learning outcomes, but they also raise concerns regarding scale and inclusivity.

Zhai et al. (2021) and Chen et al. (2022) each offer longitudinal reviews of 20 years of AI in education. These studies demonstrate increasing refinement of AI-based tools—from early, tutor-like programs to increasingly sophisticated, responsive learning environments—and suggest the need to match these technologies to educational objectives.

Kizilcec (2024) contributes an important dimension by centring the locus of discussion on educators and asserting the need for any AI integration to be grounded in teacher understanding and trust. Dimitriadou and Lanitis (2023) also provide a critical review of smart classroom technologies and discuss technical, social, and institutional challenges that hinder AI uptake.

Suryanarayana et al. (2024) connect AI to educational sustainability and propose that AI can help manage learning systems in a manner that promotes quality and access over time. Similarly, Luan et al. (2020) and Panigrahi et al. (2021) emphasize the potential of AI to enhance teaching efficiency and improve learner outcomes.

With promises come pitfalls, as Cain et al. (2023) offer a cautionary lens, warning against the unsupported use of conversational AI such as chatbots that are beneficial but can spread misinformation and bias, and encourage over-dependence.

The moral implications of generative AI are also highlighted by Baidoo-Anu and Ansah (2023), who investigate the ways in which a tool like ChatGPT can facilitate learning but calls for responsible use. Meanwhile, Holmes et al. (2019) and Luckin et al. (2016) provide perspective on the fundamental role that AI should play—augmenting, rather than replacing, humans in teachers' administrative tasks and in helping students to become more successful individual learners.

Global agencies such as United Nations Educational, Scientific and Cultural Organization (UNESCO) (2021) and the Organisation for Economic Co-operation and Development (OECD) (2021) provide strategic policy frameworks, promoting equitable access, ethical use, and the building of AI competencies. Selwyn (2019), as one might expect, is more cautious, warning educationalists and policymakers against allowing technology to create a blind spot to the relational, emotional, and social nature of education.

Finally, Baker and Inventado (2014) highlight the technical backbone of AI in education through data mining and analytics, which enable systems to detect patterns and personalize learning paths effectively.

3. AI Techniques for Collaborative Learning

Artificial intelligence is enhancing collaborative learning experiences, allowing students to work together more effectively, communicate clearly, and receive personalized support.

Some AI approaches designed to empower students and teachers, while fostering collaborative online and in-person learning experiences, are also available, such as:

- *Language and Text Processing with Artificial Intelligence*: AI-driven NLP technology can power real-time translation, sentiment analysis, and automated feedback. It assists students in working together more productively by removing language barriers and assessing discussion quality in discussion boards and chats. NLP-based chatbots supplement this immediacy with more personalized, conversational guidance, helping students navigate complex group projects and discussions.

- *ML and Adaptive Learning*: ML offers a powerful way to analyse the ways that students are learning and interacting with their peers to develop recommendations for optimal group activities. It makes sure that students with highly complementary—but individually weak—skill sets get assigned to the same group, maximizing the efficiency of teamwork. It can help mitigate learning equity gaps. Adaptive learning platforms are increasingly using ML to determine the optimal difficulty level for a question based on the performance of the entire group so that all members remain challenged.
- *Intelligent Tutoring Systems (ITS)*: AI-driven tutors allow learners to practice collaboratively, with AI chatbots delivering hints, suggestions, and corrections in real time. Finally, ITS can track the progress of the group as a whole, find areas of weakness, and recommend aligned learning materials to optimize the overall performance of the group.
- *AI-driven Recommendation Engines*: AI recommends the most relevant study materials, discussion topics, and peer collaborations to students, based on their interests and previous engagements. These systems make sure that students are engaging with the content that is most relevant to them, maximizing the benefits of group learning.
- *Sentiment and Emotion Analysis*: AI can assess students' emotions and engagement levels in discussions with facial recognition or text-based sentiment analysis. When students start exhibiting frustration or disengagement, AI can respond with encouragement or recommend that a teacher step in and provide support.
- *Automated Moderation and Conflict Resolution*: AI tools can use these models to automatically detect when a user is behaving inappropriately, straying too far off-topic, or arguing with other members of the group, and apply moderation in an automated fashion. This helps to create a productive, positive collaborative learning community. AI empowers collaborative learning. At its best, AI strengthens collaborative learning experiences with intelligent support, adaptive feedback, and immersive interactive environments. Some important AI techniques applied in collaborative learning are given in Table 6.3.

Table 6.3 AI techniques used in collaborative learning

AI Technique	*Description*	*Applications in Collaborative Learning*
Natural Language Processing (NLP)	AI-powered language understanding for analysing and generating text	• AI chatbots facilitating discussions • Sentiment analysis for monitoring group interactions • Automated summarization of discussions
Machine Learning (ML)	AI models analysing data patterns to improve learning experiences	• Personalized group recommendations based on learning styles • Predictive analytics to identify struggling students in teams • Automated peer assessment and feedback

Contd.

Table 6.3 *Contd.*

AI Technique	*Description*	*Applications in Collaborative Learning*
Intelligent Tutoring Systems (ITS)	AI-driven virtual tutors providing real-time assistance	• AI-assisted group problem-solving • Context-aware hints and guidance • Adaptive suggestions based on group performance
Recommendation Systems	AI suggesting learning resources based on user behaviour	• Recommending relevant articles, videos, or exercises for group projects • Suggesting optimal team formation based on skills and interests
Speech and Gesture Recognition	AI interpreting voice and gestures for interaction	• Real-time speech-to-text for accessible collaboration • Gesture-based interactions in virtual group environments
Collaborative AI Agents	Virtual AI assistants that support teamwork	• AI moderators ensuring balanced participation • AI-driven brainstorming assistants • Automated conflict resolution tools
Learning Analytics	AI analysing collaboration patterns and engagement	• Tracking student contributions in group projects • Providing insights into teamwork effectiveness • Detecting disengaged students for timely intervention

By integrating these different AI approaches, collaborative learning is more dynamic, personalized, and engaging, thereby increasing the collaboration and knowledge-sharing opportunities available to students. Cutting edge AI methods such as NLP, ML, ITS, recommender systems, sentiment detection, and automated moderation are further revolutionizing collaborative learning through the pervasive power of AI. These technologies all enhance both learner engagement and communication, while providing personalized assistance, helping make learning in large groups more efficient and more effective.

3.1 NLP in Education

Natural language processing (NLP) is an AI technique that enables machines to understand, process, and respond to human language. In education, NLP plays a crucial role in enhancing communication, collaboration, and personalized learning. Key NLP applications include chatbots, automated feedback, and sentiment analysis.

3.1.1 Chatbots for Learning Assistance

The AI-powered chatbots are like virtual assistants, helping students and teachers by providing answers and guidance. Chatbots help in learning in the following ways:

- 24/7 Support: Students can ask questions and get answers anytime.
- Homework Help: Chatbots explain concepts and solve doubts.

- Group Work: Chatbots facilitate discussions, suggest resources, and help students stay on track with projects.
- Administrative Support: They help with frequently asked questions (FAQs) about schedules, deadlines, and course materials.
- Example: A chatbot like "EdTech AI" can help with math problems by explaining solutions step-by-step.

3.1.2 Automated Feedback Systems

The NLP-powered systems give real-time feedback on student work, helping them improve assignments and writing. The benefits of automated feedback are:

- Instant Grading: AI quickly grades essays, quizzes, and assignments.
- Grammar and Style Suggestions: NLP-based tools like Grammarly correct errors and suggest improvements.
- Concept Understanding: AI highlights weak areas and provides additional learning materials.
- Example: Tools like Grammarly or Turnitin use NLP to check grammar, plagiarism, and writing clarity.

3.1.3 Sentiment Analysis for Student Engagement

The AI sentiment analysis looks at student emotions and engagement levels from text inputs, voice tone, or facial expressions. Sentiment analysis helps in the following ways:

- Detecting Struggles: AI can pick up on frustration or confusion in student responses and alert teachers to intervene.
- Adapting Content: If students find content too hard, AI can suggest easier explanations or extra resources.
- Boosting Discussions: NLP can analyse discussion forums to make sure students stay engaged and get the support they need.
- Example: An AI system in an online course can detect if students are using negative words like "confused" or "stuck" frequently and suggest simpler explanations.

The NLP-powered chatbots, automated feedback, and sentiment analysis are changing education by increasing interaction and by providing real-time guidance and better learning experiences. These AI tools create a more efficient, personalized, and engaging learning environment.

3.2 Machine Learning and Deep Learning

Machine learning and deep learning (DL) are game-changing AI technologies that power personalized learning and predict student behaviour. They analyse huge amounts of data to find patterns and adapt teaching methods to improve student engagement and results. One of the main applications of ML and DL is personalized learning pathways. By analysing a student's previous performance and learning style, AI can create a study plan tailored to their needs. Adaptive learning platforms use AI to adjust lesson difficulty in real time based on a student's responses so that they are never bored or overwhelmed. AI also recommends personalized content—videos, books, exercises—that match a student's learning pace and preferences, making learning more engaging and effective. Another key application is predicting student behaviour. ML and DL analyse patterns such as student participation, quiz scores, and assignment submissions to predict performance, engagement, and even dropout risk. This allows educators to identify students who are struggling early on and intervene with targeted support. For example, AI can alert a teacher when a student is at risk of falling behind so they can provide help before the issues escalate.

3.3 Computer Vision

Computer vision, an area of AI that allows machines to interpret and comprehend visual data, is greatly improving virtual classrooms with technologies such as gesture recognition and attention tracking. Gesture recognition enables computers to understand human motions, including hand gestures or body positions, facilitating more engaging and immersive learning experiences. In virtual classrooms, learners can engage with material by using actions such as pointing at a display or raising their hands. This immediate interaction enhances learning engagement and allows students with disabilities to participate actively by utilizing gestures rather than conventional input tools such as keyboards or mice. For instance, in a virtual biology class, learners could employ hand movements to focus on particular sections of a 3D cell model or signal their responses without typing.

Conversely, attention tracking employs computer vision to observe students' involvement by examining their facial expressions, eye movements, and body posture. This enables AI systems to identify if students are focused, gazing at the screen, or losing concentration. This tracking provides teachers with immediate feedback on student engagement, allowing them to modify their teaching methods or step in when needed. Furthermore, attention monitoring can encourage AI-powered platforms to adjust the speed or material of a lesson according to student engagement.

For example, a system could notify a teacher when a student's eyes look away from the screen, suggesting a lack of engagement, prompting greater interactive participation. Collectively, these computer vision technologies—gesture recognition and attention tracking—are enhancing virtual learning environments, rendering them more interactive, inclusive, and attuned to students' requirements, thereby fostering a more efficient and captivating online educational experience.

3.4 Reinforcement Learning

Reinforcement learning (RL) is an AI technique that enables systems to learn through trial and error, receiving feedback in the form of rewards or penalties based on their actions. In education, RL is particularly effective in the development of AI tutors that can adapt to students' unique needs and learning styles. These AI tutors continually evaluate a student's responses and learning progress, adjusting their teaching methods and content accordingly. In the context of personalized learning, RL allows AI tutors to provide adaptive feedback. As students interact with the system, the AI learns which teaching strategies are most effective for them. For example, if a student struggles with a particular math concept, the AI tutor may provide more targeted practice problems, offer explanations in different formats (e.g., video, text, or interactive), or simplify the content. As the student progresses and improves, the AI tutor adjusts the difficulty of tasks to match the student's growing skill level, ensuring that they are constantly challenged without feeling overwhelmed. This adaptive learning process allows students to learn at their own pace while receiving personalized guidance that evolves with their needs. RL also enables AI tutors to provide real-time intervention. By monitoring student behaviour and performance, the system can detect when a student is falling behind or struggling with a topic and intervene promptly. This might involve offering hints, providing additional resources, or guiding the student through step-by-step solutions. Over time, as the AI tutor accumulates more data about the student's learning habits, it becomes increasingly effective at predicting and addressing gaps in knowledge, thus enhancing the overall learning experience. In summary, RL empowers AI tutors to continuously adapt to a student's needs, providing a highly personalized and dynamic learning environment that supports individual growth and achievement.

3.5 Recommender Systems

Recommender systems are AI-driven tools that analyse data about a user's preferences, behaviour, and interactions to provide personalized content suggestions. In education, recommender systems play a crucial role by tailoring learning materials, such as courses, videos, articles, and exercises, to suit the individual needs and interests of each student. By collecting data on how students interact with content—such as which topics they engage with the most, their performance in quizzes, or their learning style—these systems can suggest relevant resources that align with their current level of knowledge and learning goals. For example, if a student consistently struggles with a particular topic in mathematics, a recommender system might suggest additional practice exercises, video tutorials, or alternative explanations to help strengthen their understanding (see Figure 6.2). Similarly, if a student shows an interest in a specific subject like history or coding, the system can recommend courses or supplementary materials related to that area, keeping the student engaged and motivated. As the

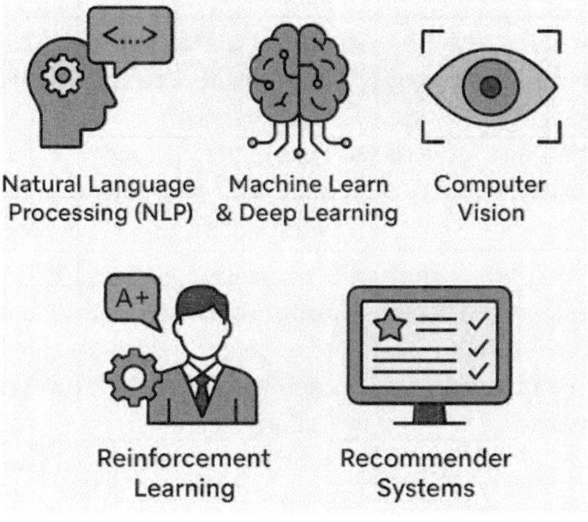

Fig. 6.2 AI techniques and their applications in education.

student progresses, the system learns from their interactions, continuously refining its suggestions to ensure they receive the most relevant and effective content for their ongoing development.

Recommender systems also help educators by providing insights into which resources are most effective for different students, allowing them to better support individual learning needs. Overall, by offering personalized content based on student preferences, learning styles, and performance, recommender systems enhance the educational experience, making learning more targeted, engaging, and effective (see Figure 6.3) text is moved to Table 6.4, Natural Language Processing (NLP) under AI Technique and their applications in education.

Table 6.4 AI techniques and their applications in education

AI Technique	*Description*	*Applications in Education*	*Examples*
Natural Language Processing (NLP)	AI that enables machines to interpret and respond to human language	• Chatbots for instant learning assistance • Automated feedback on assignments • Sentiment analysis for tracking engagement	• AI-driven chatbots answering student queries • Grammarly for grammar correction • Sentiment analysis tools detecting student frustration
Machine Learning (ML) & Deep Learning (DL)	AI techniques that allow systems to learn from data and make predictions	• Personalized learning pathways based on performance • Student behaviour prediction for early interventions	• Adaptive learning platforms adjusting difficulty • AI systems predicting student performance or at-risk behaviour
Computer Vision	AI that enables machines to interpret visual data (images, videos, etc.)	• Gesture recognition for interactive learning • Attention tracking to monitor student engagement	• Virtual classrooms using gestures for interaction • AI tracking student focus in video lectures
Reinforcement Learning (RL)	AI where systems learn by receiving feedback from actions, optimizing future decisions	• AI tutors that adapt to students' needs and progress • Personalized feedback and real-time interventions	• AI tutoring systems adjusting difficulty based on student performance • RL-based educational games for skill development
Recommender Systems	AI systems that suggest personalized content based on user behaviour and preferences	• Personalized content recommendations (e.g., courses, videos, exercises) • Tailored learning resources for students	• Coursera recommending courses based on completed modules • YouTube suggesting educational videos based on watch history

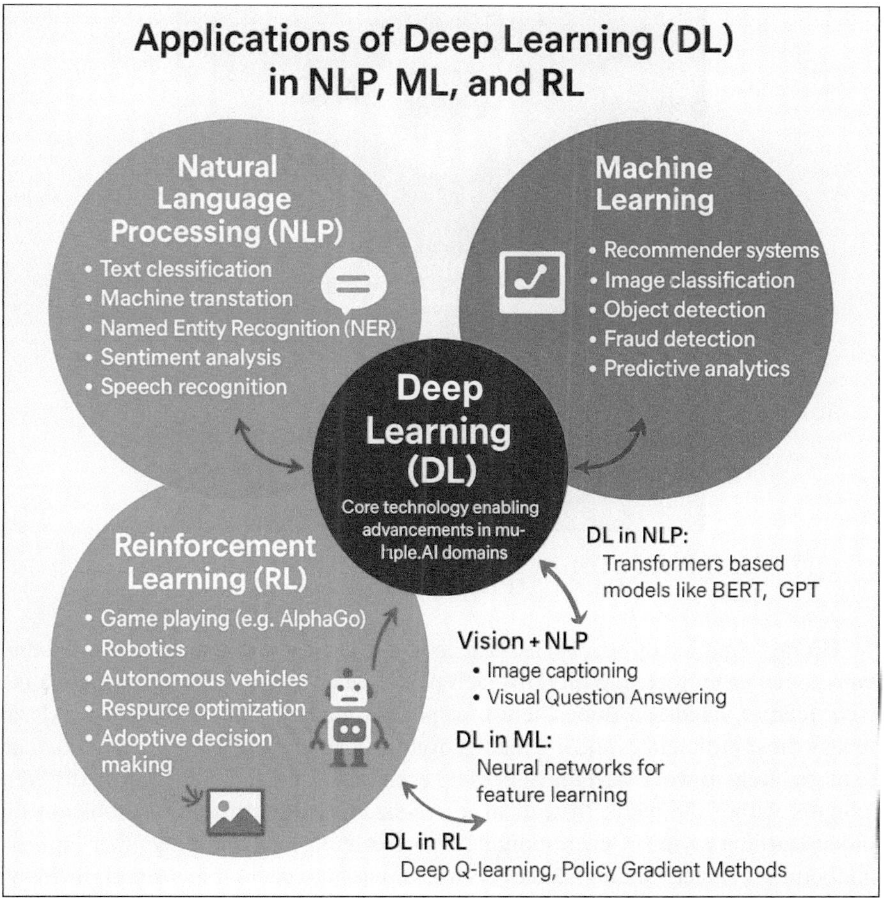

Fig. 6.3 NLP in education.

4. AI in Face-to-Face Collaborative Learning

4.1 AI-assisted Group Formation based on Learning Styles

When it comes to group projects or team-based learning, putting the right people together can make a big difference. That is where AI comes in. With AI-assisted group formation, the system can look at how each student learns best—whether they prefer visual aids, hands-on activities, discussions, or reading and writing—and use that information to build well-balanced teams. Instead of leaving group selection to chance, AI helps match students whose learning styles complement each other. This way, everyone gets a better chance to contribute in their own way, learn from different perspectives, and grow together as a team. It is like having a smart helper behind the scenes, making sure the group dynamic works in everyone's favour. Figure 6.4 illustrates key AI technologies used in education, including NLP, recommender systems, learning analytics, and robotics.

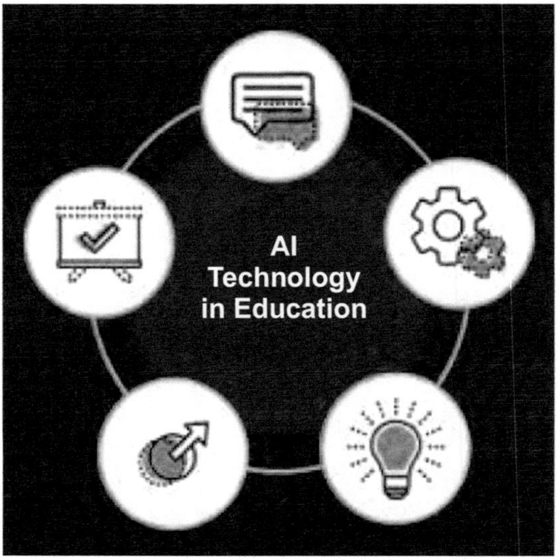

Fig. 6.4 AI Technologies in education.

Group formation through AI assistance takes the guesswork out of putting student teams together by focusing on how each person learns best. Whether someone is a visual learner, hands-on (kinesthetic), or prefers reading and listening, the AI can identify these preferences and use them to form balanced groups. This means students are more likely to work well together, help each other out, and benefit from different strengths within the team. Instead of a one-size-fits-all approach, AI looks at the unique learning styles of each individual to create custom groupings that improve collaboration, understanding, and overall learning outcomes. It is a smart, inclusive way to make sure everyone gets the most out of the group experience.

Figure 6.5 shows how AI can play a helpful role in forming student groups by understanding their unique learning styles. Each person in the circle represents a different way of learning—visual, creative, expressive, or reflective. Instead of randomly assigning people into teams, AI looks at these learning preferences and creates groups where members can complement each other. This approach helps ensure everyone gets a chance to shine and learn from one another, making teamwork more balanced and effective. It is like having a digital guide that understands how we learn best and uses that to build stronger, smarter groups.

4.2 AI-powered Real-time Feedback and Assessment Tools

Imagine turning in an assignment and getting helpful feedback almost instantly—no more waiting days for results—that is the power of AI in real-time feedback and assessment. These smart tools can quickly review your work, spot mistakes, highlight what you are doing well, and even suggest ways to improve. Whether it is writing, quizzes, or problem-solving tasks, AI can analyse your progress on the spot and give you insights that help you learn faster. It is like having a personal coach who is always available, helping you grow as you go instead of after the fact. For teachers,

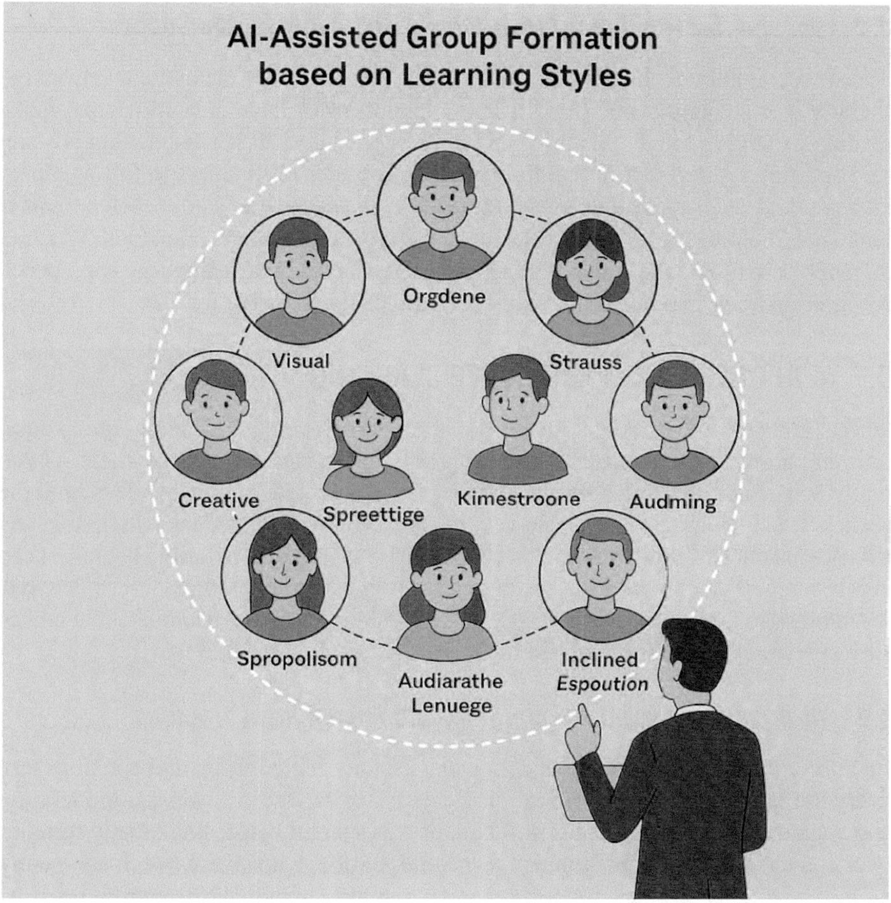

Fig. 6.5 AI-assisted group formation based on learning styles.

it also means they can better understand each student's needs and adjust their support in real time.

4.3 Intelligent Classroom Monitoring and Analytics

Think of intelligent classroom monitoring as a quiet helper in the background, keeping an eye on how the class is going. Using smart tools and data, it can track things like student engagement, participation, and even how well lessons are being understood. It is not about spying—it is more like giving teachers a clearer picture of what is working and what is not. For example, if some students seem to be struggling or losing focus, the system can flag it early, so the teacher can step in with extra support. Over time, analytics can show patterns that help improve teaching strategies and make learning more personalized. It is all about creating a classroom that adapts to students' needs and helps everyone stay on track.

4.4 Speech Recognition for Multilingual and Accessibility Support

Speech recognition technology is like giving a voice to everyone in the classroom—literally. For students who speak different languages or have disabilities that make reading or writing tough, this tool can be a game-changer. It listens to what is being said and turns it into text in real time, making it easier for learners to follow along, take notes, or even communicate their thoughts. It can also translate speech into other languages, helping multilingual students feel more included and understood. Whether it is supporting someone with a hearing impairment or bridging a language gap, speech recognition helps make learning more accessible and inclusive for all.

5. AI IN ONLINE COLLABORATIVE LEARNING

Online group work can sometimes feel a bit messy—people in different time zones, uneven participation, or just not knowing how to get started. That is where AI can really help. In online collaborative learning, AI can act like a smart guide behind the scenes. It helps organize tasks, suggest roles based on each student's strengths, and even keeps track of how everyone is contributing. If someone is falling behind, it can gently nudge them or alert the teacher. AI can also recommend resources or spark discussions to keep the group on track. It is like having a digital teammate that makes sure everyone's voice is heard and the learning stays focused and fair.

5.1 AI-driven Discussion Moderation and Engagement Tracking

In online classes or forums, keeping discussions meaningful and making sure everyone is involved can be a real challenge. That is where AI-driven moderation and engagement tracking comes in. Think of it as a smart helper that quietly watches over conversations—keeping things respectful, on-topic, and inclusive. It can gently guide discussions, flag inappropriate content, or highlight thoughtful contributions. At

Fig. 6.6 AI-driven discussion moderation.

the same time, it keeps track of who is actively participating and who might be drifting away, so teachers can step in and offer support if needed. It is not about policing students, but about creating a safe, balanced space where everyone feels encouraged to speak up and stay engaged. Figure 6.6 shows the AI-driven discussion moderation and engagement tracking.

5.2 Smart Tutoring Systems for Remote Learning Support

Learning from home can sometimes feel isolating, especially when you are stuck on something and there is no one around to help. That is where smart tutoring systems come in. These AI-powered tools act like a personal tutor, available 24/7 to guide you through lessons, explain tricky concepts, and even give you practice problems tailored to your level. They adjust as you go, offering more help when you are struggling and stepping back when you're doing well. It is like having a patient, always-available tutor who understands how you learn and helps keep you motivated and on track—even when you are learning remotely.

5.3 AI-powered Peer Assessment and Feedback Systems

Giving and receiving feedback from classmates can be super valuable, but it is not always easy to know what to say—or how fair the feedback is. That is where AI-powered peer assessment systems step in. These tools help guide students through the process of reviewing each other's work by offering prompts, checking for quality, and even spotting bias or overly harsh comments. They make sure the feedback is helpful, respectful, and focused on learning. For students, it means getting more balanced input, not just from the teacher but from peers too. And for teachers, it saves time while still building a strong, supportive learning community.

5.4 Virtual Learning Assistants and AI-powered Adaptive Quizzes

Virtual learning assistants are like friendly study buddies who are always there when you need help. They can answer questions, explain tough topics, and even remind you about deadlines—it is like having a super organized friend by your side. Paired with AI-powered adaptive quizzes, learning becomes even more personalized (see Figure 6.7). These smart quizzes adjust in real time based on how you are doing—if you are facing the questions, they will challenge you more, and if you are struggling, they will slow down and offer easier ones to help you catch up. Together, they make learning feel less overwhelming and way more tailored to your pace and needs, helping you grow with confidence.

6. AI-Powered Learning Analytics and Personalization

Every student learns differently—some move fast, others need more time, and everyone has their own strengths and struggles. AI-powered learning analytics helps make sense of all that. It quietly tracks how students are doing—like which topics they breeze through and where they tend to get stuck—and uses that information to personalize the learning experience. Instead of a one-size-fits-all approach, the system

Fig. 6.7 Virtual learning assistants.

tailors lessons, resources, and even the pace to fit each student's needs. It is like having a teacher who really *gets* you, adjusting the plan so you can learn in the way that works best for you. It helps students stay motivated and gives teachers helpful insights to better support each learner.

6.1 Predicting Student Progress and Performance

Imagine if teachers could get a heads-up when a student might be falling behind—or even when someone's about to shine. That is exactly what AI can do by predicting student progress and performance. By looking at patterns in things like quiz scores, participation, and study habits, AI can spot early signs of struggle or success. It is not about labelling students, but about helping them. With this kind of insight, teachers can step in with support *before* a student starts slipping, or challenge them more when they are ready for it. It is like having a crystal ball that helps create a smoother, more supportive learning journey for everyone.

6.2 Identifying Struggling Students for Timely Intervention

Sometimes students fall behind quietly—maybe they are too shy to ask for help or do not even realize they are struggling. That is where AI can make a real difference. By keeping track of things like missed assignments, low quiz scores, or lack of participation, AI can spot patterns that suggest a student might need extra support. It is like a gentle alert system that helps teachers catch problems early, so they can step in with encouragement, resources, or one-on-one help before things get worse. Instead of letting someone slip through the cracks, this kind of timely intervention makes sure no student gets left behind.

6.3 Learning Behaviour Analysis and Adaptive Course Recommendations

Everyone has their own learning habits—some like to dive in and explore, while others prefer to take things step by step. AI can observe these behaviours, like how long you spend on a topic, what kinds of resources you use, or how often you revisit material. This is called learning behaviour analysis, and it helps create a clearer picture of how you learn best. Based on that, the system can suggest courses or learning paths that really fit your style and goals. It is kind of like having a personal academic advisor who actually *knows* how your brain works, guiding you towards the right content at the right time to keep you motivated, challenged, and on track.

7. Ethical Considerations and Challenges

As amazing as AI can be in education, it is important to pause and think about the ethical side of things. For starters, there is the question of privacy—students' data is being collected and analysed, so we need to make sure it is handled safely and respectfully. Then there is the issue of fairness: if the AI is not trained properly, it might unintentionally favour some students over others. Plus, we do not want to rely so much on technology that we forget the human side of teaching. Striking the right balance between smart tools and real human connection is key. So, while AI opens up exciting new possibilities, we have to use it thoughtfully, keeping students' well-being, rights, and dignity at the centre of it all.

7.1 Data Privacy and Security Concerns in AI-driven Learning

With AI playing a bigger role in education, it is natural to wonder where all that student data goes and who gets to see it. Every time a student takes a quiz, asks a question, or interacts with a learning platform, bits of personal information are collected. While this helps AI customize the learning experience, it also raises serious concerns about privacy and security. What if that data gets into the wrong hands? Or what if it is used in ways students or parents did not agree to? That is why it is so important to have strong protections in place—clear policies, secure systems, and full transparency—so that students can learn with confidence, knowing their information is safe and respected.

7.2 Bias in AI-based Recommendations and Assessments

Artificial intelligence is smart, but it is only as fair as the data it learns from. Sometimes, the systems behind AI-based recommendations and assessments can unintentionally carry biases—like favouring certain groups of students over others, or making unfair judgements based on past patterns. For example, if the data used to train the AI mostly comes from one type of learner or background, then the recommendations it gives might not work well for everyone. That means some students could be pushed towards the wrong materials or judged more harshly, even if they are doing their best. It is a big reminder that while AI can be super helpful, we still need real people keeping an eye on things to make sure it stays fair, inclusive, and supportive for *every* learner.

7.3 The Balance between AI Automation and Human Interaction

Artificial intelligence can do a lot—grade quizzes in seconds, offer personalized feedback, and even suggest what you should study next. But as helpful as all that is, it cannot replace the warmth, empathy, and understanding that comes from real human connection. Teachers do more than just deliver content—they listen, motivate, and build relationships that make learning meaningful. So, while AI can take care of the repetitive tasks and help personalize learning, it is important to keep people at the heart of education. The best approach is finding that sweet spot where AI handles the techy stuff and humans bring the heart—working together to create a learning experience that is smart *and* supportive.

7.4 Digital Divide and Accessibility Issues

As education becomes more digital, it is easy to forget that not everyone has the same access to technology. Some students may not have a reliable internet connection, a personal device, or even a quiet place to study—and that creates a real gap, known as the digital divide. While AI-powered learning tools can be incredibly helpful, they can also leave some students behind if they do not have the right resources to use them. On top of that, students with disabilities may struggle if platforms are not designed with accessibility in mind. That is why it is so important to make sure that digital learning tools are inclusive, affordable, and accessible for everyone—because every student deserves the same chance to learn and succeed, no matter where they are or what challenges they face.

8. Future Trends and Innovations

The future of learning is looking pretty exciting, thanks to the rapid growth of AI and other smart technologies. We are talking about virtual tutors that feel almost human, learning platforms that adapt like a real teacher would, and even immersive experiences using technologies like VR to bring lessons to life. Imagine walking through ancient cities in history class or conducting science experiments in a virtual lab—all guided by AI that knows how you learn best. We will also see smarter ways to track progress, spot learning challenges early, and offer support that feels personal and timely. As these innovations grow, the goal is not to replace teachers, but to empower them—and to make learning more engaging, inclusive, and tailored to each individual. The classroom of the future is not just digital, it is also dynamic and built around the learner.

8.1 AI and Metaverse in Collaborative Learning

Imagine stepping into a virtual classroom where you and your classmates—no matter where you are in the world—can meet, explore, and learn together as if you are in the same room. That is what happens when AI meets the Metaverse in collaborative learning. In this virtual space, you can work on group projects, conduct experiments, or even tour historical landmarks—all through your digital avatars. AI helps make the experience smoother by guiding interactions, offering instant support, and even adjusting the environment to suit different learning styles. It is more than just cool tech—it is a new way of connecting, learning, and creating together in a space that feels engaging, immersive, and personalized. It is like the group study session of the future, where distance does not get in the way of teamwork or learning.

8.2 AI-powered VR and AR Applications

Learning is more fun—and effective—when you can actually *experience* it, and that is exactly what AI-powered VR and AR make possible. With VR, students can explore places they would never reach in real life—like diving into the ocean for a science lesson or walking through ancient civilizations in history class. AR, on the other hand, brings learning to life right in front of you, layering information or 3D objects onto the real world using a phone or tablet. Now, when you mix in AI, these experiences get even smarter. The system can adapt to how you are interacting, guide you through tasks, and offer personalized feedback along the way. It is like stepping into a lesson that knows exactly how to help you learn—making education more hands-on, immersive, and unforgettable.

8.3 AI-driven Emotional Intelligence and Affective Computing

Learning is not just about facts and grades—it is also about feelings. Sometimes, students feel frustrated, bored, or overwhelmed, and that can really affect how they learn. That is where AI-driven emotional intelligence, also known as affective computing, comes into play. These smart systems can pick up on emotional cues through things like facial expressions, voice tone, or typing patterns, and respond with empathy. For example, if a student seems stressed, the system might offer encouragement, adjust the difficulty of a task, or alert a teacher to check in. It is like giving technology the ability to care—not just track progress, but actually understand how a student *feels* in the moment. This kind of emotional awareness helps create a more supportive, human-centred learning environment where students feel seen, heard, and cared for.

8.4 The Role of Generative AI in Content Creation and Assessment

Generative AI is like having a super-creative assistant in the classroom—it can help teachers and students alike by creating content on the fly. Need a quick quiz, a reading passage, or a creative writing prompt? Generative AI can whip one up in seconds, saving time and sparking new ideas. It can also help with assessments by generating practice questions, summarizing complex topics, or even giving personalized feedback based on a student's work. For teachers, it means more time to focus on what really matters—connecting with students and supporting their growth. And for students, it means getting learning materials that are tailored, engaging, and just right for their level. It is a powerful tool that adds flexibility and creativity to both teaching and learning.

9. Conclusion

In the end, advanced AI technologies are not just reshaping how we teach, they are also transforming how we *connect*, *support*, and *inspire* learners. From smart tutoring systems and real-time feedback to personalized learning paths and virtual assistants, AI is helping make education more adaptive, inclusive, and engaging. But as we embrace these innovations, it is just as important to stay mindful of the challenges—

like data privacy, digital access, and the need for human connection. When we find the right balance between technology and empathy, we can build an education system that not only keeps up with the times but also truly empowers every learner to thrive. As discussed in earlier sections, technologies such as NLP, recommender systems, and generative AI are transforming how we design, deliver, and personalize education.

References

Baidoo-Anu, D. and Ansah, L.O. (2023). Education in the era of generative artificial intelligence (AI): Understanding the potential benefits of ChatGPT in promoting teaching and learning. *Journal of AI*, *7*(1), 52–62.

Baker, R.S.J.d. and Inventado, P.S. (2014). Educational data mining and learning analytics. In J.A. Larusson and B. White (Eds.), *Learning Analytics: From Research to Practice* (pp. 61–75). Springer.

Cain, C.C., Buskey, C.D. and Washington, G.J. (2023). Artificial intelligence and conversational agent evolution—A cautionary tale of the benefits and pitfalls of advanced technology in education, academic research, and practice. *Journal of Information, Communication and Ethics in Society*, *21*(4), 394–405.

Chen, L., Chen, P. and Lin, Z. (2020). Artificial intelligence in education: A review. *IEEE Access*, *8*, 75264–75278.

Chen, X., Zou, D., Xie, H., Cheng, G. and Liu, C. (2022). Two decades of artificial intelligence in education. *Educational Technology & Society*, *25*(1), 28–47.

Dimitriadou, E. and Lanitis, A. (2023). A critical evaluation, challenges, and future perspectives of using artificial intelligence and emerging technologies in smart classrooms. *Smart Learning Environments*, *10*(1), 12.

Holmes, W., Bialik, M. and Fadel, C. (2019). *Artificial Intelligence in Education: Promises and Implications for Teaching and Learning*. Center for Curriculum Redesign. https://doi.org/10.54675/PCSP7350

Kizilcec, R.F. (2024). To advance AI use in education, focus on understanding educators. *International Journal of Artificial Intelligence in Education*, *34*(1), 12–19.

Luan, H., Geczy, P., Lai, H., Gobert, J., Yang, S.J., Ogata, H. and Tsai, C.C. (2020). Challenges and future directions of big data and artificial intelligence in education. *Frontiers in Psychology*, *11*, 580820.

Luckin, R., Holmes, W., Griffiths, M. and Forcier, L.B. (2016). *Intelligence Unleashed: An Argument for AI in Education*. Pearson Education.

Miao, F., Holmes, W., Huang, R. and Zhang, H. (2021). AI and education: Guidance for policy-makers (Paris, UNESCO; approx. 45 pp.; ISBN 978-92-3-100447-6). UNESCO.

OECD. (2021). *AI and the Future of Skills, Volume 1: Capabilities and Assessments*. Organisation for Economic Co-operation and Development.

Panigrahi, R., Srivastava, P.R. and Sharma, D. (2021). Applications of artificial intelligence in education: Current trends and future prospects. In *AI and Big Data for Innovation in Higher Education* (pp. 53–72). IGI Global.

Pedro, F., Subosa, M., Rivas, A., Valverde, P. (2019). Artificial Intelligence in Education: Challenges and opportunities for sustainable development. 46 pages. UNESCO (United Nations Educational, Scientific and Cultural Organization), Paris. ttps://www.gcedclearinghouse.org/sites/default/files/resources/190175eng.pdf

Selwyn, N. (2019). *Should Robots Replace Teachers? AI and the Future of Education*. Polity Press.

Suryanarayana, K.S., Kandi, V.P., Pavani, G., Rao, A.S., Rout, S. and Krishna, T.S.R. (2024). Artificial intelligence enhanced digital learning for the sustainability of education management system. *The Journal of High Technology Management Research*, *35*(2), 100495.

UNESCO. (2021). *AI and Education: Guidance for Policy-makers*.

Zhai, X., Chu, X., Chai, C.S., Jong, M.S.Y., Istenic, A., Spector, M., ... and Li, Y. (2021). A review of artificial intelligence (AI) in education from 2010 to 2020. *Complexity*, *2021*(1), 8812542.

CHAPTER 7

AI-driven Collaborative Learning: Enhancing Group Dynamics, Participation, and Emotional Awareness in Hybrid Education

Karimunnisa Shaik[1*], *Dudekula Siddaiah*[2] and *B. Niveditha*[3]

1. Introduction

The growing number of changes in education has led to collaborative approaches being recognized as a main way to stimulate critical thinking, teamwork, and useful learning. Whether students are working together in-person or online, they become more involved with what they are studying and share ideas through collaboration. It improves a person's knowledge and also teaches important skills like talking, listening, appreciating others, and team play (Rosé & McLaughlin, 2021; Suthers & Verbert, 2020).

Artificial intelligence (AI) has become more important in transforming the way students and teachers work together because of technological changes. Besides undertaking administrative work or personal tutoring, AI is now being used to improve classes for groups, whether taught in-person or online (Zinelli, 2025; Luckin, 2023; Roll & Wylie, 2021). By automating some of the tasks, it now becomes important to figure out the role of AI alongside humans in cooperative activities (Ahmed & Karim, 2023; Rosé & McLaughlin, 2021).

This chapter examines how AI helps with learning by group work, encouraging involvement, and resulting in better learning outcomes. AI today ranges from helping with discussions to forming learning teams, quietly but productively backing up what teachers and students do in school (Tzirides et al., 2025; Rosé & McLaughlin, 2021;

[1] Department of Information Technology, Marri Laxman Reddy Institute of Technology and Management, Telangana, India. https://orcid.org/0009-0007-8160-252X

[2] Department of Computer Science, Pulla Reddy Institute of Computer Science, Telangana, India. https://orcid.org/0009-0009-4158-2688

[3] Department of Computer Science and Engineering, Marri Laxman Reddy Institute of Technology and Management, Telangana, India. https://orcid.org/0009-0005-5813-8082.

* Corresponding author: karimunnisa.sk03@gmail.com

DOI: 10.1201/9781003639367-7

Heffernan & Heffernan, 2021; Dillenbourg, 2022). When we look at these changes, we also discuss the moral, practical, and educational aspects of using AI in team-based learning (Lee & Zhang, 2024; Kukulska-Hulme, 2024; Tan & Lajoie, 2021).

2. Understanding Collaborative Learning

Collaborative learning is not only about giving students access to each other in teams. It mainly involves learners coming up with shared meanings by talking, questioning one another, and interactively forming new insights (Rosé & McLaughlin, 2021; Wise & Schwarz, 2021). An effective group has open talks, shows respect to everyone, and each person plays a part in reaching a shared aim. Be it face-to-face or online, working together in an effective way does not continue on its own; it needs specific planning, a well-structured environment, and serious scaffolding to help both the academic and social aspects (Suthers & Verbert, 2020; Looi & Wong, 2022).

However, collaborative learning needs some special attention to succeed. Educators come across common problems like unequal involvement by students, arguments and conflicting attitudes among classmates, blurry division of roles, limited feedback given on time, and unsuitable groupings (D'Mello & Graesser, 2021; Kukulska-Hulme, 2024; Goggins & Valetto, 2020). Such challenges may block students from getting a good interaction and reaching their learning goals. Difficulties in virtual environments are usually greater than they are in real surroundings. Because people lack face-to-face and daily social connections, such as those found in the office, opportunities for misunderstanding and feeling cut off from others may increase, and on top of that, managing time differences and electronic networking is not easy (Johnson & Lester, 2020; Dillenbourg, 2022).

Artificial intelligence can step in to make a big difference in this area. By breaking down how students interact, adjusting to learner needs, and giving immediate support, AI enhances equity, makes collaboration more responsive, and helps classes succeed no matter if they meet face-to-face or online (Zinelli, 2025; Ahmed & Karim, 2023; Chen & Looi, 2022).

3. Role of AI in Collaborative Learning

Artificial intelligence in education (AIED) offers the potential to address key barriers in collaborative learning by providing personalized support, adaptive feedback, real-time analytics, and dynamic group management. This section discusses the primary ways in which AI is being applied.

3.1 Intelligent Group Formation

An important phase of collaboration is selecting teams that have differences in both abilities and views and also share similar ways of interacting and prefer a similar working style. Not forming groups properly can cause some members to join in less, see differences arise among participants, or just become uninterested, which ruins what teamwork can do (Kukulska-Hulme, 2024; Wise & Schwarz, 2021). Usually, instructors would create groups by using methods like arranging by seating, choosing in alphabetical order, or simply deciding randomly. Even though they are easy to use, these approaches frequently omit the important factors that affect how well a group works (Rosé & McLaughlin, 2021; Heffernan & Heffernan, 2021).

Artificial intelligence uses a clever and exact method for organizing people into groups. With information about students' old grades, how they learn, their cognitive performance, approaches to communication, and personality, AI generates groups that support students' success (Wang & Lin, 2022; Noroozi & Hatami, 2023). Such systems collect past historical data to identify typical behaviours among students and anticipate those groups that are likely to be successful (Zinelli, 2025; Holstein & Aleven, 2020).

For example, an AI system can pair a student leading the way with others who are able to analyse things or come up with ideas, despite not being as active socially. It could also think about learning goals and ensure that students with different talents help each other out in useful ways. Using data, the approach makes it possible to build balanced teams in which all members can help and gain advantages from each other's skills (Tzirides et al., 2025; Zhou et al., 2025; Goggins & Valetto, 2020).

Moreover, smart algorithms for dividing students into groups can be developed to fit certain learning objectives. Homogeneity may be chosen if the task is about covering certain material, but heterogeneity is used to include multiple views on the matter. AI plays a role in making sure students experience different social groups throughout their time at school, which encourages everyone to take part in the learning process (Ahmed & Karim, 2023; Roll & Wylie, 2021; Tsai & Gasevic, 2020).

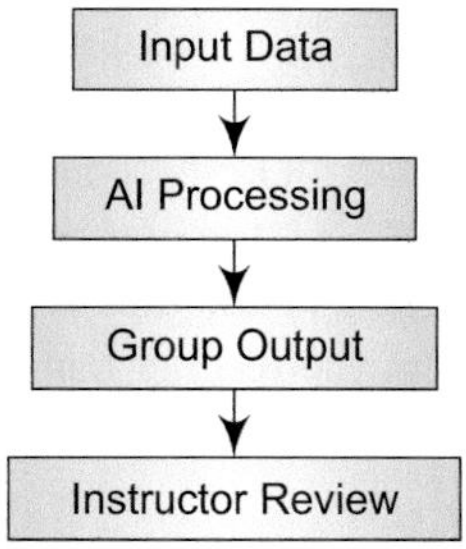

Fig. 7.1 As shown in Fig. 7.1, the AI-supported group formation process involves input data, AI processing, group output, and instructor review (Ahmed & Karim, 2023; Roll & Wylie, 2021; Tsai & Gasevic, 2020).

As AI systems become more adaptive, they can learn from the outcomes of past groupings. By analysing peer evaluations, engagement levels, and collaborative outcomes, these systems can refine their grouping strategies over time, improving accuracy and effectiveness. In this way, AI becomes not just a tool for group assignment, but also a dynamic facilitator of evolving learning communities.

Ultimately, intelligent group formation powered by AI moves beyond convenience; it becomes a strategic element of instructional design, helping to foster collaboration that is purposeful, balanced, and conducive to deeper learning.

3.2 AI-supported Facilitation: The Role of Intelligent Agents in Guiding Collaboration

A successful collaborative learning process depends mainly on effective teaching. In traditional classrooms, the teachers and facilitators lead the discussions, help everyone take part, sort out conflicts, and encourage students to learn more deeply. But with educational environments becoming more dependent on computers and online approaches, instructors are unable to help as many people at one time as they used to. At this point, intelligent agents or virtual facilitators become a promising help because of AI (Johnson & Lester, 2020; Heffernan & Heffernan, 2021; Roll & Wylie, 2021).

3.2.1 Grasping the Benefits of Using AI

An AI-supported facilitation means that AI systems are involved in or manage different learning activities in a group. AI facilitators support learners by taking part in the lessons, giving helpful feedback, encouraging students to work together, and demonstrating useful behaviour (Rosé & McLaughlin, 2021; Holstein & Aleven, 2020; Wise & Schwarz, 2021). Many times, they use natural language processing (NLP), machine learning (ML), and dialog systems, which help them make sense of student comments, look out for teamwork, and give appropriate answers. If they are created in a certain way, they act like chatbots, extra learning companions, or teachers on computers (D'Mello & Graesser, 2021; Heffernan & Heffernan, 2021; Chen & Looi, 2022).

3.2.2 Key Functions of AI Facilitators in Collaborative Learning

1. Encouraging Participation: Uneven participation is one of the biggest difficulties teams usually face. A few students take part actively in class, while most others are less involved or not involved at all. AI assistants can review who is taking turns in discussion, check the number of responses, and notice people who are not contributing much to the discussion. When it appears that some students are not participating, the teacher can offer them positive and inviting questions to help them take part (Ahmed & Karim, 2023; Rosé & McLaughlin, 2021; Brom & Hlosta, 2023).
2. Focusing and Adapting Discussions: It is easy for group discussions to take a turn away from the topic in any setting. AI can review the context of a communication and easily match its relevant meaning to the main goals. If people begin to stray, the system may remind them softly or pose some thoughtful questions (Heffernan & Heffernan, 2021; Holstein & Aleven, 2020; Wise & Schwarz, 2021).
3. Guiding Students to Think Critically: AI facilitators can be created to have strategies that encourage users to focus more on the information. They sometimes ask detailed questions, offer their own viewpoints, or try to make you say more by acting like an experienced person in charge of a discussion (Rosé & McLaughlin, 2021; Chen & Looi, 2022).
4. Taking Care of Disputes and Following Rules: Sometimes, people in a group disagree or fail to understand each other. Whilst some conflict helps students, it can also get in the way of learning. AI developers are designing advanced

systems with features that recognize any negative emotions or situations that are becoming tense. These agents are able to suggest ways to relax and let instructors know when necessary (D'Mello & Graesser, 2021; Rodrigo & Baker, 2021; Tan & Lajoie, 2021).

5. Demonstrating How to Communicate Well: When students or those with different native languages are in a group, they may be uncertain of how to collaborate well with others. AI agents can act as if they were summarizing comments, asking for further understanding, and showing respect when disagreeing with someone (Kukulska-Hulme, 2024; Looi & Wong, 2022).

3.2.3 Examples

- Bazaar is a conversational agent designed for online STEM teamwork. It prompts students to debate and align on solutions. Studies show students paired with Bazaar engaged more deeply and performed better (Rosé & McLaughlin, 2021; Heffernan & Heffernan, 2021).
- AutoTutor and DeepTutor, originally designed for individuals, are now adapted for small-group tutorials (D'Mello & Graesser, 2021; Heffernan & Heffernan, 2021; Chen & Looi, 2022).

3.2.4 Why Using AI is Beneficial for Facilitation

- AI agents can efficiently support large-scale environments such as MOOCs and blended classrooms (Kiesler et al., 2025; Ahmed & Karim, 2023; Luckin, 2023).
- Since AI follows consistent methods, it supports students fairly and scales easily.
- With each interaction, AI systems become more tuned to learners' needs, improving support and responsiveness (Holstein & Aleven, 2020; Roll & Wylie, 2021).

3.2.5 Challenges and the Benefits of Teamwork

Still, AI does not fix every possible problem. AI has difficulties with feeling empathy, making discerning decisions, and understanding cultural differences, while people do these things naturally (Lee & Zhang, 2024; D'Mello & Graesser, 2021; Tan & Lajoie, 2021). Using both AI and human educators offers the greatest benefit, since AI is fast and can help many students, while humans can handle more delicate, emotional, or moral matters (Johnson & Lester, 2020; Kukulska-Hulme, 2024).

In brief, there are several essential points about learning. The help of AI makes it easier to guide cooperation that is both balanced and aims to achieve set goals (Tzirides et al., 2025; Ahmed & Karim, 2023; Wise & Schwarz, 2021). Intelligent agents can instantly guide, change the focus, and back up students in their learning (Rosé & McLaughlin, 2021; Holstein & Aleven, 2020).

Even though AI is strong, it should work together with human facilitators (Zinelli, 2025; Heffernan & Heffernan, 2021; Luckin, 2023).

Table 7.1 highlights the primary functions of AI in collaborative learning environments. It includes examples of AI tools and references that support each role's effectiveness in enhancing teamwork and engagement.

Table 7.1 Key Roles of AI in Collaborative Learning

Role	*Description*	*Example Tools*	*Supporting Reference*
Intelligent Group Formation	Uses student data to form balanced and effective groups	AI algorithms	Rosé & McLaughlin (2021)
AI-supported Facilitation	Virtual agents guide collaboration and participation	Bazaar, AutoTutor	Heffernan & Heffernan (2021)
Real-Time Analytics	Monitors group dynamics and provides instant feedback	Dashboards, NLP	Brom & Hlosta (2023)
Emotion Recognition	Detects emotional states via text, speech, and video	Emotion AI systems	Tan & Lajoie (2021)

3.3 Real-time Analytics and Feedback

Group learning depends a lot on when each activity takes place. While taking part in group work, no matter if it is face-to-face or through an online platform, the process of interacting is usually just as important as the result. However, many obstacles including miscommunication, disengagement, and different involvement of students appear in the classroom and call for quick action. Most of the time, feedback given after an activity or observed by the instructor does not address things that appear while learning happens. This is the place where real-time use of AI for analytics and feedback can change things for the better (Ahmed & Karim, 2023; Brom & Hlosta, 2023; Tsai & Gasevic, 2020).

Such systems can always observe students as they work together and store this information for review. They receive inputs in the form of text messages in chats or documents, what is spoken on audio platforms, and information about how users behave, move the cursor, or spend time working. Some advanced tools are able to understand a person's emotions based on their facial expressions and how they speak. By analysing all the data, AI forms a changing picture of how every student and group is functioning (Rodrigo & Baker, 2021; Gašević & Dawson, 2020).

The main benefit of this process is that it supplies timely and useful information to guide activities. Whenever a group member has not participated for a long time, the system could encourage them to take part again. At the same time, it could inform the teacher so they can help right away. A similar thing can happen if the discussion begins to veer off the assigned topic; in this case, AI might offer hints or even guide the group back to discuss the main idea. Since there is always an immediate response to their questions, the learners continue to move ahead and meet their objectives (Zinelli, 2025; Ahmed & Karim, 2023).

Besides merely checking how much time people spend on the site, AI can also study how they communicate with others. These tools can check which statements students make, such as fact sharing, debating, asking, or offering analysis. When the system notices that a group's talk is not thoughtful, it could try to get group members to talk about their ideas deeper, mention alternative beliefs, or support their statements. With these prompts, the teacher does not have to closely watch the students' discussions all the time (Rosé & McLaughlin, 2021; Wise & Schwarz, 2021; Tsai & Gasevic, 2020).

Since students do not interact face-to-face online, a fast analysis helps even more. AI has the ability to detect early signs of disengagement by spotting brief responses, longer breaks between talks, or a sharp decrease in contact. Should AI sense frustration or confusion in someone's emotional tone, it can let them know about helpful things, break time, or say something uplifting. Although empathetic AI is still being perfected, it helps the teacher understand students' feelings and encourage them in online learning (Lee & Zhang, 2024; D'Mello & Graesser, 2021; Tan & Lajoie, 2021).

Real-time learning analytics help both the teacher and the learner. It is better for instructors to watch group activities through dashboards than to revisit notes or transcripts later (Kay et al., 2022; Brom & Hlosta, 2023; Tsai & Gašević, 2020). They give a picture of the groups that are more involved, the members making the biggest contributions, and the teams facing challenges. Being able to see their students' progress, a teacher can help each learner with targeted guidance when necessary. Instead of only watching or solving problems as they occur, the teacher directs students using information from data (Brom & Hlosta, 2023; Tsai & Gasevic, 2020; Gašević & Dawson, 2020).

As an example, take a class where students, seated in groups, work on a design assignment. As learners are working, the AI system works behind the scenes, checking their files and listening to what they are saying. It detects that one student has been quiet and has not made any progress in a while, and that the team has not given any reasons from sources to back up how they made their decisions. The AI recommends making decisions with the help of evidence and gently suggests the silent student to come with ideas. At the same time, the teacher spots the patterns appearing on their dashboard and takes part in the group to give more help. At this point, technology teams up with people's understanding to give all students the best and most equal learning opportunities (Zinelli, 2025; Ahmed & Karim, 2023; Chen & Looi, 2022)

Yet, while real-time AI is very effective, it can also cause some major concerns. If students receive too much feedback for how many times they mess up, it can seem overbearing (Ahmed & Karim, 2023; Brom & Hlosta, 2023). Privacy has to be respected, so systems that observe our voices, movements on camera, or emotions should work openly and be approved by users (Lee & Zhang, 2024; Kukulska-Hulme, 2024; Rodrigo & Baker, 2021). It is also hard for AI to understand sarcasm, the meaning behind cultural differences, and how context works. For that reason, using AI feedback should never be taken as a replacement for human decision-making (D'Mello & Graesser, 2021; Johnson & Lester, 2020; Tan & Lajoie, 2021).

If established and implemented in the right way, real-time analytics and feedback can improve group learning a lot. They deliver prompt information, encourage everyone to be involved, and make it easy for teachers to respond to students' needs determined as a group (Zinelli, 2025; Ahmed & Karim, 2023; Tsai & Gasevic, 2020). Thanks to digital progress and growing data in education, such tools help create kinder and more interactive learning environments (Luckin, 2023; Chen & Looi, 2022; Gašević & Dawson, 2020).

3.4 Content Recommendation and Personalization

The way groups are set up at the beginning is among the most vital, though usually overlooked, part of good collaborative learning. The arrangement of students by age or grade largely affects their interactions, communication, and the success they observe. Usually, in traditional classrooms, the instructor would arrange the groups himself or herself, often based on what is most comfortable or random. Of course, these methods are simple, but they usually ignore the way a person's characteristics can shape the way a group works together. This is one of the advantages of AI, since it makes use of data to create groups that are well-balanced, productive, and varied (Wang & Lin, 2022; Heffernan & Heffernan, 2021; Noroozi & Hatami, 2023).

In intelligent group formation, to create student teams AI looks at several things like cognitive skills, styles of communication, prior experience, emotional abilities, and personality. Composition of groups should focus on sharing necessary roles and talents to ensure optimal collaboration. Usually, AI systems use student academic history, the way they act online, their manner of studying, and a survey about their own abilities to predict how they will interact in various groups (Zinelli, 2025; Rosé & McLaughlin, 2021; Goggins & Valetto, 2020).

So, for example, AI on an online learning site may review student profiles, what they have handed in, and their posts in discussion forums to gauge who usually takes charge, asks difficult questions, and helps others. For that reason, it helps to make sure that every group has leaders, creative people, and supporters. It is possible for the algorithm to bring together a student with logical skills, one with strong verbal communication, and another with creativity and uncommon ideas. With this approach, students' skills and abilities in different subjects and ways of thinking are represented (Zhou et al., 2025; Roll & Wylie, 2021; Chen & Looi, 2022).

The reason AI is powerful in this situation is that it can learn and adjust as time goes by. Should specific groups always result in successful cooperation and studying, the system notices and uses this information to arrange groups in the future. Alternatively, if certain students have problems with some groups, the system should keep them from being grouped together again. As a result, AI moves beyond the role of former alone and keeps helping by learning from past experiences as well as new data (Ahmed & Karim, 2023; Holstein & Aleven, 2020; Tsai & Gasevic, 2020).

An illustration could be based on a business course that combines learners in class and has them come up with marketing campaign ideas. This AI can access students' writing abilities, creativity marks from previous endeavours, logs recording their videos, and reviews given and received by peers. It discovers that a few students work efficiently when they share the words with a partner, whereas the others join groups where structure and organization are key. With this information, it creates groups that are diverse and at the same time suited to each other. As time passes in the project, feedback and level of engagement among participants are evaluated, and grouping is changed to encourage a better connection among peers (Tzirides et al., 2025; Wise & Schwarz, 2021; Gašević & Dawson, 2020).

Naturally, it is important to handle the use of AI in group making with attention. Smart data usage can easily turn into dependable profiling. Clear algorithms are required, as students or teachers should be allowed to change them when the situation calls for it. It is important for students that they have a say when choosing their teachers. People running courses must know how the information is being used and avoid groupings that may strengthen unjust divisions (Lee & Zhang, 2024; Luckin, 2023; Kukulska-Hulme, 2024).

Still, AI can help groups come together in many significant ways. Recognizing and appreciating students' skills in their group is likely to make them feel more a part of the team and excited to participate. When people are grouped with intentional diversity and similar interests, they usually function better as a team. Instructors are given the opportunity to save time and ensure all the students show the necessary skills and knowledge from the start (Dillenbourg, 2022; Looi & Wong, 2022).

Sometimes the difference between insight and intuition is clear in the end results. Teachers are able to work together harmoniously, every person in the discussion adds something useful, and peers can help each other due to their complementary abilities. In this sense, AI is a way to help and become a valuable asset in shaping fulfilling classroom activities (Heffernan & Heffernan, 2021; Suthers & Verbert, 2020; Holstein & Aleven, 2020).

3.5 Sentiment and Emotion Recognition

Since most collaboration is now online, one main problem encountered by educators is difficulty in understanding their students' emotions. When teaching in a traditional classroom, teachers are able to sense whether a group is unsure, unhappy, tuned out, or upbeat from their facial work, body postures, and intonation. Such signals usually help teachers choose when to act and how to help their students. Unfortunately, virtual classrooms may eliminate these human signals, leading to students' emotional isolation and quiet problems without anyone realizing. To overcome this issue, AI is now applying methods that can analyse student emotions and sentiments to give fresh ideas about how students are doing (Lee & Zhang, 2024; D'Mello & Graesser, 2021; Tan & Lajoie, 2021).

Artificial intelligence studies student emotions in group work by using languages, tracking their face movements, understanding what they say, and watching how they type. It uses both words and other physical actions to guess if they are confused, bored, curious, stressed, or enthusiastic. If embedded into learning software, student response data can adjust the learning experience, give helpful advice, or tell teachers about any emotional hindrances students have (Rodrigo & Baker, 2021; Chen & Looi, 2022).

Artificial intelligence can spot the emotions in written comments when working in collaborative text apps. Repeated signs of irritation or apathy in a student's language, for example, saying, "I don't get this", can be detected by the system. It could suggest that the instructor reaches out or sends an encouraging message automatically. Additionally, if someone says something like "I like this a lot" or "This is an excellent plan", then it can tell us what to keep doing and pass it on to other groups (Rosé & McLaughlin, 2021; Noroozi & Hatami, 2023).

Special software in collaboration environments for video communication can quickly understand people's facial expressions and general tone of speech. If computer vision is combined with AI, it may identify behaviour such as furrowed brows, sighing, or not moving the eyes a lot as signs of someone being tired or having confusion. At the same time, it may listen to your voice to know if your stress or excitement is reflected in changes in pitch or speed. Combining all these signs makes it possible to see the group's atmosphere clearly, even if they are not present together in person (Lee & Zhang, 2024; Gašević & Dawson, 2020).

This technology can be put to use in intelligent tutoring systems that oversee students' discussions in small groups. When the system detects that group members are facing difficulties, it may interfere by giving a hint, by making the instructions clearer, or by advising a short break. Sometimes, if a student feels socially excluded or stressed, the system can recommend that the instructor offers private care over an automated message. Through this approach, the humanity in education is kept while making use of what AI can achieve well (D'Mello & Graesser, 2021; Heffernan & Heffernan, 2021; Tan & Lajoie, 2021).

The capacity to perceive and react to emotions is very important for forming inclusive and fair learning spaces. Members of the class with various backgrounds and ways of communicating can feel differently about working together as a group. Some find music exciting, yet it could also make some people feel nervous or separated. By being more aware of emotions, AI can highlight issues faced by quiet students so that educators can address each student's needs in a better way. Students now use language to support their studies and also their personal and social health (Kukulska-Hulme, 2024; Looi & Wong, 2022).

It goes without saying that using emotion recognition in schools involves major ethical obligations. People have important concerns regarding privacy, getting people's agreement, and interpreting what data means. People's emotional states depend on their personality and the details of the situation. A silence in the middle of a speech may look awkward, but at times it suggests real consideration. While yelling might seem like a sign of frustration, it could just as well suggest a positive feeling. AI tools, in spite of their progress, should be handled cautiously and with the knowledge that they are not capable of fully reading students' emotions. It is very important that the collection and use of emotions are clear to learners, and they should always have the freedom to refuse (Lee & Zhang, 2024; Luckin, 2023; Tan & Lajoie, 2021).

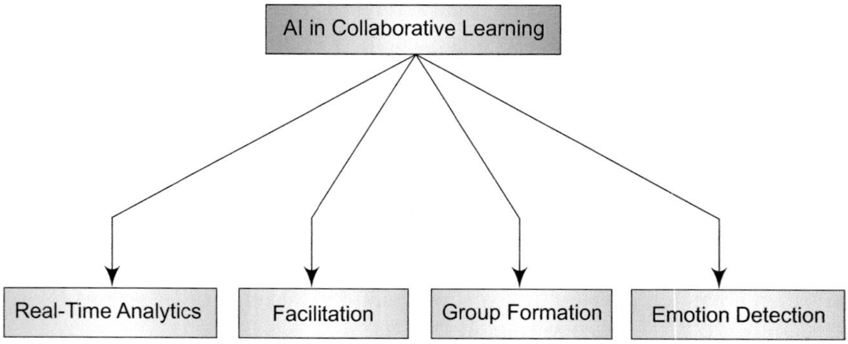

Fig. 7.2 AI in collaborative learning: Key functions.

Despite these caveats, emotion-aware AI has the potential to profoundly improve collaborative learning by bringing empathy and responsiveness into digital spaces. It can bridge the emotional gap in online education, help students feel seen and supported, and empower teachers with timely insights that might otherwise remain hidden. When used ethically and thoughtfully, sentiment and emotion recognition becomes more than a technological feature; it becomes a pathway to more human-centred education, even in the most technologically advanced environments.

4. AI in Face-to-Face vs Online Collaboration

4.1 Face-to-Face Contexts

In traditional schoolrooms, AI may be used with smartboards, sensors that pick up voices, and classroom surveillance. They gather data in the background and give information about learning that does not interfere with teaching. Indeed, AI can examine conversations among people by listening to the recordings and finding out if they are collaborating by, for example, asking questions, sharing details, and summarizing things (Johnson & Lester, 2020; Suthers & Verbert, 2020; Dillenbourg, 2022).

4.2 Online Environments for Learning

Working with other students using online tools gives AI more input, yet it is less personal than face-to-face classes. AI gets around this by keeping logs of keyboard presses, mouse movement, and when users interact with a computer system. Professional dashboard systems apply these inputs to display the team's interaction and give tips on how to increase their engagement (Ahmed & Karim, 2023; Brom & Hlosta, 2023; Gašević & Dawson, 2020).

Artificial intelligence is influencing the way people learn together, and we can see how some things tie them together and others are very different when it is used at the same time for face-to-face and online learning. While AI's main goal of helping communication, engagement, and education is the same, its methods and problems differ a lot depending on whether teaching takes place at school or on the internet (Zinelli, 2025; Roll & Wylie, 2021; Tsai & Gasevic, 2020).

In such learning settings, AI tends to blend in and be used throughout the classroom environment. Camcorders and microphones in modern classrooms collect data used to study the interactions happening in the classroom. They may keep records of which students are talking, how many take part in the class, and if the groups organized for collaborative learning are focused. Today's tools are able to pick up on gestures like a nod, eye signal, or body position, indicating if a student understands the lesson. Real-time dashboards provide instructors with information about times of active student participation or moments when the students start off track. As a result, teachers can instantly respond to the situation, letting the more reserved students join in or checking if everyone is following the discussions (Heffernan & Heffernan, 2021; Brom & Hlosta, 2023; Chen & Looi, 2022).

An important plus to this form of collaboration is the availability of various forms of data and information. If students sit together in a classroom, their nonverbal actions plus their tone of speech and facial expressions give AI useful hints to analyse. As a result, group activities and the emotions present among participants can be examined more accurately. An AI can tell if a group has stopped talking by checking their expressions and gaze to determine the reason. It is easier for activities to move between a human and an AI facilitator in this environment. After the alert, a teacher can move to the group, ask a question, and fit comfortably back into the discussion (Lee & Zhang, 2024; D'Mello & Graesser, 2021; Tan & Lajoie, 2021).

Meanwhile, online collaboration learning comes with a unique setting where the part AI plays become clearer and sometimes preferred. Being apart, students use online forums, video conferencing, and group documents to talk with other students. In this kind of setting, AI plays a major part in handling, understanding, and assisting with social interactions. It looks at messages sent in writing, follows activity in online shared directories, times the responses, and watches students' faces if cameras are installed. Since online instructors do not physically see their students, it is much harder for them to identify those who pay attention and those who do not. AI observes digital information and converts it into helpful advice for teaching (Ahmed & Karim, 2023; Rosé & McLaughlin, 2021; Gašević & Dawson, 2020).

Besides, students in virtual classes might face extra challenges because of technical problems, everyday chores at home, or loneliness. Online collaboration AI programs are meant to identify gaps in attention, including inactive users or shallow, quick messages, and then deal with the situation if necessary. Sometimes, they give students hints to motivate them, suggest how to handle complicated problems, or motivate them to engage with their peers through discussion. AI helps students in schools here by providing more than support; it links them socially and emotionally because meeting in person is not always possible for them (Lee & Zhang, 2024; Kukulska-Hulme, 2024; Noroozi & Hatami, 2023).

Still, there are specific issues that appear only in the digital setting. The kind of data used with AI is usually not as wide and comprehensive as human knowledge. As one example, the wrong tone can be taken from written or typed words, and silence can be misunderstood. If a student is typing slowly, it might be because they are really thinking or because the internet is not very strong. Therefore, AI systems ought to be planned with care, and students should also have a way to report their situation when the context is not clear. Also, since AI watches activities on multiple devices in the digital world, people are increasingly concerned about the effect on their privacy, freedom, and choice (Luckin, 2023; Rodrigo & Baker, 2021; Tan & Lajoie, 2021).

All of these differences aside, AI is able to improve learning in both types of schools by offering personalized content, encouraging an equal amount of participation, and providing timely interventions. What makes these approaches different is not how powerful AI can be, but how AI is actually applied. In classrooms with people present, AI backs up the teacher with data-based inputs that are not noticeable to the students. When education happens online, it helps out by adding structure, advice, and emotional support because there is no teacher present in person (Zinelli, 2025; Johnson & Lester, 2020; Roll & Wylie, 2021).

Table 7.2 Differences between face-to-face and online AI-supported collaboration

Aspect	*Face-to-Face*	*Online*	*Reference*
Data Collection	Cameras, audio sensors, classroom tools	Logs, chat analysis, video conferencing	Johnson & Lester (2020)
Instructor Role	Real-time intervention with physical cues	Relies on dashboard inputs and alerts	Wise & Schwarz (2021)
Emotional Awareness	Direct observation of student behaviour	AI infers emotions from digital traces	Lee & Zhang (2024)

Table 7.2 compares how AI technologies operate differently in face-to-face and online collaborative learning settings. It outlines variations in data collection, instructor involvement, and emotional awareness.

Figure 7.3 shows that AI supports collaborative learning by monitoring participation and providing feedback in both face-to-face and online settings. This leads to better engagement, balanced group work, and quicker support.

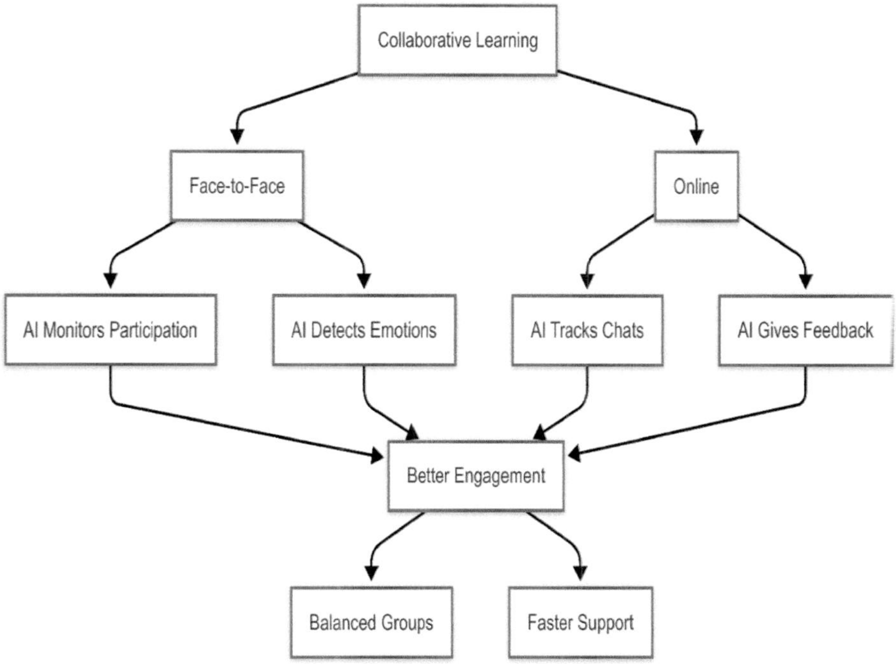

Fig. 7.3 AI in collaborative learning.

Ultimately, the value of AI in either setting depends on how well it is integrated into the pedagogical fabric of collaboration. The most effective use of AI is not one-size-fits-all but context-aware. It respects the unique dynamics of in-person interaction while also embracing the flexibility and scale of online learning. Whether students are huddled around a table or separated by time zones and screens, AI can help bring them closer not just to the content, but also to each other when applied with care, transparency, and a deep understanding of the learning experience.

5. Challenges and Ethical Considerations

Even though AI can improve the process of working together in groups, its use should be managed with care and respect for ethics. Ensuring that data is safe is one of the biggest issues. AI depends on using private information about students such as text, how they speak, behaviour, and their feelings. With weak security, this data may be misled or exposed to others outside the company. All the data should be protected, stored in a safe way, and always collected with the clear approval of users (Lee & Zhang, 2024; Luckin, 2023; Tan & Lajoie, 2021).

Deliberation is needed as algorithms may be biased too. AI learns using historical data, and these data might be affected by current injustices. If these approaches are not correctly watched, they may end up repeating stereotypes, which can leave out or hurt specific groups of students. Getting data from a diverse range of cases and keeping an eye on it through auditing helps ensure there is fairness (Zinelli, 2025; Kukulska-Hulme, 2024; Tsai & Gasevic, 2020).

It is also very important that the process is clear. A lot of AI tools do not explain their decision-making processes. Everyone should know the purpose behind the grouping that happens or the behaviours that draw attention. Clearly sharing information with users builds trust and means their input can be important (Roll & Wylie, 2021; Wise & Schwarz, 2021).

One more concern is that over-relying on automation can become a problem. Although AI may handle the monitoring and giving feedback part, it does not have the creativity and spontaneity that human involvement offers. Educators are still very important in learning, and they should use AI to make their tasks easier rather than having it replace their role (Johnson & Lester, 2020; Heffernan & Heffernan, 2021; Holstein & Aleven, 2020).

Table 7.3 Ethical challenges in AI-enhanced collaborative learning

Challenge	*Description*	*Mitigation Strategies*	*Reference*
Data Privacy	Sensitive student data is collected and stored	Consent mechanisms, secure storage	Zinelli (2025)
Algorithmic Bias	AI may perpetuate existing biases	Diverse datasets, regular audits	Kukulska-Hulme (2024)
Transparency & Oversight	Lack of clarity in AI decisions	Explainable AI, user control	Tzirides et al. (2025)

Table 7.3 outlines key ethical concerns associated with AI in collaborative learning, including privacy, bias, and transparency. It also suggests strategies for responsibly addressing these challenges.

All in all, AI can improve group learning a lot when handled well, just as long as it is used in support of people, not against them (Ahmed & Karim, 2023; Rosé & McLaughlin, 2021; Looi & Wong, 2022).

6. Future Directions

In the years ahead, AI-assisted team learning may become more closely connected with emerging technologies to give learners bettermore responsive, and personalized experiences (Bura, 2025; Zhou et al., 2025; Chen & Looi, 2022). One of the biggest areas is Extended Reality (XR), in which AI will help people work together in virtual and augmented spaces. Students here can take part in simulated labs, collaborate over the internet, or participate in 3D activities, while AI actively helps the group and ensures each student is engaged (Zhou et al., 2025; Chen & Looi, 2022).

Multimodal learning analytics is also creating major changes in the field. AI systems in the future will rely on several sources of information, such as video, audio, written data, or even signals of a person's eyes or heart. The use of multi-level observation gives a clearer and truer view of how students behave emotionally and socially, and how they think while working with others (Rodrigo & Baker, 2021; Gašević & Dawson, 2020; Tan & Lajoie, 2021).

The use of AI will probably bring us *adaptive learning communities* that create groups of students and keep improving them. If the goals and past experiences of people are similar, and they get along, their group shapes their learning and relationships over time (Wang & Lin, 2022; Roll & Wylie, 2021).

When we look at these innovations together, the future envisions AI teaming up with teachers to change collaboration, forming intelligent communities that welcome everyone and take care of each learner more and more (Tzirides et al., 2025; Luckin, 2023; Kukulska-Hulme, 2024).

7. Conclusion

Artificial intelligence is about to make a big impact on how people learn together. AI is able to build strong groups, direct discussions, comment on activities, and detect emotions, so it offers many new benefits for students in both face-to-face and online classes. With the help of these tools, educators can deal with old problems like unbalanced participation, low engagement, and various organizational issues since these tools offer help that is efficient, flexible, and data-based (Zinelli, 2025; Ahmed & Karim, 2023; Rosé & McLaughlin, 2021; Tsai & Gasevic, 2020).

For example, AI-driven platforms can track which students are contributing and which are falling behind, allowing teachers to intervene early and fairly. These systems can offer tailored feedback, suggest group pairings that balance strengths and weaknesses, and adapt in real time to the evolving needs of learners. This dynamic support allows more inclusive group dynamics, promoting equity among students of different backgrounds and skill levels.

Regardless of AI's potential, we should be careful about its role in collaborative learning. The way these tools function greatly relies on their design as well as the ethical, teaching, and human settings they are used in. Privacy needs to be respected, steps should be made to address biases, and honesty should guide everything (Lee & Zhang, 2024; Luckin, 2023; Tan & Lajoie, 2021). Data collected to enhance collaboration should be stored and processed responsibly, with transparency about how it is used and who has access to it. If not carefully handled, the use of AI in learning environments could unintentionally reinforce stereotypes or exclude certain voices.

Above all, empathy, being creative, and having a sense of curiosity are what make any joint learning experience especially valuable (D'Mello & Graesser, 2021; Roll & Wylie, 2021). While AI can help structure and enhance collaborative work, it cannot replace the depth of human interaction, emotional understanding, and spontaneous creativity that occur in group learning environments. These human qualities must remain at the core of any educational effort.

Artificial intelligence is strongest not by replacing human contact, but by helping to improve it. If applied thoughtfully, AI can encourage learning that is more accepting, more closely connected to its users, and captures their interest. AI offers a way to build better friendships among learners whether they are working side by side or remotely. By using AI to reduce friction and personalize support, educators can create space for more meaningful dialogue and cooperation. It is not AI itself that matters, but what it is meant for that will play a role in guiding how learning will evolve (Tzirides et al., 2025; Holstein & Aleven, 2020; Kukulska-Hulme, 2024). The future of collaborative learning lies in blending human insight with AI support to create richer, more connected educational experiences.

References

Ahmed, A. and Karim, N. (2023). *Computers & Education*, *184*, 104560.

Baker, R.S. and Yacef, K. (2020). *Journal of Educational Data Mining*, *12*(1), 1–17.

Brom and Hlosta (2023). *Journal of Learning Analytics*, *10*(1), 44–59.

Bura, C. (2025). Generative AI in learning: Empowering the next generation of education. Preprint. https://doi.org/10.20944/preprints202501.1234.v1

Chen and Looi (2022). *Science Education*, *106*(2), 345–370.

Dillenbourg (2022). *Educational Research Review*, *18*, 100280.

D'Mello, S. and Graesser, A. (2021). *Emotion Review*, *13*(1), 48–60.

Gašević and Dawson (2020). *Internet and Higher Education*, *45*, 100722.

Goggins and Valetto (2020). *IEEE Transactions on Learning Technologies*, *13*(2), 285–298.

Heffernan, N. and Heffernan, C. (2021). *AIED Proceedings 2021*, 302–311.

Holstein, McLaren and Aleven (2020). *ACM TOCHI*, *27*(1), 1–34.

Johnson, W.L. and Lester, J.C. (2020). *Educational Psychologist*, *55*(4), 217–231.

Kay et al. (2022). *Learning Analytics Review*, *12*(2), 56–71.

Kiesler, N., Smith, J., Leinonen, J., Fox, A., MacNeil, S. and Ihantola, P. (2025). The role of generative AI in software student CollaborAItion. Proceedings of the 30th ACM Conference on Innovation and Technology in Computer Science Education (ITiCSE 2025), 1, 1–7. ACM. https://doi.org/10.1145/3724363.3729040.

Kukulska-Hulme (2024). *British Journal of Educational Technology*, *55*(1), 22–40.

Lee, M. and Zhang, Y. (2024). *AI in Education*, *45*(2), 78–95.

Looi and Wong (2022). *Journal of Computer Assisted Learning*, *38*(5), 1032–1045.

Luckin, R. (2023). *Pearson Research Series*.

Noroozi and Hatami (2023). *Educational Media International*, *60*(3), 204–221.

Rodrigo and Baker (2021). *Affective Computing*, *12*(4), 854–864.

Roll and Wylie (2021). *AIED*, *32*(2), 175–187.

Rosé, C.P. and McLaughlin, E.A. (2021). *International Journal of Artificial Intelligence in Education*, *31*, 476–512.

Suthers and Verbert (2020). *Computers in Human Behavior*, *104*, 105512.

Tan and Lajoie (2021). *Ed Tech Research & Development*, *69*(3), 1251–1273.

Tsai and Gasevic (2020). *Computers & Education*, *150*, 103839.

Tzirides, Anastasia Olga, Galla, Michele, Cope, Bill and Kalantzis, Mary. (2025). Thinking Through AI: Advancing Cognitive and Collaborative Research for AI in Education. Preprint. January 2025. DOI: 10.35542/osf.io/s8hqe

Wang, T. and Lin, C. (2022). *Educational Technology & Society*, *25*(1), 23–34.

Wise and Schwarz (2021). *Learning Sciences*, *30*(3), 401–432.

Zhou, X., Smith, Y. and Patel, Z. (2025). Title of the Article. Journal Name, *23*(4), 123–145.

Zinelli, H. (2025). Title of the Article. Journal Name, *45*(2), 123–145.

Chapter 8

Shaping the Future of Education: Exploring the Potential and Consequences of AI and ChatGPT in Educational Settings

S. Durga Devi[1], *R. Ravinder Reddy*[2*], *Satyanarayana Nimmala*[3], *Ravi Uyyala*[4] and *M. V. Krishna Reddy*[5]

1. Introduction

Artificial intelligence (AI) has developed and established itself as a disruptive element in education that has changed how knowledge is delivered, accessed, and assessed (Isotani et al., 2025). In the last decade (2015-2025), AI technologies have evolved the most from minimal automation tools to systems that provide personalized learning, assist teachers, and deliver applied and real-time feedback to students (Ali et al., 2025). As educational institutions worldwide seek to increase learning outcomes and address gaps in achievement, the prospect of AI integration is rapidly growing in primary, secondary, and post-secondary education Alharbi et al. (2023).

One of the more revolutionary developments is that of large language models like ChatGPT. Using deep learning and natural language processing, ChatGPT is capable of chatting in ways that resemble humans, generating and producing designs or content, as well as answering questions and providing assistance to pedagogic activities (Alharbi & Alsubaie, 2023). Due to its versatility and ease of access, ChatGPT serves as a powerful paradigm of what AI in education can offer, including

[1] Department of Computer Science and Engineering, Chaitanya Bharathi Institute of Technology, Hyderabad, India. https://orcid.org/0000-0002-5163-8886
[2] Department of Computer Science and Engineering, Chaitanya Bharathi Institute of Technology, Hyderabad, India. https://orcid.org/0000-0001-6488-9935
[3] Computer Science and Engineering, Hyderabad, India. https://orcid.org/0000-0002-9433-1489
[4] Department of Computer Science and Engineering, Chaitanya Bharathi Institute of Technology, Hyderabad, India. https://orcid.org/0000-0002-6913-7529
[5] Department of Computer Science and Engineering, Chaitanya Bharathi Institute of Technology, Hyderabad, India. https://orcid.org/0000-0002-7247-6580
* Corresponding author: rravinderreddy_cse@cbit.ac.in

DOI: 10.1201/9781003639367-8

developments such as personalized tutoring, automated grading, and interactive learning on a larger scale. The intent is to provide a critical assessment of the possible benefits and possible negative consequences of using AI in education, in particular ChatGPT. The assessment uses some current meta-analyses, primary research studies but also some studies from the field, as a way to provide a sufficiently neutral lens for educators, administrators, and policy developers (Muñoz et al., 2024). Here we consider some of the real-world scenarios for understanding purposes through a discussion of the case studies.

1.1 Real-world Case Studies

i. *Case Study 1: Personalized Learning in K-12 Mathematics*: In a public school district in the US, teachers implemented ChatGPT-powered tutoring into their math curriculum. Students that worked with ChatGPT for personalized problem-solving help showed 15% greater improvement in standardized test scores than peers who used traditional resources. Teachers shared that the AI tutor helped them spend more time with students that needed extra help which, overall, increased student engagement in the classroom.

ii. *Case Study 2: Academic Writing Support in Higher Education*: A university context used ChatGPT as an academic writing support system for undergraduate students based in the UK. Students could receive instant feedback regarding their grammar, structure, and argumentation. Students overall improved their essay grades by an average of one letter grade overall, and students who engaged with ChatGPT, overall, improved more, the non-native fluency students being among the most likely to improve. Faculty were also concerned about a serious increase in originality and academic integrity issues, therefore introducing stricter checks on plagiarism.

iii. *Case Study 3: Advancing STEM Education in Under-developed Communities*: In a rural region of India, a pilot programme provided ChatGPT-enabled chatbots on low-cost tablets to support science education. Students from very remote areas, with little or no access to qualified teachers, interacted with chatbots to enrich their understanding of scientific constructs, conduct virtual experiments, and seek clarification on questions they had pursued in their exploration. In addition to advancing science literacy, the programme has achieved lower dropout rates by bringing experiential learning to these students. These cases exemplify the ways AI in education has documented potential as well as documented challenges. As the technology expands opportunities for learning while promoting equity and improving learning outcomes, it raises important and complex questions regarding ethics, equity, and access, and the professional work of educators. This chapter serves as a critical examination of these complexities with the purpose of demonstrating evidence-informed imperatives for action when embedding AI as an education resource.

As we look forward, the emerging world of AI technologies will continue to create new possibilities and challenges for education. Multimodal immersion through AI systems, advanced personalization, and deep learning analytics may afford use with AI that impacts the landscape of education to a greater extent. And yet, the primary considerations that have emerged from this study focusing on pedagogy, building AI literacy, responding to equity issues, and maintaining academic integrity will likely be central to any successful technology use that is developed.

2. Related Work

The intersection of AI and education has garnered considerable scholarly attention over the past decade (2015-2025). As AI progresses through its maturation process, the phase of experimenting with AI will yield to AI projects that move towards more scalable implementations. This has led to in-depth investigations into the effectiveness of AI across pedagogical (learning), ethical (morals and etiquette), and structural (infrastructures and frameworks) modes. Recent research argues that AI technology can foster personalized and student-centred learning experiences. For example, Holmes et al. (2019) argue that AI systems that include dynamic content adaptation, formative feedback, and learning analytics enable in-service teachers to diagnose students' misconceptions and adjust instructional decisions. This aligns with a socio-constructivist pedagogical approach that emphasizes active learning and scaffolding experiences by teachers.

Notably, large language models (LLMs), like ChatGPT, are gaining momentum for their natural language abilities. Zhai (2023) explored how ChatGPT can act as a writing assistant, cognitive partner, and brainstorming tool. Zhai (2023) pointed out the utility of ChatGPT in two varying levels of writing tasks such as surface levels (i.e., mechanical, such as grammar correction) as well as deeper level (i.e., gestural, such as constructing an argument). Zhai also exemplified caution regarding over-reliance, as well as the complex boundaries between assistance and automation. Kong et al. (2024) provided a cross-national perspective, highlighting how ChatGPT and similar tools are trained in three nations (China, Australia, and Germany). Their conclusion suggests that cultural and curricular context matters with respect to AI adoption, and variations of acceptance by teachers, policy, and levels of digital literacy among students shaped their situational response to AI as well.

Another area of interest is student engagement and motivation. Dai et al. (2023) argue that students interacting with ChatGPT showed higher rates of cognitive engagement and less reluctance to ask questions, an advantage that seems to be an even bigger benefit among more introverted students or those studying in a second language. The 24/7 availability of the chatbot also helps make asking for help more palatable, helping with self-directed learning. While these benefits may be promising, concerns are still valid. As McGee (2024) points out, there is an academic integrity risk associated with AI tools. In other words, while ChatGPT, for example, can assist students in figuring out what their assignments want them to do, there is also the risk that students may use it to generate full answers, effectively bypassing higher order thinking. Thus, institutions seem to be wrestling with how to determine sensible balance between the benefits that come from using AI tools and student success, while at the same time attempting to promote originality and integrity.

In addition, unequal access to AI creates significant issues of equity. The Organisation for Economic Co-operation and Development describes that learners in under-resourced schools or rural areas may be lacking the infrastructure needed to engage with AI-tools, and this may further entrench digital divides. Addressing these inequities includes purposeful policy measures providing devices, teacher training, and developing content that is locally relevant. ChatGPT has also been examined as a means for lifelong learning and adult education. Its conversational interface has been used to engage learners in simulations of workplace contexts, explain complex technical concepts, and provide coaching in non-traditional learning contexts, such as in corporate training and online micro-credentials (Bryant & Singh, 2024). It is worth noting that these applications help move AI beyond traditional education and exemplify how versatile AI has become. Overall, the studies cited above illustrate that while ChatGPT and AI systems continue to offer promise to support teaching and learning, their effectiveness will be a factor of how they are intentionally situated into pedagogy. Educators need to be trained to see AI not as a replacement but to enhance cognitive and metacognitive learning processes.

2.1 The Effectiveness of ChatGPT in Education

In recent meta-analyses, researchers have found that ChatGPT has a significant positive effect on student learning performance, overall moderate positive effects on learning perceptions, and even higher-order thinking, with a large effect size ($g = 0.867$). These consistent results across all primary, secondary, and higher learning educational settings influenced students across all subjects, with the strongest effects in skill-based and science, technology, engineering, and mathematics (STEM) fields. Other contributors to the effectiveness of ChatGPT in those academic settings could be related to its ability to deliver immediate, personalized feedback and the ability to scaffold challenging problem-solving tasks.

2.2 Generative AI and ChatGPT in Academic Settings

Generative AI, and especially LLMs like ChatGPT, exists as a disruptive technology in education. Researchers emphasize that ChatGPT is able to support interactive learning by facilitating tutoring sessions, producing academic material, and providing assistance with language barriers (Zhang & Wang, 2023). Studies performed by Patel et al. (2024) show that students using ChatGPT for writing assistance reported improved clarity and structure in essays, while also recognizing potential risks such as reliance on AI-produced content. Faculty views are mixed. While some educators hailed ChatGPT as a way to improve productivity (Johnson & Kumar, 2022), others were concerned about plagiarism and effect on the development of critical thinking (Rivera & Thomas, 2021). In a recent survey of 300 post-secondary instructors, 65% of respondents indicated their desire to explore and implement ChatGPT tools with constraints or guidelines while a further 25% wanted to pause or ban ChatGPT tools until ethical guidelines develop (Alharbi & Alsubaie, 2023).

2.3 Challenges and Ethical Considerations

While there are many benefits of using ChatGPT, there are also challenges relating to academic integrity (including potential plagiarism and over-reliance on AI-generated content) as well as equity considerations, including the potential to widen existing gaps in education due to differences in access to AI technologies. Ethical implications of ownership (data privacy), algorithm bias, and the changing landscape of work for educators mean that there is also a need for policies and regulations to manage responsible use of AI in education (Muñoz et al., 2024). Some of the key highlights from this review are listed here.

i. *Educational Effectiveness*: ChatGPT displays potential to significantly enhance learning due to its ability to provide personalized instruction, to provide immediate feedback and to re-teach, and to adapt to students as they learn. Plus, the round-the-clock accessibility and individualized support capabilities of ChatGPT help to overcome many barriers to educational support.
ii. *Pedagogical Fit*: If used as intended, ChatGPT fits into the best practice of formal learning theories and instructional practice. Examples include constructivist theories that emphasize hands-on learning, scaffolding, and personalized instruction. Chat GPT's ability to operate in (or below) the student zone of proximal development is meaningful pedagogically.
iii. *Barriers to Implementation*: There are considerable barriers to successful ChatGPT implementation such as preserving academic integrity, the need for extensive professional development, availability of technological infrastructure, and equity with access and digital literacies.
iv. *Best Practices*: Effective implementation will come from practices that stress gradual implementation, explicit instruction in AI literacies, blended learning, and an evaluation of outcomes. Institutions that have successfully leveraged ChatGPT adoption typically leverage comprehensive frameworks that stress considerations of policy, pedagogy, technology, and professional development.

This chapter examines theoretical underpinnings, constructivist learning theory, social learning theory, and cognitive load theory that provide important considerations for how to enhance, not replace, traditional learning approaches with ChatGPT. Given that ChatGPT has the potential to minimize extraneous cognitive load while still aiding in schema formation and metacognitive development, it holds the greatest promise for compounding, expanding, enriching, and improving learning effectiveness. Despite this potential for learning enhancement, some challenges, particularly those related to academic integrity, atrophying skills, and inequitable access, merit careful and ongoing consideration. A major concern relates to instances where students may rely on ChatGPT too often, taking away from their critical thinking and reasoning skills, which should be prevented through purposeful pedagogy and assessment patterns.

The use of ChatGPT in education is not just a technological upgrade but also a true re-imagining of how learning can be supported, personalized, and scaled in practice. It will be as much about how the education community (teachers, schools, systems) can thoughtfully develop responses to the opportunities and challenges that these powerful generative AI tools provide, while still honouring educational excellence, equity, and integrity. While education systems are still understanding how and where to use AI, the experiences and insights compiled here provide a starting point for making well-informed decisions about where we want to take education through AI integrations. Our aim is not to replace teachers, but to meaningfully augment learning and experiences for all learners.

3. Theoretical Framework and Foundations for Pedagogy

ChatGPT's implementation in education can be viewed and understood through several theoretical frameworks that can help us better understand the potential ramifications of its use in learning. In this section of the chapter we engage with the pedagogical foundations that inform AI-enhanced education, and the theoretical frameworks that can provide insight into the implications of its implementation.

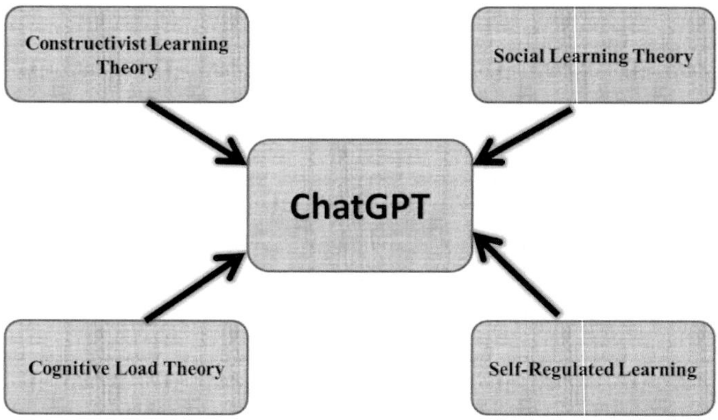

Fig. 8.1 Framework for AI-supported learning by ChatGPT.

The outline of the fundamental theoretical frameworks and foundations for pedagogy that inform ChatGPT's implemented use in education is displayed in Figure 8.1. At the centre is ChatGPT, which represents its central role in AI-supported education. The arrows lead to well-known learning theories.

3.1 Constructivist Learning Theory

Constructivist learning theory, established by Piaget and Vygotsky, suggests when learners are given opportunities to interact with the environment and their social circumstances, they are able to construct knowledge (Piaget, 1954). ChatGPT supports constructivist principles in its role as an interactive partner for students. ChatGPT can react to student questions, support students through scaffolding, and help note taking that allows students to construct knowledge through conversation. Additionally, the adaptive programming of ChatGPT allows for very personalized learning experiences in real time that respects personal learning paths and prior knowledge structures.

There are also elements of Vygotsky's zone of proximal development (ZPD) that has importance in regard to educational uses of ChatGPT. The difference between what a learner can do without assistance or support and what a learner can do when working with a more knowledgeable other (i.e., an adult or peer) makes the boundaries of a ZPD (Vygotsky, 1978). ChatGPT has the ability to offer anecdotal evidence to any learner in terms of a "more knowledgeable other", in the moment, to allow for just-in-time support which can make the difference to get to a learning goal. Additionally, the way ChatGPT works, it provides just-in-time references, based on each individual student, which can help maximize the efficiency of learning in terms of scaffolding and individual needs.

3.2 Social Learning Theory

Social learning theory, developed by Bandura, emphasizes that both observation and modelling have value as learning processes in conjunction with one's social environment (Bandura, 1977). While it is clear that ChatGPT is not human or social, it is able to model good problem-solving, provide models of academic writing, and demonstrate good expert thinking. Students could curate their own models for effective learning. However, it raises the question whether the social aspect of learning still holds genuine meaning, and that students benefit from having real human mentors on the other side of the virtual process instead of AI systems employing Socratic questioning.

3.3 Cognitive Load Theory

Cognitive load theory, developed by Sweller, provides a useful lens on how ChatGPT could also potentially optimize student learning performance by reducing cognitive load (Sweller, 1994). The theory breaks cognitive load down to state three types: intrinsic cognitive load (necessary to learning), extraneous cognitive load (not related to the intended learning outcomes), and germane cognitive load (related to schema acquisition). ChatGPT can potentially reduce the extraneous cognitive load by performing low-level routine tasks like checking grammar and formatting, freeing students' focus and cognitive resources to focus on higher-order thinking and concepts. Utilizing ChatGPT effectively could efficiently use cognitive resources to perform (PM) increasingly meaningful learning tasks than was previously possible.

3.4 Self-regulated Learning

Self-regulated learning theory focuses on the students' ability to self-monitor, control, and self-regulate their own learning processes (Anderson & Shattuck, 2012). ChatGPT can offer self-regulated learning by providing real-time feedback, enabling students to set their own learning objectives and providing metacognitive questions which urge them to think about their learning strategies (Zimmerman, 1990). The AI's "always available" factor allows the students to learn at their own pace, ask for clarifications, if necessary, and also enables greater autonomy and accountability over their learning.

This theory emphasizes students' ability to self-organize, self-manage, and self-evaluate their own learning. By providing immediate feedback, tailored suggestions, and reflective prompts, ChatGPT provides the means for self-regulation and greater independence and self-directed learning.

4. Methodological Approaches to AI Integration

The successful integration of ChatGPT into educational environments will require systematic methodological approaches that will consider pedagogical goals, student attributes, and contextual factors from institutions. This section identifies some of the key methodological frameworks and approaches to implementation.

4.1 Design-based Research Approach

Design-based research can be defined as the iterative design and development of AI-supported educational interventions (Anderson & Shattuck, 2012). Design-based research is collaborative, working with researchers, educators, and students to design something practical that deals with educational problems we face in the real world. Figure 8.2 depicts the design-based research approach for ChatGPT-based educational interventions. It shows the four phases and how they are slipped between.

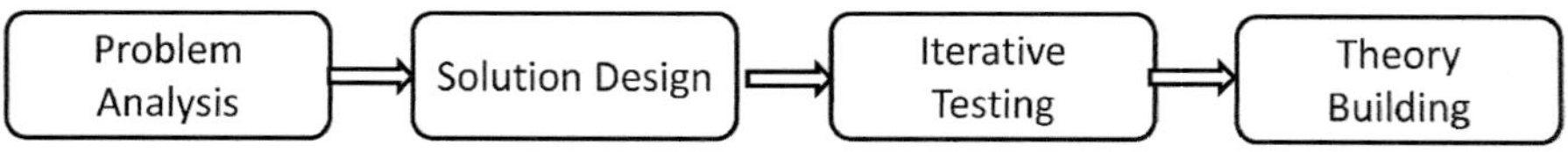

Fig. 8.2 Design-based research approach.

Design-based research protocols can support the following stages while using ChatGPT:

i. *Problem Analysis*: Identifying important educational problems that ChatGPT can solve
ii. *Solution Design*: Creating pedagogically sound AI-enhanced learning opportunities
iii. *Testing Prototypes*: Collecting user feedback, and/or altering ChatGPT-based applications
iv. *Theory Building*: Developing knowledge related to meaningful ways to integrate AI

4.2 Blended Learning Models

ChatGPT is most ideally situated as part of a blended learning model incorporating both AI support and human instruction. Table 8.1 identifies blended learning models and compatibility with ChatGPT augmentation.

Table 8.1 Blended learning models and opportunities to integrate ChatGPT

Model	*Description*	*ChatGPT Integration*	*Effectiveness Rating*
Rotation Model	Students alternate between online schooling and live schooling	AI tutoring during online stations	High
Flex Model	Content delivery could be online and teacher support as required	ChatGPT as primary content delivery with teacher checks	Medium
À la carte Model	Students are engaged in online courses and traditional front facing courses	ChatGPT to provide supplementary learning and homework-assistance tutor	High
Enriched Virtual Model	Primarily online and engaged face to face periodically	ChatGPT to provide uninterrupted support between sessions	High

Various blended learning models provide diverse opportunities to integrate ChatGPT, as outlined in Table 8.1, with the rotation and à la carte models having the highest effectiveness ratings based on the balanced approach to human-AI collaboration.

4.3 Assessment and Evaluation Strategies

When integrating ChatGPT meaningfully, powerful assessment strategies are needed to help discern AI-supported learning and AI dependence. Current assessment strategies may need to be adapted in these new contexts, but still must consider academic integrity. The important elements to consider when assessing include:

a. *Formative Assessment*: ChatGPT can facilitate formative assessment in providing timely feedback on student work, and to identify parts of the work that may need improvement. However, there must be some assurance by the educator that the AI-oriented feedback is supplementary to a human evaluation process.
b. *Authentic Assessment*: Performance-based assessment requiring students to use knowledge in real-world contexts can be helpful to assess learning outcomes while minimizing opportunities to misuse AI.
c. *Process Assessment*: Rather than outcome assessment, process assessment can provide a better understanding of the interactions students are having with ChatGPT while sufficiently developing AI literacy skills.

5. Advantages and Opportunities

The use of ChatGPT in education provides a variety of advantages that can support, augment, and improve educational outcomes, and address systemic difficulties in education. In this section, we will review the broad opportunities that ChatGPT offers students, educators, and educational institutions.

5.1 Customized Learning and Adaptive Learning

One of the key advantages of ChatGPT is its ability to provide customized learning experiences for individual students that are aligned with their needs, learning styles, and pace. Unlike the traditional one-size-fits-all approach, ChatGPT can customize its responses, explanations, and support based on students' interactions and exhibited understanding levels.

a. *Individualized Feedback*: ChatGPT can deliver custom, individualized feedback on students' work right away, allowing the learner to easily identify strengths and weaknesses in their work during the creation of the project. This feedback loop allows for a much faster turnaround time on student learning and helps students make corrections before misconceptions become too ingrained.
b. *Adaptive Difficulty*: The AI has the ability to change the difficulty level of explanations and examples of lessons, based on a student's level of understanding so that the content is not too basic, or too difficult. This allows the AI to adapt the information to suit the optimal challenge level within each student's ZPD.

c. *Learning Style Accommodation*: ChatGPT presents information in multiple formats, for example, through textual explanations, procedural steps, analogies, or examples, allowing for different learning preferences and cognitive styles.

5.2 Improved Accessibility and Inclusion

ChatGPT greatly enhances accessibility in education for students from diverse populations—including students with disabilities, English language learners, and students in resource-limited settings.

a. *Language Support*: ChatGPT can provide English language learners with specific examples and explanations in simpler language, provide translations from English and other languages, and simply allow the learner to practice the use of the language in conversational exchanges. AI has two great advantages - not judging language learners, which allows them to feel less anxious in the language learning process, and patience, which allows learners to practice and play language until they understand. ChatGPT provides a barrier-free opportunity to freely explore, learn, and practice a new language.
b. *Disability Accommodation*: For students with learning disabilities, ChatGPT can take complex concepts and break them down into manageable pieces, present multiple formats for explanation, and offer unlimited repetition without frustration or judgement.
c. *24/7 Access*: ChatGPT is always available to students—there is no time of day when it is not available. Human tutors or human teachers can only provide so much time per student, and that access is often dictated by time zones, working hours, and sometimes geographical location. This is an essential tool for adult learners and students in different time zones.

5.3 Cognitive Skill Development

ChatGPT can act as a cognitive tool that enhances the range of thinking skills and learning strategies, as shown in Figure 8.3.

Figure 8.3 shows how ChatGPT integration can concurrently enhance multiple cognitive skill domains and provide a complete learning environment with specific competencies related to twenty-first–century learning.

a. *Critical Thinking Development*: As discussed earlier, ChatGPT can ask thought-provoking questions to challenge assumptions and provide guidance through logical reasoning processes. Although some may consider it an inexpensive logical anecdote, by using the AI in Socratic dialogue with students, the students can develop analytical skills by evaluating arguments and evidence.
b. *Creative Problem-Solving*: The AI can enhance brainstorming sessions, provide an alternative approach, and assist students in discovering the creative solutions to a complex problem. The AI could serve as a reference book with the knowledge to make innovative connections across disciplines.
c. *Metacognitive Awareness*: ChatGPT can prompt students with reflections on the learning process and increase awareness of the thinking strategies and preferences when learning. To develop awareness is metacognitive development, which is essential to being a lifelong learner and self-directed student.

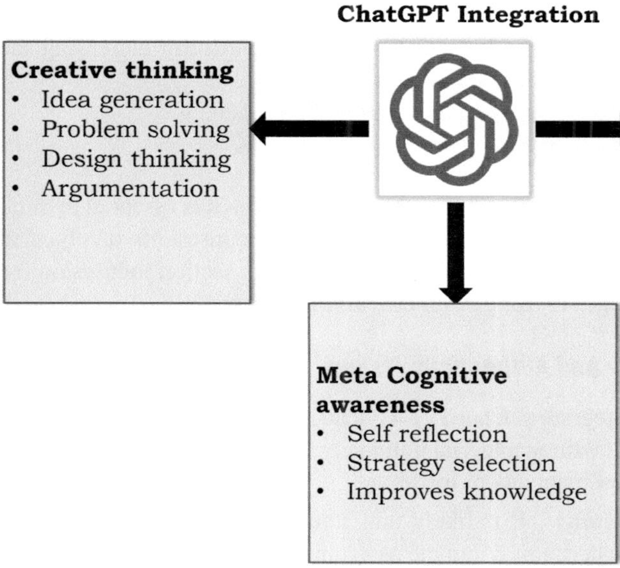

Fig. 8.3 Cognitive skills improved by ChatGPT integration.

5.4 Teacher Support and Professional Development

ChatGPT has many advantages for teachers, acting as a professional development and instructional resource to promote teaching effectiveness.

a. *Lesson Plan and Curriculum Development*: Teachers can use ChatGPT to generate lesson ideas, create assessment tools, create questions for discussion, or generate learning activities all aligned to curriculum documents.
b. *Differentiated Instruction Support*: The AI can assist educators with multiple versions of assignments, assessment, or explanations for diverse learners in a heterogeneous class.
c. *Professional Learning Communities*: ChatGPT can support professional discussions by providing summaries of research, suggesting best practices or views on pedagogical dilemmas.

5.5 Scalability and Cost-Optimization

The ChatGPT tool offers unprecedented scalability and cost optimization in providing educational services from an institutional perspective. Amazon served millions of customers from one fulfilment centre.

a. *Tutoring at Scale*: ChatGPT provides personalized tutoring support to an unlimited number of students at once, alleviating the shortage of available qualified tutors, and making quality academic support available to all students.
b. *Fewer Resource Constraints*: Educational institutions can expand their support services without proportional increases in human resources, thereby making quality education more sustainable.

c. *Access to World*: ChatGPT enables educational institutions to provide services to students in underserved markets across the globe, thereby increasing the availability of quality education in underserved areas.

6. Challenges and Limitations

While ChatGPT has lots of potential to help enhance education, it is essential to think about, and work through, some significant challenges and constraints involved in implementing it in a responsible and successful manner. This section focuses on the primary issues and challenges in using AI in educational settings.

6.1 Academic Integrity and Authenticity Issues

The primary barrier to integrating ChatGPT is academic integrity. AI's ability to generate very sophisticated written material ultimately calls into question authorship, originality, and authentic assessments of learning.

a. *Plagiarism and Cheating*: It is likely that students will use AI to generate more sophisticated material or complete assignments and submit them as their own work. Although the writing is intended to reflect their learning and skill development, use of AI-generated content diminishes meaningful assessment of learning and skill development. While a traditional plagiarism checker can often catch the student who simply copies and pastes directly from a source, it cannot catch a student who uses AI to generate content. This may require potential new ways of monitoring academic integrity.
b. *Assessment Validity*: Current assessment methods may become invalid when AI tools exist as resources where students can essentially cherry-pick their written and completed assignments or have the AI tool complete or solve processing tasks for them. This requires a fundamental re-thinking of how learning is assessed and validated.
c. *Loss of Skills*: Over-reliance on ChatGPT for writing, math, and problem solving may lead to inhibited development of important academic skills. Students may flounder in their skill development and rely on AI-dependent writing and forget how to write independently.

6.2 Pedagogical and Learning Concerns

With ChatGPT integration, there are several pedagogical concerns that can impact the quality and nature of learning experiences.

a. *Surface-level Learning*: There is a risk that ChatGPT can facilitate surface learning by providing quick answers rather than prompting students to have a more conceptual understanding and critically engage with the content. Students may be focused on getting answers instead of developing a solution process to the problem and emphasizing conceptual understanding.
b. *Reduction of Human Interaction*: Over-reliance on AI tutoring may come at the expense of interactions that students need in order to establish human connections necessary for social learning, social-emotional development, and

collaborative skills. Human teachers can provide subtle feedback and empathy that AI systems cannot.

c. *One-Size-Fits-All Limitations*: Even with its adaptive features, ChatGPT may not effectively recognize and capture the elaborate individual differences, cultural contexts, and individual learning needs that human educators can.

6.3 Technical and Infrastructure Challenges

In the less than fifteen-year period that ChatGPT is presumably going to be integrated with education, or at least accessible to educational institutions, there may well be technical challenges that bar and/or impair the successful integration of the functionality in educational settings.

a. *Digital Divide*: Not all students will have equal access to the computer technology, connectivity, and necessary digital literacy skills to properly utilize ChatGPT. The divide will likely exacerbate already established educational inequalities and produce a new disadvantage.
b. *Reliability of Services*: The reliance on AI systems introduces the risk of server downtime or site updates, and various other technical glitches that could potentially disrupt learning tasks and assessment procedures.
c. *Complexity of Integration*: Adding ChatGPT into existing learning management systems, institutional processes, and the practices of teaching implementation will likely require technical expertise and on-site capacity that none of the education systems have.

Table 8.2 provides a broad perspective of identified barriers and their associated levels of impact.

Table 8.2 Challenges and impact assessment of ChatGPT integration

Challenge Category	*Specific Challenge*	*Impact Level*	*Mitigation Difficulty*	*Timeline for Resolution*
Academic Integrity	Plagiarism/cheating	High	High	Medium-term
Academic Integrity	Assessment validity	High	High	Long-term
Academic Integrity	Loss of Skills	Medium	Medium	Long-term
Pedagogical	Surface learning	Medium	Medium	Medium-term
Pedagogical	Reduction of human interaction	Medium	Medium	Ongoing
Pedagogical	Limited personalization	Low	Low	Short-term
Technical	Digital divide	High	High	Long-term
Technical	Reliability of services	Medium	Medium	Short-term
Technical	Integration complexity	Medium	High	Medium-term
Ethical	Data privacy	High	Medium	Medium-term
Ethical	Algorithmic bias	Medium	High	Long-term
Economic	Implementation costs	Medium	Medium	Medium-term

Table 8.2 indicates that challenges will have different levels of impact and different levels of mitigation difficulty; institutional issues such as academic integrity and the digital divide appear to be the biggest challenges that will require long-term strategies to resolve.

6.4 Privacy and Ethical Issues

The potential educational use of ChatGPT might have considerable ethical implications with respect to the rights and interests of students.

a. *Data Privacy and Security*: Sensitive student information about learning disabilities and personal information might be stored in the online databases and learning challenges may exist during educational use of ChatGPT. When providing an AI-supported service, educational organizations have to ensure that they have complied with legal privacy regulations such as Family Educational Rights and Privacy Act (FERPA) and General Data Protection Regulation (GDPR).
b. *Algorithmic Bias*: Bias inherent to AI systems can continue, or exacerbate, biases concerning gender, race, culture, or socio-economics. These biases can reduce quality and equitable education support for other student populations.
c. *Consent and Transparency*: Students and parents must be well informed about the use of AI systems in educational contexts, what data are being collected, and how AI-generated feedback and recommendations are created.

6.5 Institutional and Policy Implications

Educational institutions face considerable organizational challenges with regard to successfully integrating ChatGPT.

a. *Policy Development*: Institutions will need to create framework policy documents that simultaneously address AI use, expectations for academic integrity, specify acceptable and unacceptable practices related to AI assistance. Policymakers must address innovation, educational quality, and commitment to ethical expectations.
b. *Faculty Development and Support*: Educators will need considerable professional development to effectively integrate ChatGPT into their educational practices, including understanding the capabilities and limitations of AI systems and continuing to be pedagogically effective.
c. *Quality Assurance*: Institutions need to develop new and different processes to ensure quality educational experience and learning outcomes in the integrated use of AI tools in their course designs and assessments.

6.6 Long-term Societal Impacts

The use of ChatGPT in education has the potential to lead to effects on a wider scale that shape and impact the overall future of human learning and development.

a. *Workforce Benefits*: If students rely too heavily on AI assistance, then there is a risk that they are not prepared for working environments that require them to engage in independent critical thinking, creativity, and problem solving.

b. *Cognitive Development*: If there is an over-reliance on AI to do the "thinking" for us and make intellectual decisions or conclusions, it may undermine our cognitive development and impede our ability to attend for extended periods of time, read deeply, and think independently.
c. *Educational Equity*: ChatGPT holds the promise of equalizing access to quality educational support. But there rises another challenge in terms of inequities and difference in access among those who are able to actively engage with AI tools and applications and those who cannot.

7. Implementation Strategies and Best Practices

The successful implementation of ChatGPT in schools requires strategic planning that takes into account the current opportunities and challenges discussed in the previous sections. This section offers implementable guidelines, best practices, and strategic recommendations to help guide educators, administrators, and policymakers.

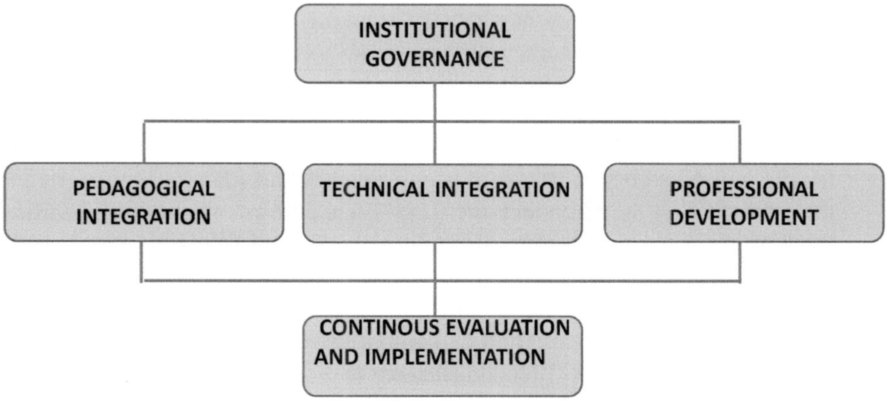

Fig. 8.4 Comprehensive ChatGPT implementation framework.

Figure 8.4 shows a complete implementation framework which attempts to create stability in the relationship between technological potential and pedagogic effectiveness.

7.1 Institutional Framework for AI Integration

Educational institutions should develop complete frameworks to ensure responsible implementation of ChatGPT and other AI tools. This framework should include the establishment of policies, developing infrastructure, and engagement with all stakeholders.

a. *AI Governance Committee*: Institutions should assist in developing multidisciplinary committees including some or all or educators, administrative personnel, IT professionals, ethicists, and students who will assist in overseeing the AI implementation plans. Committees should create policies, maintain their view of progress during implementation, and recommend solutions to challenges.

b. *Explicit Use Policy*: Detailed policies must be put in place to identify acceptable and unacceptable uses of ChatGPT, guidelines for academic integrity, and consequences for misuse. Policies should be revised periodically to correspond with technological changes and best practices.
c. *Technology Planning*: Institutions must build a solid technological infrastructure to make ChatGPT work optimally with respect to reliable internet accessibility, devices to use, and for incorporation into established learning management systems.

7.2 Pedagogical Integration Strategies

As discussed, integrating ChatGPT into practice should include consideration of pedagogical strategies to leverage the capabilities of AI, while sustaining educational quality and learning outcomes.

a. *Gradually model*: A gradual introduction to the use of ChatGPT is a strategy that could be utilized. Educators could consider low-stakes activities and then develop students' use of ChatGPT, making use of its functionality and developing students' AI literacy simultaneously. A sequence like this would help build students' confidence and competence before a major assessment task can include the use of ChatGPT.
b. *Explicit instruction in AI literacy*: Students need explicit instruction on how to AST with ChatGPT, why to evaluate AI outputs, and what the limitations and the capabilities of the AI systems are. This also includes developers using critical thinking skills when evaluating AI outputs, ethical considerations with AI usage, etc.
c. *Blended learning approaches*: Integrating ChatGPT into learning is most effective when done in addition to human instruction, rather than replacing activities led by teachers. Blended learning approaches enable efficiency from AI while retaining the human elements of education.

7.3 Assessment and Academic Integrity Approaches

To sustain academic integrity and to take advantage of the educational potential ChatGPT offers, educators will need to develop innovative assessment approaches and clear parameters around appropriate AI use.

a. *Genuine Assessment Design*: One way for educators to address the consequences of students accessing ChatGPT is to construct assessments that ask students to apply knowledge in settings similar to the real world, include reflections about the learner's own experience, and require a higher-level thinking benefit that is hard for AI to achieve. Some examples include:
    i. portfolio assessments that represent what the students learned during the learning process;
    ii. performance-based assessments that require taking action;
    iii. reflective writing that connects real-life experiences with academic course content;

iv. collaborative assessments where the students rely on the input from members of their group, and a collective goal to reach, whether academic or otherwise.

b. *Process-focused Evaluation*: Assessments as an end product should shift towards evaluating how students learn—which includes measures of their research strategies, their development of reasoning, and their methods of problem solving. Emphasis on process, instead of assessing simply an end product, will make it much more difficult for students to misuse AI while requiring cognitive engagement.

c. *AI-transparent Assignments*: Some assignments allow for explicit inclusion of ChatGPT use. In these assignments, students would be required to record the prompts they used with AI, evaluate the benefits/drawbacks of the AI responses, and provide evidence for how they built off or used the AI in some way. This promotes transparency of AI use, promotes ethical use of AI, and promotes not only the importance of the AI tool for learning but also the importance of developing critical evaluation skills for emerging technologies.

d. *Honour Codes and Academic Integrity Agreements*: Educational institutions should revise their honour codes to include AI use, for example, explaining when it is appropriate to use AI and when it is academic dishonesty. Students should sign an agreement acknowledging that they have received instruction on academic dishonesty with AI and that they understand each component of the policy.

7.4 Professional Development and Teacher Learning

Effective professional development is necessary for emerging technologies like ChatGPT to become successful because educators must be able to engage in a high-quality manner in their use of AI tools in education.

a. *Basic AI Literacy*: Teachers must have a foundational level of understanding of how AI systems operate, what AI can and cannot do, and when and how teachers may use AI systems. A foundational level of AI literacy allows educators to make informed decisions about when and how to integrate AI technologies into their pedagogical practice.

b. *Pedagogical AI Integration*: Professional learning should centre on how teachers may use ChatGPT as part of their existing curriculum, what an AI-infused learning activity may look like, and how they can meet their learning requirements while reimagining how they work with AI technologies.

c. *Support and Communities of Practice*: The effective integration of AI tools into pedagogical practice is reliant on support, professional learning communities, mentorship, and continued access to training as AI technology is rapidly evolving.

7.5 Student Support and Digital Citizenship

Students need support to use ChatGPT in a thoughtful and ethical way.

a. *Digital Citizenship Education*: Students need to understand responsible use of AI, the limitations of AI, ethical use of AI, and applying critical thinking while engaging with AI systems.

b. *Technical Skills Development*: While ChatGPT is designed to be intuitive, effective prompting, result evaluation, and AI-assisted learning in conjunction with independent learning require instruction for students.
c. *Teaching Academic Integrity*: Students need to be aware of the limits of using AI, what happens when they go over those limits, and how to use AI to be ethically responsible.

8. Future Directions and Research Implications

The integration of ChatGPT in education represents an evolving field with significant potential for future development and research. This section explores emerging trends, research opportunities, and future directions that will shape the landscape of AI-enhanced education.

8.1 Technological Changes and Improvements

Given the rapid pace of AI development, it will not be a matter of whether there will be additional developments in ChatGPT in the future, but what further advances will be made in functionality that could drive new sources of educational use.

a. *Multimodal AI Integration*: Future AI products will potentially incorporate text, voice, image, and video capabilities, allowing for a more immersive and engaging experience. Students could then interact with an AI tutor at multiple sensory levels, which could enhance their understanding and memory in unique ways.
b. *Elevated Personalization*: Advanced AI systems could gain more complex understandings of individual learner's learning styles, emotional states, and cognitive patterns, creating possibilities for personalized learning and personalized support like never seen before. In the future, with AI integrated into online learning environments, learning analytics will be continuous and will enable educators and learners to track progress, recognize challenges, and revise learning methods in real time.
c. *Real-time Learning Analytics*: In the future, the integration of AI will enable continuous learning analytics, helping educators and students keep track of their progress, identify challenges, and adapt their learning approaches in real time.

8.2 Emerging Research Areas

There are a number of research areas that provide possibilities for advancement of knowledge about AI, the role it can play in education, and the best ways to implement it.

a. *Long-term Learning Outcomes*: Longitudinal studies should investigate the long-term impact of ChatGPT integration on student learning outcomes, skills development, and academic success at different educational tiers and across disciplines.
b. *The Effect of Cognitive Development*: Research should investigate the effect of prolonged AI engagement on students' cognitive development, particularly

in developing critical thinking, creativity, and independent problem-solving capabilities.

c. *Equity and Access Studies*: More robust research is needed to examine how ChatGPT integration contributes to educational equity, particularly with regard to students' different socio-economic status, cultural contexts, and varying learning abilities.
d. *Teacher Professional Development*: There is a need for studies examining sound professional development models to prepare teachers for the use of AI tools while still navigating pedagogical effectiveness and student engagement.

8.3 Policy and Regulatory Considerations

For AI to be widely adopted in education, policies need to be carefully thought out at the institutional, national, and international levels.

a. *Educational Standards and Frameworks*: Education systems may need to take into consideration educational outcomes enhanced by AI and create new standards and frameworks that consider educational quality and integrity.
b. *Privacy and Data Protection*: Policies needs to be put in place to ensure the protection of student data and privacy of learners using AI-enhanced education experiences as AI-based systems and AI-enhanced learning environments continue to evolve.
c. *International Coordination*: There may be a need for international coordination to address issues related to educational AI that cross national borders and to facilitate consistency of ethical standards for education systems that may have different practices and ethics.

8.4 Institutional Transformation

The integration of AI in education will lead to deep changes across educational institutions as it develops increasingly sophisticated forms of integration.

a. *Organizational Structure Changes*: Institutions create new structures, roles, and responsibilities to develop a functional AI integration plan and to provide educators and students with a supported social environment or ecosystem of AI-related enhanced learning contexts.
b. *Curriculum Redesign*: Academic programmes may require wholesale redesign to incorporate AI literacy, along with pedagogical emphasis on human-only skills, to prepare students for an AI integrated workplace.
c. *Revolutionary Assessment*: Conventional assessment practices will give way to novel evaluation processes and new course content that incorporate AI capabilities to maintain academic integrity and provide a meaningful assessment of student learning.

9. Recommendations and Guidelines

The comprehensive analysis presented in this chapter offers the following actionable recommendations for stakeholders involved in ChatGPT integration in educational spaces.

9.1 Recommendations for Educational Institutions

a. *Create Universal AI Policies*: Institutions should develop concise, wide-ranging policies that govern AI use while balancing innovation versus student well-being, academic integrity, and quality of education.
b. *Invest in Infrastructure and Support*: Institutions need to provide adequate technological infrastructure to support AI utilization and provide ongoing support. They should plan to budget for necessary hardware, software, and human resources.
c. *Focus on Professional Development*: Involving teachers in AI programme integration should also include professional development as a necessary programme and it should be ongoing as educators progress in their AI experiences.
d. *Develop Evaluative Measures*: Institutions should develop strong evaluative measures of AI integration effectiveness, including those related to student learning outcomes, satisfaction metrics, and unintended consequences.

9.2 Recommendations for Educators

a. *Start Slow and Scale Carefully*: Educators should start with low-risk activities for AI integration and then begin to expand its use as their comfort ability and expertise grow.
b. *Stay Pedagogically-minded*: AI tools should be integrated in support of learning objectives, not for novelty or ease. Technology decisions should always be pedagogically focused.
c. *Build AI Literacy*: Educators should spend time to understand AI capabilities and limitations so that they may make informed decisions about integration and facilitate accurate use with students.
d. *Foster Critical Thinking*: Spend special time to develop students' critical thinking skills to analyse AI-generated content, while preserving independent rational thought and reasoning ability.

9.3 Recommendations for Policymakers

a. *Fund Research and Development*: Policymakers should make an investment into the pedagogical exploration of AI, including its efficacy and unforeseen consequences to support evidence-based policy-making.
b. *Provide Framework for Digital Equity*: Policies should ensure equitable access to AI educational tools regardless of socio-economic and geographic boundaries to avoid deepening existing social inequalities.
c. *Establish Ethical Frameworks*: Ethical frameworks must be developed for the use of AI in education to protect students as subjects of research while also promoting innovation.
d. *Nurture International Collaboration*: Global collaboration should be encouraged in order to disseminate best practices, share common challenges, and establish common standards with respect to the use of AI in education.

9.4 Recommendations for Students

a. *Build AI Literacy*: Students should create a knowledge base on the following elements of AI: capabilities, limitations, and appropriate use in order to maximize key benefits and avoid reliance.
b. *Uphold Academic Integrity*: Students will abide by ethical use of AI and understand how to draw the line when taking assistance while remaining responsible for their own learning.
c. *Foster Critical Thinking*: Students should develop and strengthen critical thinking skills for their own evaluation of AI-generated information and content and retain their ability to be independent thinkers.
d. *Seek Balance*: Students should strike a healthy balance between using AI and engaging in their own work to become effective and skilful learners.

10. Conclusion

The use of ChatGPT and AI tools in education represents a significant opportunity and a difficult challenge in the contemporary educational landscape. This thorough review has shown that there is immense potential for personalized learning, accessibility, and educational goals with AI tools, but that thoughtful consideration must be given to the implications for academic integrity, pedagogical efficacy, and social equity. The body of evidence throughout this chapter suggests that ChatGPT can be an effective learning tool when implemented deliberately and methodically. The meta-analyses and research studies cited showed substantial positive effects on learning outcomes, especially learning skills in the skill-based and STEM fields relevant to education. The effect sizes indicate a meaningful educational impact of ChatGPT as a learning tool. The case studies delivered promising findings from both formal and informal educational contexts, and demonstrated both the potential capacity of AI tools like ChatGPT, and the challenges of implementing new AI across diverse settings.

While ChatGPT can truly reshape education today and in the future, opportunities will be missed unless there is a commitment to responsible use, successful implementation, ongoing assessment, and continued improvement. If thoughtfully applied within the bounds of rigorously researched pedagogical practice and ethical principles, ChatGPT, as well as other innovative AI and IT technologies, can facilitate progress towards educational purposes and developmental needs in an increasingly complex and ever-changing twenty-first century. The journey of AI in education has just begun and whichever way it migrates, the lessons from these first implementations will fuel developments for years to come. By continuing to focus on student learning and development, educational equity, and academic integrity simultaneously with the exciting possibilities these new technologies offer, we in the education community can help navigate towards an era in which AI and AI tools enhance, and not erase, the essential human facets of what makes education meaningful and transformative.

The recommendations made in this chapter prove that a strategic, evidence-based approach is needed to integrate AI into education. The important elements for educational organizations include policy development, funding and resourcing for professional development, and measures of evaluation. The recommendations for educators address the need to facilitate AI literacy and build on pedagogical capacity to make the most of AI. Policymakers need to consider equity issues, develop ethical frameworks, and maintain supports for innovation and research.

References

Alharbi, S. and Alsubaie, M. (2023). Faculty perceptions of ChatGPT integration in higher education: Opportunities and concerns. In *Emerging Trends in Educational Technology* (Springer) are Dominik May (University of Wuppertal, Germany) and Michael Auer (CTI Frankfurt, Germany). (pp. 113–128). Springer.

Ali, Y., Wang, Y., Wang, Q. and Wang, X. (2025). The effect of ChatGPT on students' learning performance, learning perception, and higher-order thinking: A meta-analysis. *Humanities and Social Sciences Communications*, URL:https://www.nature.com/articles/s41599-025-04787-y.

Anderson, T. and Shattuck, J. (2012). Design-based research: A decade of progress in education research? *Educational Researcher*, *41*(1), 16–25.

Bandura, A. (1977). *Social Learning Theory*. Prentice Hall.

Bryant, L. and Singh, M. (2024). AI chatbots in workforce training: A study of ChatGPT in corporate learning environments. *Journal of Workplace Learning*, *36*(2), 154–170.

Dai, H., Li, Y. and Huang, P. (2023). Impact of AI-based chatbots on student engagement and learning efficacy in online settings. *Computers & Education*.

Holmes, W., Bialik, M. and Fadel, C. (2019). *Artificial intelligence in education: Promises and implications for teaching and learning*. Center for Curriculum Redesign.

Isotani, S., Millán, E., Ogan, A. and Santos, P. (2025). Artificial intelligence in education. In: Luckin, R. et al. (eds) *Artificial Intelligence in Education*. AIED 2024. vol 13966, pp. 133–143. Springer.

Johnson, D. and Kumar, S. (2022). Addressing ethical challenges of AI in higher education. In *Ethics and Governance in AI Applications* (pp. 145–160). IGI Global. Available at: https://www.igi-global.com/chapter/addressing-ethical-challenges-of-ai-in-higher-education/285679. URL was last accessed 15/6/2025.

Kong, S., Mason, A. and Wei, M. (2024). Global perspectives on the adoption of ChatGPT in education: Emerging challenges and innovations. In: R. Luckin et al. (Eds.), *Emerging Trends and Innovations in Artificial Intelligence and Higher Education* (pp. 263–273). Springer.

McGee, P. (2024). Generative AI and academic integrity: Managing the dual use dilemma. *Journal of Educational Integrity*, https://www.frontiersin.org/journals/education/articles/10.3389/feduc.2024.1328769/full this is online article link is provided

Muñoz, M., Chaudhry, A. and Stojanov, M. (2024). The use of ChatGPT in teaching and learning: A systematic review of strengths, weaknesses, opportunities, and threats. *Frontiers in Education*, *9*.

Patel, R., Chen, J. and Nguyen, T. (2024). Effects of ChatGPT on student writing quality and dependency. *Journal of Educational Computing Research*, *62*(1), 15–36.

Piaget, J. (1954). *The Construction of Reality in the Child*. Basic Books.

Rivera, L. and Thomas, G. (2021). AI-assisted learning: Effects on student critical thinking skills. In: Michael C. Smith, Deborah L. Nussbaum, *Emerging Trends in Educational Psychology* (pp. 72–88). Springer.

Sweller, J. (1994). Cognitive load theory, learning difficulty, and instructional design. *Learning and Instruction*, *4*(4), 295–312.

Vygotsky, L.S. (1978). *Mind in Society: The Development of Higher Psychological Processes*. Harvard University Press.

Zhai, X. (2023). ChatGPT in education: Opportunities, concerns, and future research directions. *International Journal of Educational Technology in Higher Education*, *20*(1), 1–9.

Zhang, Y. and Wang, H. (2023). The role of ChatGPT in transforming higher education: Opportunities and challenges. In: Moshe F. Gelman, Haggai Koiransky, Daniel R. Parker, *Advances in Educational Technology and Learning* (pp. 89–104). Springer.

Zimmerman, B.J. (1990). Self-regulated learning and academic achievement: An overview. *Educational Psychologist*, *25*(1), 3–17.

Chapter 9

Edge AI for IoT in Education: A Smart School – An Advanced Research Study

S. Sowjanya Chintalapati[1*] and *Lakkineni Swathi*[2]

1. Introduction

The world is changing as a result of emerging technology in ways that were previously unthinkable. Wireless connectivity, widespread digitization, and advancements in smart device technology have drastically altered the environments in which we work, live, and study. Disruptive technologies, which have a significant impact on how organizations operate and how society functions overall, are responsible for all of the developments. Artificial intelligence (AI), virtual/augmented/mixed reality, cloud computing, Internet of Things and Edge AI for IoT, e-services (such as e-trading, e-banking, e-learning, and e-health), block chain technology, etc., are some of the most successful disruptive technologies. Disruptive technology integration has several advantages for our economy and society in terms of everyday operations planning. The key question is how these technologies, and the data analysis that goes along with them, may be used to improve social and economic functioning, resulting in elevating individual standards of everyday life. In keeping with these trends, smart homes, buildings, cities, and even smart schools are becoming more than ideals. A developing technology called the Internet of Things (IoT) enables objects, gadgets, and other objects (such sensors and actuators) to connect with one another inside a system to perform various tasks. Sensors collect and transmit data in real time. Depending on what they are meant to be used for, sensors come in different varieties. Actuators carry out various activities in response to the input data from other networked devices.

Edge AI for IoT-enabled systems use sensors to measure, record, and monitor data. For processing, analysis, and archiving purposes, the data are sent to control

[1] Department of CSE, Sreyas Institute of Engineering and Technology, Hyderabad, India. https://orcid.org/0009-0005-5651-5438

[2] Ford Motor Company, MI, USA. https://orcid.org/0009-0006-2769-1421

* Corresponding author: soujikits@gmail.com

DOI: 10.1201/9781003639367-9

centres either on the network or in the cloud. These data are used to make decisions, which leads to the actuators receiving the appropriate commands (Abdel-Basset et al., 2019). Based on our analysis of related studies taking into account coordination with the relevant research, the applied intelligence is predicated on the utilization of numerous interconnected IoT, specially Edge AI for IoT, gadgets that can share information via networks and the cloud and send data wirelessly or through wire. By enabling improved control, management, and decision-making, the aim is to create a whole ecosystem made up of smart systems and related cutting-edge applications, resulting in efficiency, convenience, and sustainability. Users interact with these smart systems in a multitude of ways via the IoT by means of smart gadgets. Modern Edge AI for IoT gadgets enable apps to access real-time data that is appropriately collected, handled, and processed to help decision-makers, authorities, and end users make decisions that will improve daily operations and general quality of life. This chapter seeks to demonstrate how Edge AI for IoT is used to create smart environments across several domains. Specifically, a thorough examination of the idea of smart schools is conducted. According to the research, Edge AI for IoT devices can be used to create safe and energy-efficient classroom environments, preserving the right conditions for learning and facilitating effective teaching-learning process (Chan, 2002).

2. Diversified Application Areas of Edge AI in a Smart Environment

This allows for the monitoring and control of a wide range of variables that have a significant impact on our lives:

- Keeping track of energy distribution and conservation
- Tweaking traffic signals to traffic patterns
- Facilitating optimization of routes
- Avoiding gridlock
- Streamlining public transportation
- Trash collection
- Sanitation on the streets
- Irrigation systems
- Controlling parking spaces and electric vehicle charging stations
- Improving air quality and lowering pollution
- Lowering maintenance costs and expenses
- Enhancing sustainability and overall quality of life

Smart homes are equipped with IoT networks that connect appliances, security systems, virtual assistants, utility meters, and other devices to enable the automation of household chores. More conveniently and effectively, users can operate appliances remotely while keeping an eye on how they are being used. In a same vein, intelligence encompasses many facets of life, including education, where automation and optimization are brought about by smart campuses, institutions, and schools. Security and access control are offered in the framework of smart campuses and schools, together with energy-efficient heating and cooling systems.

Additionally, the astute application of digital technologies improves learning opportunities and educational services. Therefore, in order to better satisfy the requirements and preferences of the present digital age, educators should use student-centred approaches and individualized learning to make better use of developing technology. The smart environment enables dynamic, captivating, and collaborative learning, in contrast to traditional teaching methods. It makes use of a number of cutting-edge technologies that make it possible to modify the teaching process to fit the needs, interests, and ability levels of the students, thereby boosting motivation. Intelligent decision-making for the entire educational system is also made possible by the integration of information and communication technology (ICT)–based educational applications.

3. Edge AI for IoT in Education

The IoT has a lot to offer the education industry. A review of the literature demonstrates the variety of uses of Edge AI for IoT in education for educators, students, and administrators. This section provides a summary of potential Edge AI for IoT applications in educational settings that have been highlighted in the literature. Although not new, the idea of "smart schools", or ICT-driven schools, is still in its infancy. This section focuses on how Edge AI for IoT can help make the idea of a "smart school" a reality.

3.1 Edge AI for IoT for Smart School Management

- *Energy management*: An intelligent energy management system needs to be both interoperable and connected. Cost-effective energy management and conservation would be very beneficial to school administration in a smart school scenario. A school organization can increase its smart energy utility and save hundreds of dollars a month. Electrical energy, for instance, can be used whenever needed. Sensors in school buildings have the potential to enable on-demand electrical energy use and to switch off unnecessary appliances, reducing the cost and carbon footprint associated with each unit of power utilized. An Edge AI for IoT-based communication framework can relay energy-related information to the management system and users in a school setting. A platform for Edge AI for IoT-based energy management has been built by a study (Choi, 2012). The creation of a demand response (DR) energy management system can be aided by such a platform.
- *School Premises Security*: One crucial Edge AI for IoT-powered smart school aspect is school premises security. The majority of schools are not equipped with a safe infrastructure that can recognize and identify incidents of theft, fire, abuse, sexual assault, and other mishaps that may happen on school grounds. By using a real-time camera to record scenes linked to IoT sensors and devices, school administration can monitor the school grounds, spot potential accident scenes, and, in the event of an emergency, start an action plan right away. The research project in Hua (2012) suggested the secure system of Internet of school things (S-IoST), an Edge AI for IoT-based alert mechanism. This system makes use of intelligent transportation systems, 5G cellular technologies, sensing technologies, and IoT networks for sophisticated communication (Hua, 2012; IBM, 2012).
- *Management of students with special needs*: To support their learning, students with mental health issues, learning challenges, hearing loss, or vision impairments

require comprehensive and varied solutions. For children with particular needs, Edge AI for IoT can offer an environment that is both personalized and easy to use. For instance, a system that combines a tablet and sensor-connected gloves can produce vocal speech that is translated from sign language. Teachers can utilize this system to facilitate communication in the classroom for deaf and mute students. Various studies examined the use of Edge AI for IoT and its potential for kids with special needs (Kankaanranta, 2014; Smart SC Report, 2014). Assistive technologies for students with disabilities include, for instance: Students with mobility issues can be benefited from an on-screen keyboard and a screen reader, that can convert text-to-speech applications which can read aloud and this information is very helpful to those who are blind or visually impaired (Research and Markets, 2014).

- *Smart school transport*: An intelligent transport system for students can be operated by school administration with the help of Edge AI for IoT solutions. An IoT-based car monitoring system is described in a study (Samsung, 2013). A microcontroller and a mobile phone application make up the suggested system. The global positioning system (GPS) is used to determine the school bus's position from a gadget. An alcohol sensor and a panic switch were incorporated into the system to ensure the children's safety. The real-time bus status is provided to parents and school administration by this intelligent transportation system.
- *Student's health monitoring system*: The performance of its students is the most important thing to every educational institution. A student's health problems have a significant impact on their performance. A student's academic performance suffers when they are ill or experiencing psychological issues since they are unable to concentrate on their academics. Many health data are gathered via sensors. The cloud and Edge AI for IoT gather and analyse massive amounts of data to assess the health and happiness of students. An Edge AI for IoT-based Ambient Intelligence Assisted Health Monitoring System (AmIHMS) has been proposed for student health monitoring in a study (Uskov et al., 2015). A framework for cloud-centric Edge AI for IoT-based student health monitoring was proposed by research work in Zhu et al. (2016). A different study (Kim et al., 2014) examined the use of physiological markers and Long ShortTerm Memory (LSTM) for emotion identification. An Edge AI for IoT framework for healthcare was proposed by researchers here.

3.2 Edge AI for IoT for Teachers

- *Autonomous attendance system*: For teachers, recording every student's attendance is a demanding and time-consuming chore. Instructors must concentrate on their main responsibility, which is to instruct and support students in their learning. At the entrance of the classroom door, an automated attendance system based on biometric attendance or barcode with an identity card number can be employed. As a result, every kid coming in the door will have a registration. Parents can also be apprised of this information so that they are aware of whether or not their child is in class. Likewise, research indicates that automatic attendance systems employing Edge AI for IoT in conjunction with facial recognition technologies are possible (Middleton, 2015; MESTRK, 2011; Lee, 2015).

- *New pedagogies*: The idea of a classroom has evolved in light of technological improvements. Traditional methods and physical space are absent from modern classrooms. Teachers are enabling learners with a new set of pedagogies that are driven by digital technologies and digital learning environments because modern classrooms offer new ways to learn. Consequently, fostering digital pedagogies refers to how teachers might utilize specific technology to facilitate teaching and learning in the most efficient way possible. One such instance is the flipped classroom, which is used by teachers to educate computer networks while including Edge AI for IoT elements (Gros, 2016).
- *Feedback and assessment systems*: Quick and simple methods to examine student performance and evaluate them based on their involvement, engagement, interaction, and task completion are to use embedded evaluation technologies that are based on the IoT. A study used attention scoring assessments in e-learning to suggest an IoT-based framework for student interaction (Mayer & Cukier, 2013). In a similar vein, a different study proposed an Edge AI for IoT-based real-time data mining method for student assessment (Boulanger et al., 2015). Edge AI for IoT and raspberry pie are utilized in research work (Picciano, 2016) to measure student engagement.
- *Edge AI for IoT for STEM education*: Education in the fields of science, technology, engineering, and mathematics (STEM) necessitates practical laboratory experiences as well as measurements, observations, and experimentation. Real-time measurements and data processing are now feasible thanks to IoT sensors and devices. One other benefit of Edge AI for IoT is that STEM students can use design thinking and other STEM abilities through project-based learning with Edge AI for IoT kits (Stackpole, 2017). How Edge AI for IoT applications can be used in physics experimental classes was demonstrated in a study (Spector, 2014). A case study of Edge AI for IoT integration for an embedded system course was covered in research in Hwang (2015), which also demonstrated how Edge AI for IoT may be used in STEM education.

3.3 Edge AI for IoT for Learners

- *Distance learning*: The COVID-19 pandemic has demonstrated the efficacy and concreteness of online learning platforms. Students from throughout the world attended school in their homes during the global school closure, thanks to the technological developments in the field of education. Any type of distance learning, including hybrid, e-learning, online, mobile, and virtual classroom learning, is possible. Distance learning encompasses all of these formats. Edge AI for IoT has the power to successfully improve remote learning. One such instance is the study that showed how the IoT might boost learning efficiency in distant learning by up to 20%. Edge AI for IoT devices tracked the amount of brain activity throughout the video session and fed the algorithm with feedback and digital picture (Chatti, 2010).
- *Increased engagement and productivity*: Students become more interactive and engaged in smartphone-based online classes, often known as virtual

classes or e-classes. Students' motivation in participating in various tasks and actively participating in the assessment and feedback process is developed through interaction-based learning. As a result, Edge AI for IoT-based learning environments improves engagement and productivity among students. For instance, barcode-enabled e-books facilitate interactive reading for pupils. An IoT gadget called Scan Maker can swiftly scan editable text from books, papers, and other materials, and transfer them straight to a computer, tablet, or phone. Text can be translated into forty different languages using this gadget (Kopainsky et al., 2012).

- *Digital learning environments that are personalized*: A digital learning environment offers a customized learning experience. A good example of this is Blackboard (Andrade & Du, 2007), which provides interactive learning management systems (LMS), websites that inform parents, students, and others about events, news, and grades, as well as virtual classroom technology that promotes enhanced cooperation (Fig. 9.1).

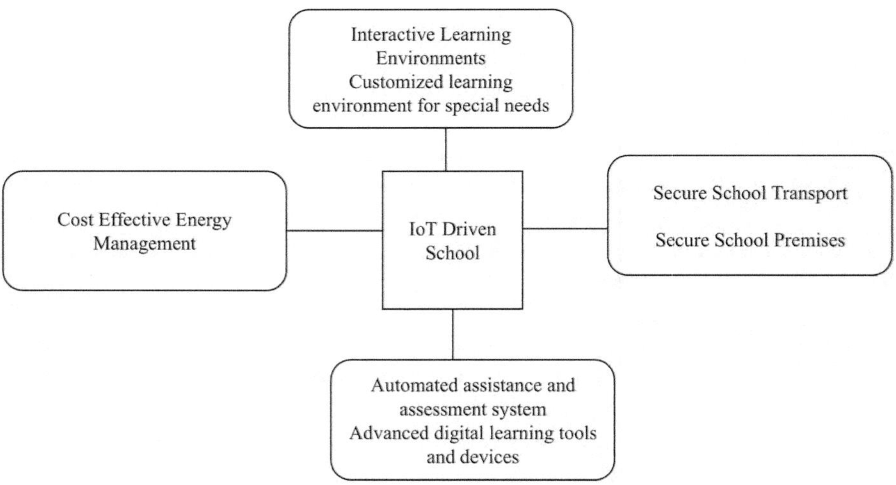

Fig. 9.1 Edge AI for IoT-driven smart school.

4. Bottlenecks in Edge AI for IoT Adaptation in Educational Settings

Edge AI for IoT integration in education has numerous advantages, but it also has significant risks and obstacles (Pons et al., 2015). The primary dangers and difficulties related to Edge AI for IoT adaption in education are covered in this section.

4.1 Security and Privacy

The existence of extremely sophisticated technologies linked to the internet presents a risk of cyber-attacks. It is simple for outsiders or students to attack educational institutions. Controlling an institution's systems has the potential to render it inoperable as well. It is also possible for a connectivity issue to prevent the system from operating as a whole. Data breaches continue to be a major security risk.

Authentication, integrity, end-to-end security, and confidentiality present security problems (Durall & Leinonen, 2015).

Massive amounts of data are collected by sophisticated Edge AI for IoT sensors and devices working with various technologies including near-field communication (NFC), radio frequency identification (RFID), 4G, and 5G. The data is a major source of concern. What happens to that data? Who gains from the control of this data? These are the privacy-related questions pertaining to internet technologies that gather vast amounts of personal data (Knight, 2013). Edge AI for IoT adoption in the education industry can only be guaranteed going forward if the security concerns around it are resolved. Since only authorized individuals should be able to access the data collected, confidentiality is crucial to Edge AI for IoT security. Users can be both humans and robots, and data leakage to unauthorized users—whether they are humans or machines—is a major concern for Edge AI for IoT secrecy.

Another security risk is large-scale data management. Since Edge AI for IoT is gathering and analysing massive amounts of data, a reliable data management solution is necessary. While being transmitted from one sender to another, data may be lost, altered, or leaked. In order to prevent data breaches and mishandling, a secure data management system is necessary.

4.2 Scalability and Reliability

Scalability is a major issue in educational environments. There will be a massive data collection process while using Edge AI for IoT apps. A significant amount of data will be produced by Edge AI for IoT integration in education. As a result, data analysis is crucial to gaining deep insights from data. As a result, one barrier to Edge AI for IoT use in educational settings is data scalability. Scalability is a topic covered in the study by Buckingham Shum et al. (2012), which also suggests a social recommender system based on Hadoop and parallel computing to offer students individualized learning materials based on massive data analysis. Another study suggested an Edge AI for IoT solution based on secret sharing that is both scalable and secure (Kruse & Pongsajapan, 2012). The study also proposed that distributed Edge AI for IoT infrastructure deployment can lead to increased scalability and dependability of Edge AI for IoT systems. Concerns about scalability arise when Edge AI for IoT-based solutions are not long-term viable. All social classes ought to have equal access to these technologies. Additionally, schools must to have sufficient cash to pay for these services and maintain a reliable system.

4.3 Dehumanization and Ethical Concerns

Although technology is a tremendous tool for our daily life, ethics and humanization should not suffer because of it. It is true that technology exists to simplify our lives in many ways, but in order for any technology to function, we must establish an operating protocol. One major problem that would occur if systems operated more autonomously and with less human interaction is dehumanization. Over-interaction among machines may reduce human oversight and give the companies that provide these services more authority and control (Durall & Toikkanen, 2013). Consequently, there is a very high chance of an unethical and unfair power distribution.

User awareness is yet another crucial consideration. Users need to know the potential for attacks and be taught self-defence techniques on the internet. Governments should plan public awareness and education campaigns about new technologies and how to use them, in addition to developing mechanisms for security requirements.

5. Role of Edge AI for IoT in Intelligent Educational System

In the context of smart schools, the intensive, pervasive ICT has recently been a driving force behind the paradigm shift towards intelligent education. Following this pattern, our team conducts research to model an intelligent learning environment based on the concept of a smart classroom. It uses many different technologies and incorporates a lot of digital gadgets into the classroom to make learning more intelligent and boosts student achievement. A concept like this may be used in both virtual and real-world settings, and it addresses the demands of the digital age as well as the difficulties that arise during world crises. The intelligent atmosphere provided by the Intelligent Education System (IES) enhances learning in a personalized, efficient, effective, and enjoyable way.

It is accomplished by making use of the entire spectrum of modern educational technologies. They include networked interactive devices (including laptops, cell phones, computers, and tablets) and wearables, sensors, actuators, cameras, RFID tags, RFID readers, and other state-of-the-art technological instruments. Additionally, all environmental and educational data is tracked, collected, processed, analysed, and managed using specialist software. Furthermore, smart decision-making based on intelligent technologies like learning analytics, AI, semantics, ontologies, neural networks, and data mining provide intelligent resource control and intelligent process management.

Within the framework of IES, Edge AI for IoT plays two roles: it facilitates creative teaching methods that optimize the learning process and it allows for intelligent control and management of classroom environmental elements. It involves the integration of Edge AI for IoT devices within IES, such as sensors, actuators, wearable devices, and cameras, in order to allow the educational system to remotely track and regulate the physical characteristics of the surroundings as well as the factors related to the educational procedure (students' behaviour, engagement, and concentration). One essential requirement is that IoT devices be able to connect to a network and communicate with one other.

They collect, store, retrieve, and synchronize data via the internet in addition to performing remote monitoring and control. The physical aspects of the school environment, such as humidity, lighting, temperature, and oxygen levels, all have an impact on learning. To provide the best possible learning environment, they are appropriately observed and managed. An enormous amount of data about the classroom environment is gathered and analysed by several sensors. Through the use of cameras, motion sensors, wearable technology, RFID readers, and other devices, the learning process flow is monitored. They monitor students' participation, keep an eye on their presence, identify inappropriate behaviour, and keep tabs on how students communicate within the learning system (Terzieva et al., 2022).

6. Edge AI for IoT Applications in Teaching–Learning Pedagogy

The networking of devices other than desktop and laptop computers, or the "internet of things", will fundamentally alter our way of life in the next years. Among those is the use of Edge AI for IoT to transform the education industry. Every day, this transition is happening faster and faster. Every student connects to their school or any other educational institution via a smart device. The IoT (Edge AI for IoT) has made it possible for educational institutions to monitor enrolment, enhance campus safety, and relieve teachers from abundance of paperwork. Approximately 82% of children between the ages of 12 years and 17 years own smartphones, and they utilize them for both leisure and educational purposes, according to a survey from emarket retail. The majority of the upcoming generation's activities rely on mobile phones, and they are already glued to them. Edge AI for IoT has the ability to significantly improve education, even while user safety and protection remain points of worry.

Two viewpoints exist regarding Edge AI for IoT applications in educational institutions: those of teachers and personnel, as well as those of students. Instructors are able to monitor their pupils' attendance. Teachers can measure student attendance based on the location or information that students provide using sensors installed in classrooms, libraries, or canteens. Teachers can virtually observe their classes thanks to the integration of Edge AI for IoT in schools. We examine the different ways that Edge AI for IoT is being used in the field of education. The aforementioned uses make it abundantly evident that Edge AI for IoT will be highly profitable in the education industry in the future.

6.1 Energy Management

a. Growing concerns about global climate change necessitate changes to the educational system in order to safeguard the finite energy supplies.
b. Publishing books and course manuals in vast volumes results in paper wastage.
c. By utilizing Edge AI for IoT framework, books are accessible in the electronic mode on gadgets for learners to peruse and make adjustments.
d. The fitting of sensors and actuators in lights and faucets to cause them to turn on right away without a person's assistance.
e. Edge AI for IoT reduces the amount of harmful gases in the atmosphere, such as CO_2 and CO.

6.2 Attending Lectures Online and Remotely

a. Edge AI for IoT helps to provide a secure and safe learning environment for students.
b. Edge AI for IoT enabled the majority of establishments worldwide to transition to remote learning methods.
c. On-campus sensors keep an eye on student activities and react to incidences of bullying or property damage.
d. RFID chips embedded in student ID cards capture pupils' daily attendance.

6.3 Monitoring the Health of Students and Personnel

a. An educational institution can be more productive when its staff and students are well. To be more engaged in class, students need to take better care of their health and get enough sleep every day.
b. Student health is identified and monitored by physiological signs. Since this data is kept in databases, medical professionals may easily evaluate the health of students.
c. Vital data such as blood pressure, medical history, allergies, and prescriptions for emergency care in case of accidents can be saved via Edge AI for IoT.

6.4 Applications for Improving Interactive Learning

Education may be made much more enjoyable and fascinating with the help of Edge AI for IoT. It keeps pupils interested in learning and, in general, contributes to the development of their personalities and mental abilities.

a. Virtual classrooms on TVs and mobile phones enable students to continue learning even while unwell.
b. Edge AI for IoT-based graphic classrooms use 3D images, colours, and sounds to enhance learning experiences.
c. Voice-to-text converters improve note-taking efficiency and convenience.

6.5 Smartboards

Edge AI for IoT technology, such as electronic boards and smartboards, simplifies teaching and learning for students. It is user-friendly and reliable. It enables a user to display visuals on the board and communicate through the subject matter.

a. The board's interactive display enables personalized learning by allowing for image manipulation, drawing, and other features.
b. Colours and voice can be used by smart pens to generate 3D creations. Voice-to-text sensors convert teacher lectures into direct observations given to students.

6.6 Monitoring Disabilities

Developing instructional tools can sometimes ignore students with difficulties, including dyslexia. Special software tools can be created using Edge AI for IoT to help kids with disabilities learn.

a. Transcripts translate voice to text, allowing students with hearing disabilities to comprehend their notes.
b. Students who are visually impaired benefit from Text-to-voice apps.
c. The way disabled youngsters learn is changed by specialized tablet apps. These applications are looked at to enhance education.

6.7 Attendance and Evaluation of Tests

A student must adhere to certain regulations in order to pass, such as maintaining an attendance rate of 80% throughout the semester, among others.

a. Internet of Things apps monitor student attendance and send staff reports.
b. Students who have lower attendance rates receive a notice, and they can monitor their progress via Edge AI for IoT-developed applications.
c. Students can now take tests remotely thanks to the IoT. Devices equipped with sensors can monitor a student's movements to make sure they are not cheating.

6.8 Improvement of Safety Measures

In schools, technology and education go hand in hand. Schools utilize a variety of Edge AI for IoT solutions to protect their children.

a. For example, sensors for fire detection. In the event of a threat, sensors notify everyone on the school grounds and can even notify the fire department.
b. Internet behaviour tracking to make sure pupils are not accessing offensive material while on school property.
c. Video surveillance to monitor what students are doing during lunch breaks or after classes.
d. Additional Edge AI for IoT applications to guarantee school safety include visitor management, digital signs, and access control.

6.9 Augmented Reality in the Classroom

The technology that makes audio, video, and images come to life and encourages in-person engagement is called augmented reality (AR). Students learn and comprehend the concepts more effectively as a result.

a. Project mapping ideas in AR display physical distances so that students can better understand subjects.
b. AR expands learning beyond books and textbooks by fostering interactive learning in the classroom.

6.10 Building Temperature Control

The required temperature in classrooms and gymnasiums varies.

a. IoT applications can sense human presence and turn on or off the energy in response.
b. These applications can also adjust the temperature and light levels based on the surroundings.

6.11 Acquiring Data in Real Time

a. The fact that the majority of the curriculum and teaching methods used in the current educational system are outdated is one of the main causes for concern. In other words, the subjects are not appropriate for today's students.
b. The IoT (Edge AI for IoT) makes it easier to gather and analyse data in order to create courses that emphasize student participation, sharpen attention, and progress learning on a much larger scale.

6.12 Bus Safety in Schools

Due to reckless drivers and a high number of school bus accidents worldwide, parents are hesitant to send their kids to school. However, Edge AI for IoT can help with this issue.

a. Edge AI for IoT applications allow drivers to keep an eye on the well-being of the students they are transporting. From school property, schools are able to follow both the driver and the bus.
b. Parents can monitor their children's location to ensure they get to school safely and on time. Edge AI for IoT can also be used to check speed limits and tire punctures, raising alarms during danger.

6.13 Ensure that Kids Return Home Securely

Many parents are concerned that their children will arrive home late or become lost on the way from school. Edge AI for IoT is benefiting children and parents by guaranteeing that children reach home on time.

a. School shoes or backpack sensors, for instance, can notify parents of a child's whereabouts and other facts.
b. In this way, parents may monitor their children from a distance. When a youngster is in danger, these Edge AI for IoT apps sound an alarm.
c. As soon as a child leaves the designated routes while heading home, these applications notify parents and the local police departments of their whereabouts.

6.14 Instructors

In the educational system, instructors, tutors, and professors are just as significant as pupils.

a. The IoT makes teaching easier.
b. IoT applications collect student attendance data, monitor their academic progress, and rank the pupils according to various attributes like grade, sport, and so on.
c. It bridges the knowledge gap between educators and students by assisting teachers in better understanding the needs of their students.

6.15 Exam Monitoring

Exams are an essential component of a student's success. Therefore, it is critical to design a system that guarantees fair and appropriate exam administration.

a. Edge AI for IoT is assisting educational institutions in maintaining order during tests.
b. Examiners are receiving notifications and are being able to notice faint noises thanks to Edge AI for IoT-incorporated LED screens.
c. Students can also see warning messages shown on these gadgets.

7. Pros and Cons of Edge AI for IoT in Academia Field

7.1 Pros of Edge AI for IoT in Education Field

The aforementioned applications make it abundantly evident that Edge AI for IoT will be highly profitable in the education industry in the future.

- *Effective administration of the educational system*: IoT is assisting in cutting down on time and expenses associated with manual labour, including paperwork, money management, and supply tracking.
- *Information gathering in real time*: Edge AI for IoT handles terabytes of data at once. In order to create a dependable and effective educational system, this data forecasts and examines the actions of both staff and students.
- *Resource management improvisations*: The IoT is assisting in the rescue and management of energy and water use. Edge AI for IoT technologies in education have led to a significant decrease in the use of paper.
- *Global connections*: Edge AI for IoT systems expose us to schools and colleges throughout the world because they depend on the internet. This provides us with a means of interacting with international systems and enhances the prominence of education in teaching field.
- *Incorporates safety concerns*: A lot of parents in remote areas are concerned about bringing their children to school. Some people might not have the money to pay for it, while others might not have enough resources to support a quality education. These worries are being addressed by Edge AI for IoT, which ensures that the youngster feels protected and secure while receiving a top-notch education.

7.2 Cons of Integrating Edge AI for IoT in Schooling

Edge AI for IoT in education has a few disadvantages too.

- *Expensive infrastructure*: Edge AI for IoT-controlled systems are costly to set up and purchase. They need a significant quantity of both hardware and software. To run the educational system, an experienced group of IT officials would need to be hired.
- *In-class ethical behaviour*: Edge AI for IoT uses knowledge-based tests to assess assignments, but there should be measures in place to stop plagiarism, cheating, and other forms of fraud.
- *Inadequate data storage infrastructure*: The IoT necessitates cloud storage of massive volumes of data processing information. Edge AI for IoT ecosystems are incompatible with outdated data storage methods.
- *Security and safety concerns*: IoT exposes personal data to thousands of users who can use it for their own gain. Access to student and staff data and details is made possible by Edge AI for IoT. This private information is accessible to hackers.

8. Considerations While Creating Smart Educational Settings

The International Association for Smart Learning Environments defines a smart learning environment broadly, stating that a learning environment is considered intelligent when it incorporates adaptive technologies or is designed with state-of-the-art features and capabilities that improve comprehension and performance. In general, a smart learning environment is one that is effective, engaging, and efficient. In their broadest meaning, smart learning environments refer to a new breed of educational systems that effectively and efficiently blend technology and pedagogy to enhance student learning (IASLE, 2017).

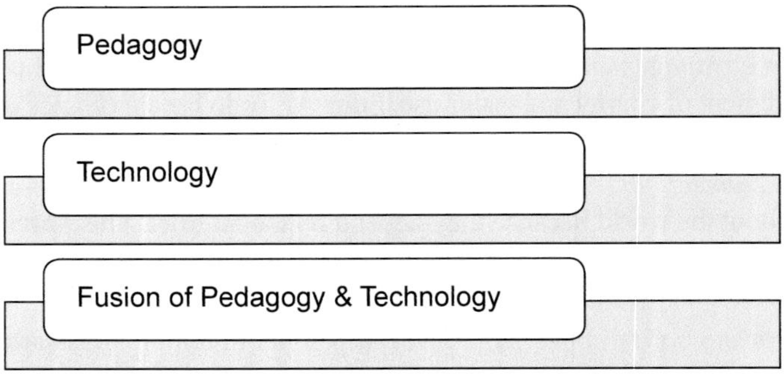

Fig. 9.2 Intelligent educational framework design.

When creating intelligent learning environments, two important factors need to be considered:

(a) involving users in the design process and
(b) offering supportive services to enable users to submit pertinent input.

The potential for smart learning is greatly influenced by the architecture of the learning environment, and participative procedures are essential in the creation of learning ecosystems. To make learning ever more flexible, contextual, and unhindered—whether interacting with others and businesses at home, at work, or in academic settings—transparently immersive experiences transmedia among people, businesses, and devices must be prioritized in smart learning (Goodyear & Retalis, 2010; Shoikova et al., 2016).

9. Conclusion and Future Ahead

The idea of learning anywhere, at any time, is not entirely uncommon. However, it is critical to consciously plan and encourage these processes when they are thought of as typical everyday activities. As was previously noted, in order to develop autonomous, adaptable learning environments that meet the needs of individual learners, smart learning environments must integrate formal and informal learning. To find meaningful learning trends, these settings must combine historical data with real-time learner location information using big data and learning analytics techniques.

Multiple virtual learning environments are integrated with a traditional classroom setting in context-aware learning environments. A fresh idea known as "Education as a Service" is making waves as a way to address the problems posed by open and global marketplaces. This strategy delivers educational resources as a service, making them easily accessible to learners worldwide.

From this vantage point, we may anticipate significant changes to the organizational structures and instructional methodologies of traditional education. Lectures, for instance, might be taught separately from the course material. An instructor other than the one in charge of the course may provide some of the lectures. Tests can also be administered separately, with an outside entity administering the exams in place of the course instructor. Services need to take the perspective and learning path of the learners into account. In an intelligent learning environment, students would have access to a variety of services during various learning phases, delivered by various academic institutions both online and offline.

Modern technology may improve economies through education. One such technology that has the potential to transform the conventional school setup and make the smart school concept possible is the IoT. This study provides an overview of Edge AI for IoT applications in educational settings. Edge AI for IoT offers a variety of advantages in learning environments. Thus, we propose that equitable and inclusive education facilitated by technology can provide high-quality results. But technology also presents obstacles. For this reason, ethical, scalable, and sustainable technology use is needed to reduce bottleneck impacts.

References

Abdel-Basset, M., Manogaran, G., Mohamed, M. and Rushdy, E. (2019). Internet of things in smart education environment: Supportive framework in the decision making process. *Concurrency and Computation: Practice and Experience*, *31*, e4515.

Andrade, H. and Du, Y. (2007). Student responses to criteria referenced self-assessment. *Assessment & Evaluation in Higher Education*, *32*(2), 159–181.

Boulanger, D., Jeremie Seanosky, Vive Kumar and Dr Kinshuk (2015). Smart learning analytics. In *Emerging Issues in Smart Learning*. Berlin Heidelberg: Springer. 289–296.

Buckingham Shum, D.G. (2012). In *Proceedings of the 2nd International Conference on Learning Analytics and Knowledge*. New York: LAK12 (ACM 2012).

Carol K.K. Chan (2002). "Co-regulation of learning in computer-supported collaborative learning environments: a discussion". Available at Smart learning environments, https://slejournal.springeropen.com /articles/10.1186/s40561-016-0026-2#CR10. Published: 23 February 2012.

Chatti, M.A. (2010). Toward a personal learning environment framework. *International Journal of Virtual and Personal Learning Environments*, *1*(4), 66–85.

Downes, S. (2010). New technology supporting informal learning. *Journal of Emerging Technologies in Web Intelligence*, *2*(1), 27–33.

Durall, E. and Leinonen, T. (2015). Feeler: Supporting awareness and reflection about learning through EEG data. *Proceedings of the 5th Workshop on Awareness and Reflection in Technology Enhanced Learning,* Durall, Eva; Leinonen, Teemu, By Aalto University (pp. 67–73).

Durall, E. and Toikkanen, T. (2013). Feeler: Feel good and learn better: A tool for promoting reflection about learning and well-being. In *Proceedings of the 3rd Workshop on Awareness and Reflection in Technology Enhanced Learning,* Publisher is CEUR-WS.org, 83–89.

Feng Li and Caohui Wang "Artificial intelligence and edge computing for teaching quality evaluation based on 5G-enabled wireless communication technology", Received Published on 23 March 2023, DOI https://doi.org/10.1186/s13677-023-00418-6

Goodyear, P. and Retalis, S. (2010). *Technology-enhanced learning*. Boston: Sense Publishers.

Gros, B. (2016). The Design of Smart Educational Environments. https://techvidvan.com/tutorials/Edge AI for IoT-applications-in-education-sector/ springer publisher.

Hwang, G.J. and Tsai, C.C. (2011). Research trends in mobile and ubiquitous learning: A review of publications in selected journals from 2001 to 2010. *British Journal of Educational Technology*, *42*(4), E65–E70.

Kankaanranta, M. (2014). Available at https://slejournal.springeropen.com/articles/10.1186/s40561-016-0026-2#CR36 https://doi.org/10.1186/s40561-016-0026-2 (2016).

Kim, S., Mims, C. and Holmes, K.P. (2006). An introduction to current trends and benefits of mobile wireless technology use in higher education. *AACE Journal*, *14*(1), 77–100.

Knight, S.S. (2013). Collaborative sense making in learning analytics. Viewing Education as a Site of Work Practice, Co-located with the 16th ACM Conference on Computer Support Cooperative Work and Social Computing (CSCW, San Antonio, 2013). San Antonio.

Kopainsky, B., Birgit Kopainsky, Pablo Pirnay-Dummer, Stephen M. Alessi's (2012). Automated assessment of learners' understanding in complex dynamic systems. *System Dynamics Review*, *28*(2), 131–156.

Kruse, A. and Pongsajapan, R. (2012). Student-centered learning analytics. *CNDLS Thought Papers*, 1–9.

Lee, A. (2015). Authentication scheme for smart learning system in the cloud computing environment. *Journal of Computer Virology and Hacking Techiques*, *11*(3), 149–155.

Mayer, V. and Cukier, K. (2013). *Big data: A revolution that will transform how we live, work, and think*. Boston: Houghton Mifflin Harcourt.

MESTRK. (2011). Smart education promotion strategy. President's Council on National ICT Strategies. Published in 2022 by the UNESCO Institute for Information Technologies in Education.

Middleton, A. (2015). Smart learning: Teaching and learning with smartphones and tablets in post compulsory education. MELSIG & Sheffield Hallam University, Sheffield. First published by the Media-Enhanced Learning Special Interest Group and Sheffield Hallam University, 2015, Quality Enhancement and Student Success, Level 1, Oneleven, Sheffield Hallam University, Howard Street, Sheffield, UK S1 1WB MELSIG is online at http://melsig.shu.ac.uk and @melsiguk

Picciano, A. (2016). The evolution of big data and learning analytics in American higher education. *Journal of Asynchronous Learning Networks*, *16*(3), 9–20.

Pons, P., Catala, A. and Jaen, J. (2015). Customizing smart environments: A tabletop approach. *Journal of Ambient Intelligence and Smart Environments*, *7*(4), 511–533.

Research and Markets. (2014). Global smart classroom market 2014–2019 Report. Available at http://www.researchandmarkets.com/reports/3063552/globa l-smart-classroom-market-2015-2019#pos-0, October 2024.

Samsung. (2013). Samsung Smart School: An interactive and collaborative learning environment for the next-generation classroom, Samsung news room.

Sharples, M., Taylor, J. and Vavoula, G. (2007). A theory of learning for the mobile age. In: Sharples, M., Taylor, J. and Vavoula, G. (2016) A Theory of Learning for *The Sage Handbook of E-Learning Research* (pp. 221–247). SAGE Publications.

Shoikova, E., Nikolov, R. and Kovatcheva, E. (2016). Conceptualising of smart education. KoMEIN Project, Conceptual Modeling and Simulation of Internet of Things Ecosystems, funded by the Bulgarian National Science Foundation, Competition for Financial Support of Fundamental Research.

Spector, J. (2014). Conceptualizing the emerging field of smart learning environments. *Smart Learning Environments*, *1*(1), 2–10.

Stackpole, B. (2017). Edge AI for IoT and augmented reality: A match made in heaven? Available at http://internetofthingsagenda.techtarget.com/feature/Edge AI for IoTand-augmented-reality-A-match-made-in-heaven?

Terzieva, V., Ilchev, S. and Todorova, K. (2022). The role of internet of things in smart education. *IFAC Papers Online*, *55*(11), 108–113.

Uskov, V., Howlett, R.J. and Jain, L.C. (2015). *Smart Education and Smart e-Learning*. Springer.

Zhu, Z.-T., Yu, M.-H. and Riezebos, P. (2016). A research framework of smart education. *Smart Learning Environments*, *3*(1), 1–17.

CHAPTER 10

The Framework Design of Intelligent Assessment Tasks Recommendation System for Personalized Learning

Anirudh AVS[1], *Aditya Krishnan*[1] and *Karmel A*[1*]

1. Introduction

The education system is undergoing a deep transformation driven by advances in artificial intelligence (AI). Traditional approaches to the creation of educational content is suffering from challenges associated with producing or maintaining quality, coherence, and comprehensive cognitive coverage in technical areas, such as engineering. This research seeks to address those challenges by developing a multi-agent AI system to create and assess educational content.

Singh and Walia (2024) examine the role of AI in educational institutions around the world and argue that these institutions face a two-pronged problem of increasing demand for high-quality instruction while managing an increasingly limited pool of resources. Engineering education, presented with multidimensional technical concepts and a rapidly expanding knowledge base, is especially difficult for content creators and instructors. Building educational materials manually takes a substantial amount of time, often resulting in inconsistent quality and coverage, and is incapable of keeping pace with new knowledge within a given area of discipline.

The result is demand for educational materials that are more accurate, comprehensive, and cognitively balanced. The multi-agent system will be using a crew-based architecture with specialized components for syllabus review, content creation, assessment generation, and assessment quality evaluation. The components will work inter-dependently, relying on defined communication protocols and workflow orchestration to produce cohesive sets of educational materials aligned to the needs of all levels of Bloom's Taxonomy. The AI system will objectively

[1] School of Computer Science and Engineering, Vellore Institute of Technology, Tamil Nadu, India.
* Corresponding author: karmel.a@vit.ac.in

DOI: 10.1201/9781003639367-10

demonstrate improvement in cognitive balance, content quality, resource utilization, and processing speed, compared to using a single model.

Through the proposed architecture, the research scope is to create a new model for AI-supported content development that more accurately meets the complex needs of the engineering education context. The multi-agent approach focuses on solving existing problems in accuracy of specifics, diversity of cognitive coverage, and efficiency of the system. Ultimately, the multi-agent product will provide educators and students with a tool that produces varied comprehensive and high-quality educational materials.

Multi-agent AI systems have particular relevance for engineering education, where complex technical concepts require multifaceted approaches to teaching and assessment. Applications include comprehensive curriculum development, balanced assessment generation, adaptive learning materials, and educational resource integration. These applications address persistent challenges in engineering education, where balancing theoretical knowledge with practical application and critical thinking development is essential for student success (American Psychological Association, 2020).

2. Literature Survey

Chai and Wang (2022) provide a comprehensive review of the creation of AI systems for the development of education-related content that is predicated on a number of important fields, such as educational theory, AI architectures, and content knowledge in the domain of engineering education. The application of AI in education has evolved through several distinct phases.

a. *Early intelligent tutoring systems (1970s–1980s)*: The first generation of education AI consisted of rule-based intelligent tutorial systems that were trying to model students' knowledge and offer advice directly. SCHOLAR (geography) and GUIDON (medical diagnosis) are examples of early attempts to create adaptive learning experiences. The systems tended to depend more on expert-written rules and domain-specific representations of knowledge.
b. *Computer-assisted instruction (1980s–1990s)*: The next round saw the introduction of more sophisticated computer-assisted instruction systems that incorporated some basic natural language processing capabilities and problem-solving functions. Systems were beginning to adopt more complex student models and pedagogical strategies although they were still hampered by primitive computational power and use of AI techniques.
c. *Web-based adaptive learning (2000s–2010s)*: The advent of the internet spurred web-based adaptive learning systems that could log and analyse student performance data at scale. For example, Carnegie Learning's Cognitive Tutor and Knewton's adaptive platform began to utilize more advanced algorithms aimed at personalization, although still relying mainly on predetermined paths through content.
d. *Early LLM-based educational tools (2010s–2020)*: The development of language models with increased capabilities facilitated the creation of a new application area of content generation, automated grading, and intelligent

tutoring. Tools began to use natural language understanding to allow more flexible interaction with learners but generally in limited contexts.

e. *Current generation of AI educational systems (2020–present)*: Recent advances in LLM-based educational tools can generate many types of content, assessments, and feedback in various areas. However, these tools are still experiencing notable challenges in achieving educational coherence, cognitive coverage, and translating educational practice.

The constructivist theory gives utmost importance to learners actively constructing knowledge, through examples, applications, and opportunities to construct or synthesize knowledge within educational materials. Active construction of knowledge, prior knowledge activation, learning in context, and social construction are the main principles of constructivist learning theory.

The self-regulated learning framework is focused on metacognitive strategies to help learners monitor and control their own learning processes, including the planning, monitoring, control, and reflection phases. Educational materials designed to foster self-regulated learning provide educational features such as language that communicates learning objectives, comprehension checks, strategic or tactical support, and prompts for reflection and metacognition.

These educational theory frameworks should be understood in order to build AI systems that create and design educational materials that are truly effective. The multi-agent design allows for the explicit integration of these educational theory frameworks through agents designed specifically for one of the aspects of learning and education design.

3. Survey on Multi-Agent AI Systems

This section reviews existing research and implementations related to multi-agent AI systems, with particular emphasis on educational applications and architectural approaches.

- *Blackboard architectures*: Various studies have examined blackboard-style architectures, where specific agents contribute to a shared information space. The EDUCATOR system (Johnson et al., 2022) provided evidence of how this architecture supports flexible collaborative activities between agents that generate and assess the content, reporting a 38% increase in coherence of the student agent's content when compared to situations where agents did not collaborate. The CollabEd Framework (Wilson & Chen, 2023) took a more formal blackboard-like approach in how content was represented and its agents were designed. The content was represented in a structured way where content elements were represented as a semantic node that contains content and semantic relationships (i.e., synonyms, parent-child, etc.), some specialized agents focused on nodes (or content elements), and there was also a meta-agent supporting the overall coherence of the nodes.
- *Hierarchical systems*: The MAESTRO system (Chang & Roberts, 2023) utilized a three-level hierarchy for science education. The three components of responsive, dynamic, and engaging planning are: agents for strategic

planning, agents for tactical development, and agents for operational creation of presentations. This type of planning hierarchy afforded better coherence of content than baseline responses, as represented by a 42% improvement between stated learning outcomes in the MAESTRO and baseline approaches. The EduArch Framework (Patel et al., 2022) used a more elaborate hierarchy with domain-specific supervisor agents, cross-cutting coordinator agents, and occupationally specific worker agents. This activity also demonstrated strong technical accuracy while being effective educationally.

- *Market-based systems*: The EduMarket System (Wei et al., 2021) used a bidding mechanism to create competition among the developer agents for content development. While the market-based approach offered some advantages in terms of resource allocation efficiency, creating content was completed 27% faster than the fixed-assignment approaches being used to evaluate the model, and while the market-based bidding mechanism used in the EduMarket System introduced complexities regarding coherence of the content delivered at a global level. The AdaptEd platform (Rodriguez & Kim, 2023) sought to address the coherence issues by introducing additional logistical structures, including a hierarchy market structure, contract specifications, a reputation system, supervision, and the use of some form of alignment as a mechanism to improve coherence. The structural changes offered an opportunity to make necessary reforms to the system while still providing the advantages of a more efficient resource allocation in the system as adoptions were made.

4. Educational Approaches

- *Curriculum planning*: The CurriculAI system (Martinez & Johnson, 2023) showed that specialized planning agents could produce sequenced courses that were coherent, and expert educators favoured the learning sequences created by CurriculAI to the manually created sequences in 72% of blind comparisons. The LearningPath framework (Ahmed et al., 2022) developed further the planning agents for the curriculum planning use case, by extending it to personalized learning paths, which included learner model agents, content mapping agents, path optimization agents, and assessment integration agents.
- *Content generation*: The EdCreator system (Zhang et al., 2022) adopted a multi-agent approach to generating educational content with agents for explaining concepts, generating frameworks, producing visual representations, and connecting concepts. Results showed a 42% improvement in overall clarity when compared to generating content with a single-model approach. The MultiEd platform (Sanchez et al., 2023) continued the multi-agent approach with more specialized agents for adapting readability, anticipating misconceptions, scaffolding, and generating motivation for students. The multi-agent platform showed improvements for both the quality of the content and increased engagement and motivation for students.
- *Assessment generation*: The AssessNet system (Patel & Lee, 2023) utilized specialized agents for question generation at various cognitive levels and offered greater balance across the cognitive levels of Bloom's Taxonomy, achieving a

64% improvement in cognitive level distribution over comparable single-model systems. The EvalGen framework (Thompson et al., 2022) addressed issues of integration using content-assessment alignment agents, difficulty calibration agents, format diversity agents, and feedback generation agents, illustrating their capabilities to create assessments aligned to educational content.

5. Evaluation Frameworks

- *Quality metrics*: The EQUIP framework (Anderson et al., 2023) supplied metrics for the measurement of various dimensions of quality in educational content including: technical accuracy, conceptual clarity, cognitive level distribution, educational alignment, and engagement potential. The Cognitive Balance Assessment (Rivera & Chen, 2022) concentrated specifically on the distribution of cognitive levels and included automated classification of content along Bloom's Taxonomy, distribution analyses, progression analyses, and integration assessments.
- *Efficiency measures*: Nguyen and Patel (2022) provided a standardized approach to measuring efficiency for computational processes in educational AI, including measuring processing time, component/time ratios, costs associated, scalability requirements, and development time.
- *Educational effectiveness*: Williams et al. (2023) provided methods for identifying pedagogical effectiveness including instructional learning outcomes measurement protocols, cognitive development assessment, knowledge retention testing, transfer capability assessment methods, and approaches to assessing learner engagement.

6. Limitations of Existing Approaches

While developments in multi-agent educational AI are promising, there are still limitations in the current state of things.

a. Limited integration: Most current systems focus on specific parts of the educational content development process rather than providing full integration across the entire educational workflow. Such limitations produce inconsistencies between curriculum planning and content development, misalignment of content to assessment, and inefficient workflows.

b. Domain specificity: Many current implementations have been in limited educational domains and have not shown ability for more complex engineering topics. This challenge has its origins in the inability to draw on narrow representations of domain knowledge, an inability to handle multimodal content, and inadequate theory–practice integration capabilities. Among the cognitive levels, even multi-agent approaches struggle to provide coverage across all the cognitive levels and are weakest for the higher order thinking skills. Examples of imbalances include knowledge and comprehension levels being more evident, rather shallow treatment of analysis, limited support of evaluation skills, and artificial designs for tasks (Gettinger & Seibert, 2023).

c. *Scalability challenges*: Efforts to scalability have been limited, with a few comprehensive educational modules demonstrating both successful quality and coherence. Considerations for scaling include degradation of coherence for volume of content, intolerable computational resource demands, additional level of coordination, and maintaining terminology coherence.
d. *Limitations of workflow integration*: Existing systems frequently do not facilitate strong integration with educational content development workflows, lack adequate support for iterative improvement, fail to provide sufficient mechanisms for human oversight and assistance, impose restrictions on the process model, and do not easily integrate with existing content management systems.
e. *Limits to transparency and explainability*: There are many approaches presently in use that do not provide effective transparency of their decision-making process, that provide outputs as "black box" content generators without clear rationalization, do not provide adequate transparency around factual claims, and have poorly documented educational decisions.

This survey will identify both the important potential areas of focus in multi-agent educational AI and the gaps that the proposed study will address. This research intends to improve upon approaches that exist already and begin the next steps towards improving the current state of the art on AI-supported educational content development.

7. Challenges in Educational AI

The application of AI to educational content creation faces several significant challenges that impact the effectiveness and adoption of these technologies.

7.1 Content Quality and Accuracy

Berger (2022) explains that AI systems that are used for educational purposes need to maintain a high level of technical accuracy, particularly in engineering-related areas. Some specific challenges are as follows:

a. *Domain-specific knowledge*: The representation of technical and specialized knowledge in domain-specific areas creates a significant challenge for AI systems. Fields of engineering use specific terminology, use particular mathematical representations, and refer to related conceptual frameworks that must all be accurately represented.
b. *Currency of information*: Rapidly changing technical fields means current knowledge bases are continuously out of date if not regularly updated. Engineering disciplines are often quickly introducing new technologies, new approaches, and/or new best practices.
c. *Conceptual depth*: In an educational context, learners need depth of explanation, rather than shallow coverage, in order to learn effectively and to fully understand a concept. Often, engineering concepts are layered in complexity, ranging from simple problem premises to more compelling and nuanced applications.
d. *Consistency*: Another critical aspect of effective learning is the uniform definition of vocabulary and concepts across educational materials. Engineering

education often requires vocabulary, definitions, notation, and conceptual frameworks to be consistent across topics, modules, and the programme as a whole.

e. *Integrated practical applications*: Connecting theory with practical applications is a significant aspect of engineering education. In general, AI systems could be effective in describing theory, but struggle to connect that theory to authentic practical applications.

7.2 Cognitive Development Balance

Educational materials should respond to multiple stages of cognitive development (e.g., Bloom's Taxonomy). Current AI systems exhibit some struggles with:

a. *Over-representing lower cognitive levels*: The single-model AI systems focus nearly exclusively on knowledge-inconvenience and simple comprehension, tending to provide careful factual explanations while providing low-complexity-level responses.
b. *Limited development of higher-order thinking*: Creating a content area that is able to develop analysis, evaluation, and creation skills presents the AI with significant challenges. The higher cognitive levels require a somewhat better nuanced understanding of how components of the learning results can be deconstructed, evaluated, and recombined.
c. *Assessment reference*: To demonstrate significance of backmatching assessments to achieve the cognitive level of the learning expectations requires sophisticated understanding of all aspects of the content and different types of assessments.
d. *Metacognitive contribution*: To understand how students develop awareness of their learning processes to identify and make sense of their own learning contributes to metacognitive awareness as well, but metacognition has not been achieved. Good AI engineering education needs metacognitive elements to help students learn how to think about and approach real world complex problems (Barr & Massa, 2023).

7.3 System Efficiency and Scalability

a. The effective use of AI in education encounters some resource challenges.
b. *Computational requirements*: The high demand for LLMs imposes a real barrier to implementation. Engineering education seeks to be comprehensive in covering complex, technical domains and LLMs are often essential to achieve that level of comprehensiveness.
c. *Processing time*: The time involved to generate learning content prevents practical implementation, which is a real issue in education contexts. Current AI learning methods often need considerable time to generate a "good" quality learning content.
d. *Costs*: Funding in education limits AI implementation possibilities. The computational resources, API charges, and development of these systems are typically out of educational budgets in the current educational AI landscape.

e. *Scalability*: The lack of ability to scale coverage across a comprehensive curriculum adds to the barrier of implementation. Engineering programmes often have dozens of classes across multiple areas of specialization, each needing considerable learning content.

7.4 Integration with Education Practice

For AI educational systems to be successful, they must mesh well with existing education practices:

a. *Pedagogical alignment*: It is key for successful deployment that the content aligns with pedagogical practices that we already use in education. Engineering programmes use a range of different pedagogical approaches ranging from traditional lectures to a problem-based approach, and everything in between.
b. *Instructor adoption*: There is value in readying these systems to educators for impact; this requires practitioners to embrace systems that complement their expertise rather than replace it. Typically, AI systems are designed and promoted as self-sufficient generators of content, and this invites rejection from instructors and educators.
c. *Learning process integration*: Operational and instructional utilization of AI should also mesh with the larger learning process, which implies a coherent understanding of the larger educational ecosystem. For instance, engineering education occurs in lectures, labs, project-based contexts, independent study, and numerous assessments.
d. *Customization capacity*: For practical deployment, consideration of customizing the system to requirements at the institutional or, more broadly, practitioner level should be anticipated. Engineering programmes attended to varying levels of biomedical engineering, Kramer connection to institutional industry, or disciplinary educational philosophies, and each of these aspects impact curriculum considerations.

These challenges are the impetus for various innovative approaches to the landscape of education like multi-agent systems in order to support multifaceted needs of a meaningful application of educational content.

8. Emerging Trends and Future Directions

Emerging trends in multi-agent educational AI could delineate future research directions in this field.

a. *Hybrid human and AI collaboration models*: Researchers are increasingly focusing on further developing collaborative models wherein multi-agent AI systems are strategically positioned to support human educators instead of acting as autonomous content creators. Some collaborative models we can employ include interactive cycles of refinement for co-created content, role-based collaborative approaches, augmented creativity tools, and expertise amplification systems.
b. *Knowledge graph integration*: Multi-agent systems are progressively relying on structured representations of knowledge in order to produce coherence and accuracy in educational content. These approaches might cover modulations of

knowledge graphs that are domain-specific, educational relationships models, cross-domain knowledge integration, and knowledge update mechanisms.

c. *Adaptive personalization*: New research is beginning to explore how multi-agent systems can continue to personalize and facilitate learning via integrating learner models, content adaptation, personalized learning paths, and adaptive scaffolding based on individual learner needs.
d. *Multimodal content generation*: Multi-agent systems are starting to advance with their capabilities to integrate different modalities of content production, from coordinated text and visual imagery (e.g., developing something like an interactive simulation), multimedia integration to managing cross-modal coherence.
e. *Ethical and responsible AI integration*: This research highlights the ethical issues raised by AI in education through efforts related to bias detection and mitigation, the development of cultural sensitivity agents, transparent attribution of sources of information, and privacy-preserving system architectures.

These emerging trends situate the current research amid, and suggest potential avenues for future developments beyond, the evolving field of educational AI research. The current review of multi-agent AI systems in educational settings informs the present research in two ways, identifying both opportunities for future developments and challenges to navigate. The crew-based architecture developed in the present research is a culmination of all these insights, specifically in the context of the challenges presented by engineering education, learning, and teaching and what can be learned (often indirectly) from the emerging core trends in educational AI.

9. Crew-based Agent Architecture

9.1 Architectural Overview

The crew-based agent architecture represents a paradigm shift from single-agent systems to a coordinated multi-agent framework designed around specialized teams. This architecture adheres to several core design principles.

a. *Modularity*: The system is organized into separate functional units (crews), each of which is responsible for specific parts of the educational content development pipeline. Modular systems allow for the evolution of a component independently from the rest of the system, maintaining the integrity of the system.
b. *Specialization*: Each crew consists of agents performing discrete tasks and roles, which allows for depth of capability and depth of role within specialized circles of knowledge, rather than depth of role across the entirety of the educational content development tasks. In the specialization of roles, higher order tasks are interchanged without redundancy and the system utilizes the most capable agent in a timely manner.
c. *Hierarchy of coordination*: The workload management system creates a logical flow where (output) from one crew informs the (input) of another crew. The integrated system of multiple agents stays on task and on time.
d. *Extensibility*: The architecture is designed to allow the addition of new crews or agents as and when there is a need for change. The degree to which the

architecture can be added to is significant, and new crews can be added without a complete systemic redesign.

e. *Redundancy and resilience*: Critical functions are performed by more than one agent in the field, creating capacity for resilience when individual agents are delayed in tasks or unable to deliver effectively in the context.
f. *Continual feedback*: Metrics of each evaluation and indicators of quality of assessment are consistently fed back into the system, developing a self-improving mechanism.

Overall, the consistent and quantified feedback creates potential for improved quality of output over time.

9.2 Inter-Crew Communication

The effectiveness of the crew-based architecture depends heavily on robust communication protocols between crews. The system implements several communication mechanisms.

a. *Structured data interchange*: Crews exchange information using standardized data formats defined in YAML configuration files, ensuring semantic consistency across the system. For example:

```
output:
format: structured_content
schema:
- chapter_title: string
- learning_objectives: list
- topic_outlines: list
- prerequisites: list
- estimated_duration: integer
```

b. *Message passing protocol*: Each crew communicates through a message broker that maintains delivery guarantees and preserves message ordering, essential for maintaining the integrity of the workflow.
c. *Context preservation*: Communication includes not just the primary data but also metadata about its origin, intended use, and relationship to other information, ensuring that receiving crews have full context for processing.
d. *Dependency management*: The inter-crew communication system tracks dependencies between tasks, allowing the system to manage complex workflows where output from one crew may be required by multiple downstream crews.
e. *Error handling*: Communication includes standardized error reporting that enables crews to recognize and respond to issues in upstream processing, either by requesting clarification or implementing fall-back strategies (Jensen et al., 2024).

9.3 Workflow Orchestration

The orchestration layer acts as the "conductor" of the crew-based system, managing tasks and actions within and across all crews and supporting the collaborative development of educational content.

a. *Timely coordination of task management*: The orchestrator manages a directed acyclic graph (DAG) of tasks, scheduling those tasks in coordination with dependencies, priority, and available resources. This is demonstrated in the tasks. yaml configuration:

```
tasks:
analyze_chapters:
priority: 1
develop_content:
depends_on: ["analyze_chapters"]
priority: 2
```

b. *Resource allocation*: Computational resources are assigned to crews on a task basis, and the resources can be reassigned based on the complexity of each task and overall system load.
c. *Progress tracking*: The orchestrator provides visibility at all times into the status of all tasks across crews, including tracking progress for the full production pipeline and identification/use of bottlenecks or delays.
d. *Exception management*: In the case of unexpected issues, the orchestration layer will enforce some level of recovery strategy, including options such as reassigning tasks, taking alternative workflow paths, or requesting human intervention.
e. *Optimization*: The orchestrator uses mechanism to record and analyse workflow patterns and identify additional potential improvement techniques or processes, including options such as parallelizing independent tasks or re-ordering workflow tasks to reduce idle time.

10. Chapters Crew

The Chapters Crew is the starting point of the educational content development process, accountable for crafting the structural design that all subsequent content will be based on (see Figure 10.1).

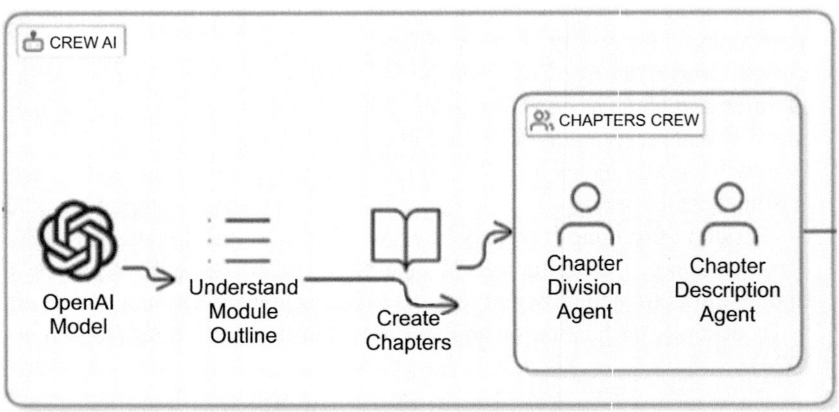

Fig. 10.1 Chapters Crew.

11. Content Crew

The Content Crew leverages advanced retrieval-augmented generation (RAG) techniques to extract, process, and synthesize educational content (see Figure 10.2).

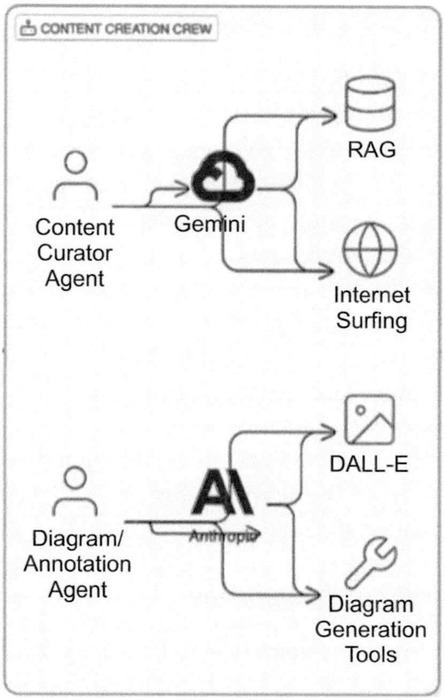

Fig. 10.2 Content Crew.

12. Testing Crew

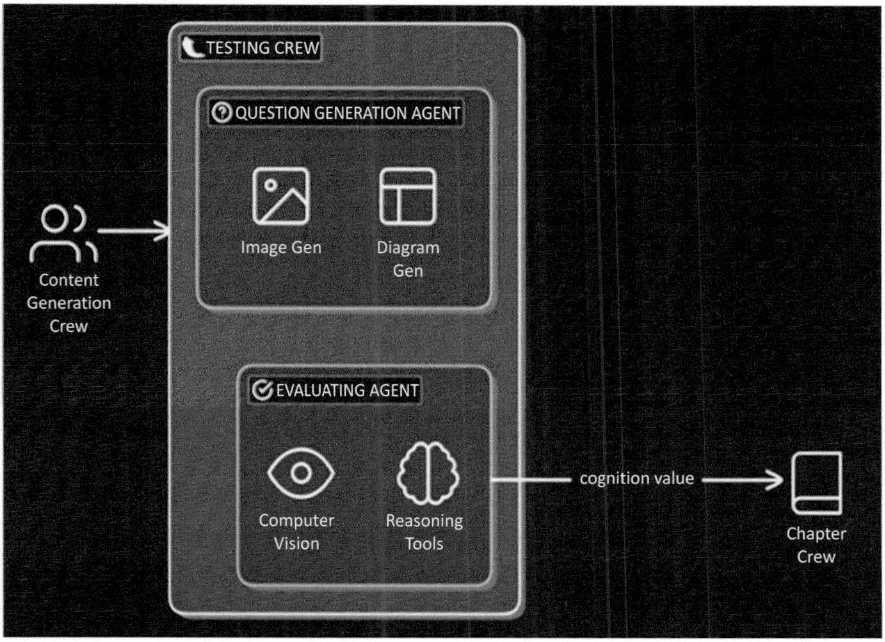

Fig. 10.3 Testing Crew.

12.1 Educational Assessment Principles

The Testing Crew operates according to established principles of educational assessment.

a. *Validity*: Assessments are utilized to assess what they are purported to assess and there is strong correspondence between test items and learning outcomes.
b. *Reliability*: Assessment tools generate dependable results regardless of time or person, creating test fairness and accuracy.
c. *Authenticity*: Assessments encourage evidence of knowledge being applied rather than straightforward accumulation of facts, demonstrating real competency.
d. *Constructive alignment*: Assessment tasks are clearly aligned with intended learning outcomes and teaching and learning activities to provide a cohesive learning experience.
e. *Formative function*: Assessments have a dual purpose of not only evaluating but also contributing to the learning process, allowing learners to receive constructive information that can help improve future assessments.

12.2 Question Generation Methodology

The Testing Crew employs a systematic methodology for generating assessment questions.

a. *Clear learning objective alignment*: Each question has been intentionally crafted to assess a specific learning objective, thus aligning the question's construction to the framework that was intended.
b. *Question distractors*: The incorrect answer choices (distractors) in the multiple-choice questions are intentional to promote a response based on a specific misconception, thus allowing the assessment to yield an informative instructional decision.
c. *Scenario construction*: All complex application questions are developed from real-world scenarios and require students to integrate multiple constructions to problem solve in a real-world application.
d. *Language precision*: All questions are written in precise and unambiguous language to measure the intended content and not the learners' interpretation of language meaning.
e. *Rationale for difficulty*: All questions are developed with an intended difficulty level that is optimal to facilitate the students' learning and within their zone of proximal development.

12.3 Cognitive Level Targeting

The Testing Crew agrees that the tests engage and assess all cognitive levels.

a. *Bloom's taxonomy application*: The questions are designed to target specific cognitive levels. For example, it demonstrates from basic recall to evaluating and creating.
b. *Lower-order thinking skills (LOTS)*: Foundational questions measure prior essential knowledge, comprehension, and application. This establishes a baseline.

c. *Higher-order thinking skills (HOTS)*: Higher-order questions expect students to analyse, synthesize, evaluate, and create. This is critical in fostering critical thinking and problem-solving skills.
d. *Metacognitive assessment*: Certain questions may prompt the students to consider their own thought processes. Therefore, students develop awareness of cognitive strategies and self-regulation.
e. *Cognitive progression*: The test instruments contain a reasonably even proportion of cognitively weighted tasks that range from a basic understanding to higher level application.

12.4 Assessment Variety and Balance

The Testing Crew ensures diverse assessment approaches to accommodate different learning styles and provide comprehensive evaluation.

a. *Assessment format*: Assessments can take different formats (multiple choice, short answer, essay, problem-solving, project-based) to assess different elements of understanding.
b. *Timing*: Assessment tasks can be from an in-the-moment formative, to a long summative task, providing feedback at various points in the learning journey.
c. *Assessment of individual and group work*: Some tasks will assess individual learning, while others will assess the process of learning collectively, and how to add to the work of others.
d. *Assessment of product and product process*: Assessment includes the evaluation of the product created (solution, essay, project) and the processes that created the product (problem solving, researching).
e. *Assessment of self and peer*: Assessments, where applicable, support self-assessment and peer feedback to develop evaluative judgement.

13. Evaluation Crew

13.1 Quality Assurance Processes

The Evaluation Crew implements comprehensive quality assurance processes.

a. *Content accuracy verification*: Educational content is systematically checked against authoritative sources to ensure factual accuracy and current understanding.
b. *Pedagogical soundness review*: Content is evaluated against established principles of effective instruction, ensuring appropriate scaffolding, explanation, and application.
c. *Accessibility compliance*: Materials are assessed for compatibility with accessibility standards, ensuring equitable access for learners with diverse needs.
d. *Language and readability analysis*: Content undergoes linguistic analysis to ensure appropriate complexity, clarity, and consistency for the target educational level.
e. *Curriculum alignment verification*: All content is checked for alignment with stated curriculum objectives and standards, ensuring comprehensive coverage without scope creep.

14. Cognitive Models and Bloom's Taxonomy in Education

14.1 Bloom's Taxonomy: Theoretical Framework

Bloom's Taxonomy, created by Benjamin Bloom in 1956 and later revised in 2001 by Anderson and Krathwohl, D. R. (2022), is a hierarchical model of cognitive processes for learning. The cognitive skills are divided into the six levels of Remember, Understand, Apply, Analyse, Evaluate, and Create. These six levels are a continuum from the mere recall of the knowledge to high-level thinking ability (Bloom et al., 2023).

In engineering education, Bloom's Taxonomy plays a crucial role in curriculum and assessment design that encourages problem-solving and innovation. At the lower levels, students memorize basic principles of physics, mathematics, and computer programming. At higher levels, they apply these principles through problem-solving, analysis of system behaviour, comparison of different solutions, and finally producing innovative designs or software.

An effective curriculum guarantees congruence between instruction objectives and assessment. The engineering courses need to organize students' learning in a manner that progresses from understanding basic theory to critical thinking and innovative problem-solving. Bloom's Taxonomy can be used as a benchmark for test development that probes the students more and offers balanced cognitive skills development.

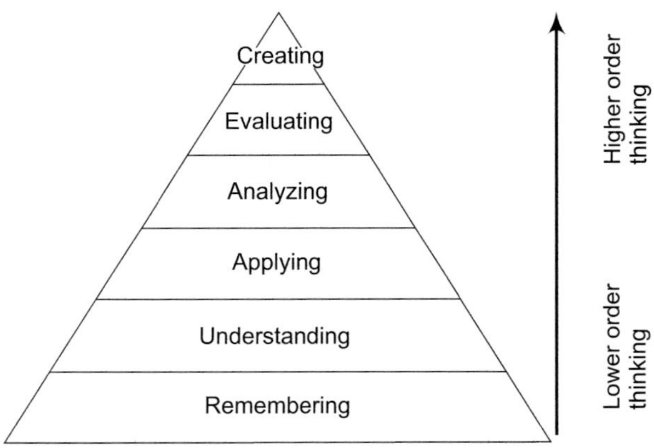

Fig. 10.4 Bloom's taxonomy pyramid.

14.2 The Importance of Bloom's Taxonomy in Education

Bloom's Taxonomy is the foundation of today's education in the sense that it offers a systematic framework of instruction and learning. It helps teachers to plan instructional materials, tests, and learning activities that promote deeper learning and critical thinking. By matching instruction to levels of cognition, instructors can facilitate effective transfer of knowledge. For instance:

1. Recall: Flashcards, quizzes, and recall exercises
2. Comprehension: Support explanation, summarizing, and discussion

3. Assigning: Offer problem-solving exercises and real-life applications
4. Analysing: Have students compare different theories or models
5. Evaluating: Conduct peer review and critical analysis
6. Creating: Invite students to make projects, research, or develop solutions

14.3 Other Cognitive Models and Comparison with Bloom's Taxonomy

There are a number of cognitive models that simplify learning, but one of the most used models is Bloom's Taxonomy. Some of the other cognitive models are also utilized.

The Structure of Observed Learning Outcomes (SOLO) Taxonomy analyses learning into five levels:

1. Pre-structural: No understanding or meaningless information
2. Uni-structural: Summary of a single idea
3. Multi-structural: Understanding of several concepts but not integration
4. Relational: Capacity to connect several ideas
5. Extended abstract: Ability to generalize and apply to new fields

While SOLO Taxonomy is more concerned with depth of knowledge, Bloom's Taxonomy is more concerned with breadth of application and creativity and is therefore more suitable for teaching engineering and AI. Anderson and Krathwohl modified Bloom's original taxonomy by replacing the noun-based categories with action verbs, i.e., "analysing" instead of "analysis". The new version also places "creating" in the topmost level instead of evaluation, highlighting the importance of innovation in the education of today (see Figure 10.4).

Bloom's Taxonomy continues to be a widely accepted model because it is practical and easy to apply throughout the curriculum. It can close the gap between theoretical learning and real problem-solving. Its hierarchical structure enables students to develop progressively higher-order thinking skills, which is especially helpful in disciplines such as AI, engineering, and computer science. In contrast to some inflexible cognitive models, Bloom's Taxonomy can be applied in most fields of study, such as sciences, humanities, business, and technology. Teachers can easily adapt their instructional approach while having a structured learning experience. Bloom's Taxonomy–created assessments ensure an exhaustive assessment of student learning. Teachers can achieve a balance between lower-order recall questions and higher-order problem-solving and creative tasks, ensuring assessments become more effective in measuring student competency. With the advent of AI-based learning systems, Bloom's Taxonomy offers a concrete framework for the design of adaptive learning models. AI-based educational tools can leverage Bloom's principles to adapt content delivery so that students are given challenges that match their level of cognitive development.

14.4 Bloom's Taxonomy and its Impact on Critical Thinking

Bloom's Taxonomy, developed by Benjamin Bloom and his team in 1956, is a systematic hierarchy of intellectual skills that is useful in the educational community, particularly in encouraging critical thinking. The taxonomy categorizes educational objectives into six levels of increasingly complex intellectual functioning: Remember, Understand, Apply, Analyse, Evaluate, and Create. The model enables teachers to guide the construction of curricula and tests that increasingly build students' ability to think critically and solve problems.

Critical thinking is the application of reasoning processes, analysis, assessment of evidence, and the drawing of informed conclusions. Bloom's Taxonomy offers a framework for the creation of these higher-order thinking skills. The lower levels (*Remember*, *Understand*) form the basis of learning whereas the higher levels (*Analyse*, *Evaluate*, *Create*) are most applicable to critical thinking. For instance:

- Reading a work of literature enables students to recognize underlying arguments and themes.
- Assessing a scientific assertion prompts students to decide whether it is true from empirical evidence.
- Creation of new solutions to real problems increases creativity and integration of knowledge.

Teachers who create questions and assignments that tap into these higher levels of Bloom's Taxonomy allow students to transcend mechanical learning and, instead, question, link, and reflect. The application of Bloom's Taxonomy allows educators to create learning activities that support an orderly sequence of cognitive development in students. For example:

- Begin with recall questions ("What are the steps of photosynthesis?")
- Advance to conceptual understanding ("Why is photosynthesis important for life on Earth?")
- Proceed to application and analysis ("How would a plant grow if there was not enough sunlight?")
- Conclude with assessment and development ("Devise an experimental framework to examine the influence of light on the growth of plants.")

This systematic approach allows students to develop their reasoning skills step by step.

Children typically transition from concretizing to abstract thought as they develop. Bloom's levels reflect this developmental pattern:

- Younger children (4–7 years) will normally function at the lower levels (*Remembering* and *Understanding*).
- Older kids (8–12 years) start *Applying* and *Analysing* concepts.
- Adolescents tend to be able to *Assess* and *Build* based on what they know.

By examining students' reactions to questions at various levels of the taxonomy, students and psychologists can determine their level of cognition and adjust instruction accordingly to serve their purposes.

Bloom's Taxonomy may be employed diagnostically to determine what cognitive processes a child has acquired and where he or she needs support. For example, a child who has difficulty with application but is good at memorization may require extra opportunities to work on applying knowledge to new situations. On the other hand, students who are skilled at evaluating arguments or developing creative ideas can gain from working with open-ended tasks and inquiry-based learning experiences.

This customized understanding supports differentiated pedagogical practices and maximizes the learning potential of each child (Lee & Seshia, 2023). Cognitive frameworks, such as Bloom's Taxonomy, are most essential in organizing education, especially in AI-based learning environments. By connecting teaching material to cognitive levels, with increasingly escalating difficulty, and adding metacognitive methods, AI education can be better made effective and personalized. Such methods not only enable students to learn, but also help in developing higher-order thinking, which is needed to solve real-world engineering and AI problems.

15. System Implementation for Embedded Systems Course

15.1 Case Study: Embedded Systems Module Creation

The initiative sought to build an all-encompassing educational module about embedded systems directed towards engineering students Heath, S. (2021).. The system's requirements and user stories were to create curricular content in a structured manner for learning with logical progression. Further requirements asked that the content be useful, incorporating applied examples, relevancy to industry, and an analysis of current needs in job ads and employer descriptions. The system would also be structured to analyse current syllabi, content required for embedded systems course development, and enrich workplace relevance with decisions to add more content to the generated syllabi, which was expected to assist instructors in gauging where to assess students at appropriate cognitive levels of Bloom's Taxonomy.

15.2 Module Structure and Progression

The implemented system established a unified module structure for Wolf, W. (2022) embedded systems education, starting from essential concepts and moving towards more advanced content. The module was designed in a way that constant learning pathways build off of fundamentals and progressively develop knowledge, so students will always have a firm grasp of topics before progression. The structure consisted of five main chapters on definitions and characteristics of an embedded system, the aspects of designed challenges, processor technology, Catsoulis, J. (2023) hardware basics, and microcontroller architecture, each organized Yiu, J. (2023) into an appropriate time allotted to its complexity (Marwedel, 2022).

15.3 Implementation Challenges

Several challenges were encountered during implementation, including:

- maintaining content coherence across chapters while ensuring each builds logically on previous material;
- balancing theoretical concepts with practical applications to enhance student understanding;
- integrating diverse educational resources without overwhelming students with excessive information;
- ensuring accurate extraction of relevant content from source documents; and
- developing assessment questions that accurately target specific cognitive levels.

15.4 Content Generation Process

The system uses a specialized syllabus analysis agent that evaluates existing curriculums and identifies key topics, learning outcomes, and content organization. The agent evaluates the logical flow of topics, identifies required content relationships, and creates an optimal sequence for learning. The method consists of:

- pattern recognition to identify chapter headers and content boundaries;
- semantic analysis to understand topic relationships and dependencies;
- learning progression mapping to ensure coherent knowledge building; and
- time allocation assessment based on topic complexity and importance.

15.5 Content Extraction and Enrichment

Content generation follows a multi-stage process:

- Extraction of core content from source documents using a specialized RAG expert agent
- Content organization according to the established chapter structure
- Enrichment with explanatory text, examples, and illustrations
- Integration of practical applications and case studies
- Addition of learning aids such as summaries, key points, and review questions

The system ensures content depth and breadth by maintaining minimum word counts per chapter and incorporating diverse explanatory approaches to accommodate different learning styles.

15.6 Educational Resource Integration

To enhance the educational value of the generated content, the system integrates various external resources:

- Relevant web-based tutorials and articles
- Video demonstrations of key concepts and applications
- Interactive simulations and exercises
- Supplementary reading materials for deeper exploration
- Industry case studies that illustrate real-world applications

These resources are curated by a specialized web researcher agent that evaluates content quality, relevance, and educational value before integration into the module content.

16. Assessment Generation

16.1 Question Development Across Cognitive Levels

The system incorporates a sophisticated question development mechanism that ensures comprehensive coverage of all cognitive levels in Bloom's Taxonomy:

- Remembering: Basic recall questions to test fundamental knowledge
- Understanding: Questions that require explanation of concepts and principles
- Applying: Problem-solving questions that involve implementation of knowledge

- Analysing: Questions requiring breakdown of concepts into component parts
- Evaluating: Questions that necessitate judgement and critical assessment
- Creating: Design-oriented questions that require synthesis of knowledge

Each assessment includes a balanced distribution of questions across these levels to ensure comprehensive evaluation of student learning.

16.2 Assessment Variety Implementation

The system generates diverse question types to accommodate different assessment needs and learning styles:

- Multiple-choice questions for efficient assessment of broad knowledge areas
- Short-answer questions to test concise explanation abilities
- Essay questions for in-depth analysis and evaluation
- Diagram-based questions to assess visual understanding and representation skills
- Code implementation questions to test practical application abilities
- Design challenges to evaluate creative problem-solving capabilities

Each question is tagged with its corresponding Bloom's Taxonomy level to facilitate targeted assessment of specific cognitive skills.

16.3 Feedback Integration

The assessment framework incorporates feedback mechanisms designed to enhance the learning process:

- Model answers for self-assessment
- Common misconception identification and clarification
- Targeted learning resources for areas of weakness
- Progressive difficulty adjustments based on performance
- Formative assessment opportunities throughout the learning process

17. Technical Implementation Details

17.1 System Architecture

The system employs a modular architecture with specialized agents handling distinct aspects of the content generation and assessment process:

- Syllabus Analyser: Processes curriculum documents to extract structure and requirements
- RAG Expert: Retrieves and processes relevant content from source materials
- Web Researcher: Identifies and evaluates supplementary educational resources
- Content Creator: Generates comprehensive, well-structured educational content
- Topic Analyser: Identifies suitable topics for assessment across cognitive levels
- Question Developer: Creates diverse, level-appropriate assessment questions

These agents operate within a coordinated workflow managed by a task orchestration system that ensures proper sequencing and dependency management.

17.2 Agent Communication Protocols

The system implements a structured communication protocol that facilitates efficient information exchange between agents:

- Task-based messaging with clear input/output specifications
- Dependency management to ensure sequential processing where required
- Standardized data formats for consistent information exchange
- Error handling and recovery mechanisms to maintain workflow integrity
- Feedback loops for continuous improvement of generated content

17.3 Resource Management Strategies

To ensure efficient system operation, several resource management strategies are implemented:

- Parallel processing for independent tasks to optimize execution time
- Content caching to minimize redundant processing
- Resource allocation based on task complexity and priority
- Load balancing across processing nodes for optimal performance
- Incremental content generation to manage system memory requirements effectively

These strategies enable the system to handle complex content generation and assessment tasks while maintaining performance and reliability.

18. Limitations of Existing AI Education Platforms

Artificial intelligence has begun to influence education through intelligent tutoring systems, adaptive systems, and automated content presentation. Existing AI educational systems are based on single-agent architecture, limiting their ability to address advanced cognitive processes and diverse educational tasks. Mishra and Dubey (2024) explore the limitations that affect learning performance, scalability, content trustworthiness, and compliance with educational theory.

18.1 Single-Agent Bottlenecks

Traditional AI-based learning interfaces tend to use one monolithic agent or a general-purpose model. Although these may generate superficial response and adaptation, they lack the ability in case of multitasking, deeper understanding of the subject matter, and heterogeneity in students. The single-agent model does not support specialization and functions as a bottleneck for higher-level cognitive processes.

Moreover, the single-agent approach restricts concurrent task execution. For instance, generating an assessment, answering a query, and identifying cognitive level simultaneously burden a general-purpose AI. This results in slow responsiveness and poor user experience.

18.2 Cognitive-level Coverage Gaps

Most of the existing platforms address the breadth of cognitive levels in Bloom's Taxonomy. They excel at performing lower-order activities such as recall, straightforward explanation, and procedural support but are weak in stimulating creativity, critical thinking, and synthesis of ideas.

Higher-order tasks like the assessment of algorithmic effectiveness, model comparison, or the generation of novel research concepts require domain-specific intelligence and elastic reasoning. One-agent systems tend not to offer proper support to students in these stages (Noergaard, 2023).

18.3 Content Coherence Challenges

Due to their general nature, current AI tutors sometimes produce text that is not logically connected, thematically cohesive, or conceptually pertinent. They produce abbreviated learning experiences in which lessons or exercises are disconnected. Students receive explanations or comments that are technically correct but pedagogically unconnected.

These contradictions erode students' comprehension, especially in those subjects that require multiple levels of understanding such as programming, data structures, and AI algorithms. Lacking logical sequence, students end up developing shallow comprehension and misconceptions.

18.4 Assessment Quality Issues

Test generation by current platforms also tends to favour surface-level questions. Although the systems are also capable of generating multiple-choice or fill-in-the-blank versions of questions, they are not able to generate more abstract tests based on more complex cognitive demands or learning objectives.

Besides, tests rarely change in real time according to student performance and lack metacognitive questioning. This leads to failure to offer personalized feedback and loss of potential for authentic reflection and improvement.

19. Novel Contributions of the Multi-agent Approach

A multi-agent system is an AI education paradigm shift. Rather than trusting one model with everything, a multi-agent system breaks tasks up among specialized agents, each optimized for particular cognitive tasks or educational outcomes.

- *Specialized Agent Cognitive Advantages*: In a multi-agent system, the agents are designed to be proficient at different levels of Bloom's Taxonomy. For example:
    - A recall agent specializes in delivering definitions, facts, and memory aids.
    - An application agent provides contextual practice and real-world examples.
    - An evaluation agent judges students' work and recommends modifications.
    - An imaginative agent directs open-ended projects and facilitates innovation.

This ability allows the system to produce more advanced, contextually attuned, and cognitively aligned responses that mimic skilled human tutoring.

- *Workflow Optimization Innovations*: Multi-agent systems permit parallel computation and cooperative effort. For instance, when one questions a pupil, multiple agents are able to work on the question in parallel from different points of cognition. A single agent provides a solution, another an evaluation, and another a prompt for reflection. This labour division conserves response time, enhances the quality of content, and ensures wide cognitive coverage in every interaction.

- *Educational Theory Integration*: One of the greatest contributions of the multi-agent approach is the ability to embed learning theory within the functionality of each agent. Agents have the ability to be programmed to apply the tenets of constructivism, scaffolding, cognitive load theory, or metacognitive strategies. For example, a scaffolding agent can have questions ordered by prior response so complexity is only added after mastery. An engagement agent can apply gamification or motivation theory to maintain learner interest and confidence.
- *Content Quality Improvements*: Since every agent is being trained for a particular task, educational content as a whole is greatly enhanced. Explanations are more targeted, examples more pertinent, and the feedback more helpful. Rather than canned responses, students are provided with answers written for them by agents with a cognitive science, pedagogy, and subject matter–based understanding.

20. Quantitative Improvements

The effect of multi-agent systems can be quantified by applying various quantitative measures to compare with traditional platforms.

- *Learning Outcome Comparisons*: Initial pilot research and experimentation with multi-agent systems indicate better test scores, increased rates of completion, and increased conceptual understanding. Through enabling varied cognitive processes effectively, students attain greater learning outcomes throughout Bloom's hierarchy.
- *Content Quality Metrics*: Coherence score, thematic coherence, and pedagogical correctness are just a few of the factors that can be utilized for evaluation of generated content quality. Multi-agent systems generate content with:
    - greater readability,
    - improved logical flow, and
    - more alignment with curriculum standards.
- *Development Efficiency Gains*: By using task allocation, multi-agent systems accelerate learning module development. Agents may be upgraded or improved separately, reducing time-to-release for new content or capability. Modularity also facilitates debugging and ongoing optimization more easily. Single-agent systems, however, would involve retraining or fine-tuning the whole model to implement adjustments.
- *Cognitive Coverage Analysis*: Head-to-head comparison between single-agent and multi-agent systems reveals the latter to have greater cognitive balance. Single-agent platforms are more inclined towards knowledge and understanding activities whereas multi-agent systems are more responsive to assessment, synthesis, and creativity-oriented activities. This balance is so that the learners are not merely learning but are also developing critical and creative thinking.

21. Qualitative Advantages

Apart from quantitative results, multi-agent system enhances learning process using a wide spectrum of qualitative gains.

- *Content Coherence Improvements*: Because agents collaborate, material is filtered and refined through various cognitive prisms. This collaboration achieves more continuity between lessons, smoother transitions between topics, and more natural language explanations. Students will be less prone to suffer from cognitive dissonance or mental confusion caused by unrelated modules or volatile vocabulary.
- *Learning Progression Optimization*: Multi-agent systems naturally facilitate progression-based learning. Agents can monitor learner skill at various levels and lead the learner to the next lesson or activity of suitable difficulty. This prevents frustration or boredom for the learner and encourages a steady, engaging trajectory of learning.

22. Evaluation and Results

A comparative analysis across various dimensions underscores the critical advantages Heller, A. (2023) of agentic AI over traditional AI (see Table 10.1).

Table 10.1 Comparison of traditional and agentic AI

Dimension	*Traditional AI (Non-agentic)*	*Agentic AI (Modular Multi-Agent)*
Modularity	Low; monolithic designs	High; distinct task-oriented agents coordinated centrally
Flexibility	Limited adaptability	Highly adaptive to new contexts and requirements
Interpretability	Limited; black-box	High; transparent with clear attribution of tasks
Adaptability	Static; low responsiveness	Dynamic; real-time adaptive learning and refinement
Context Management	Restricted context window	Enhanced; efficient context splitting and memory management
Performance	Limited for complex tasks	Improved; task-specific optimization by individual agents

23. Agentic AI in Educational Metrics

Agentic AI significantly transforms educational platforms through enhanced modularity, specialization, and adaptability across various aspects.

- *Curriculum Structuring*: Curriculum structuring agents analyse syllabi and learning outcomes to systematically develop structured course outlines, effectively handling large, complex educational materials and optimizing learning sequences.

- *Content Generation*: Dedicated content generation agents produce rich, diverse educational content, leveraging domain-specific knowledge and multimedia resources. These agents ensure quality, relevance, and pedagogical effectiveness, enhancing learner engagement.
- *Assessment and Testing*: Assessment agents craft precise and pedagogically aligned tests that address varying cognitive levels, effectively integrating Bloom's Taxonomy. Testing agents ensure consistency, comprehensive topic coverage, and adaptability to learner performance. Evaluation agents analyse learner responses, identifying strengths and areas for improvement. Feedback agents provide personalized recommendations, thus enabling continuous improvement and adaptive learning experiences tailored to individual learners (Valvano, 2023).

24. Comparative Analysis of Agentic AI and Single LLM Systems

24.1 Core Evaluation Metrics

The core evaluation metrics graph measures key performance indicators including reasoning accuracy, task completion, response coherence, context retention, and factual accuracy (see Figure 10.5).

Figure 10.5 Comparing key performance indicators

- These metrics are crucial for evaluating AI performance in educational settings, as they determine the reliability and effectiveness of content delivery and assessment processes.

24.1.1 Task Modularization Analysis

Figure 10.6 breaks down token usage by task, comparing agentic AI (multi-agent system) against a single LLM system.

This efficiency arises from task specialization where each agent is optimized for specific tasks, thereby avoiding token wastage due to unnecessary generalizations.

24.1.2 Token Distribution by Model

The token distribution graph (Figure 10.7) illustrates token allocation across different models within the agentic AI versus single LLM systems. The agentic AI effectively utilizes multiple specialized models such as GPT-4 Turbo, Gemini 1.5 Flash, Mistral, and DALL-E 3, allowing it to distribute tasks according to each model's strengths. This balanced distribution optimizes token usage and reduces costs significantly, as indicated by the total costs and cost per token.

Conversely, the single LLM system solely relies on GPT-4 Turbo, resulting in increased total tokens and higher operational costs due to inefficiencies stemming from task generalization.

24.1.3 Token Efficiency by Task

The token efficiency graph (Figure 10.8) provides a detailed comparative analysis of total token usage across tasks like data analysis, content generation, and code

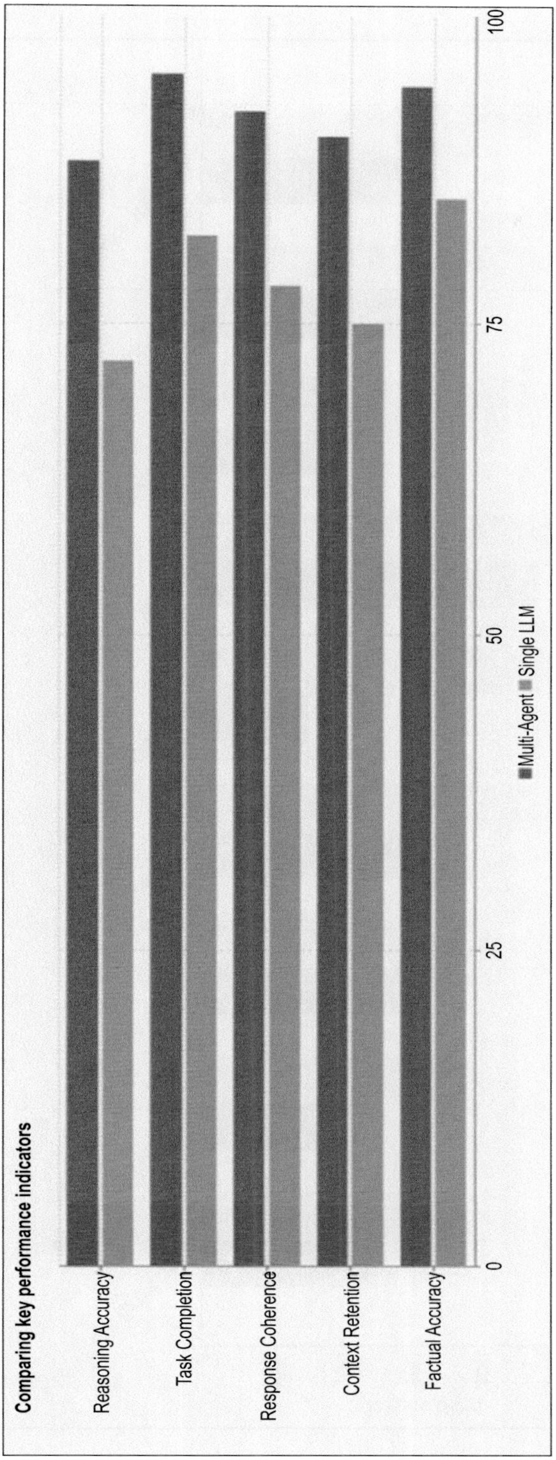

Fig. 10.5 Comparing key performance indicators.

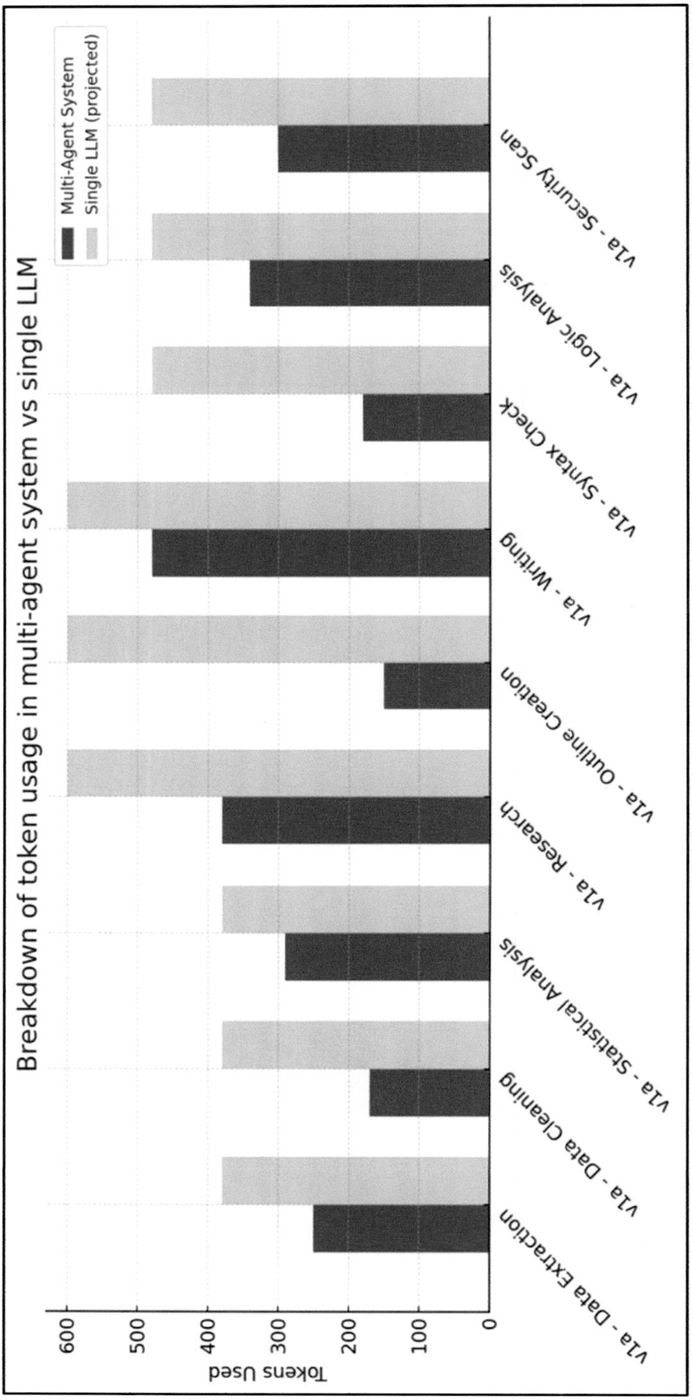

Fig. 10.6 Token usage in various tasks.

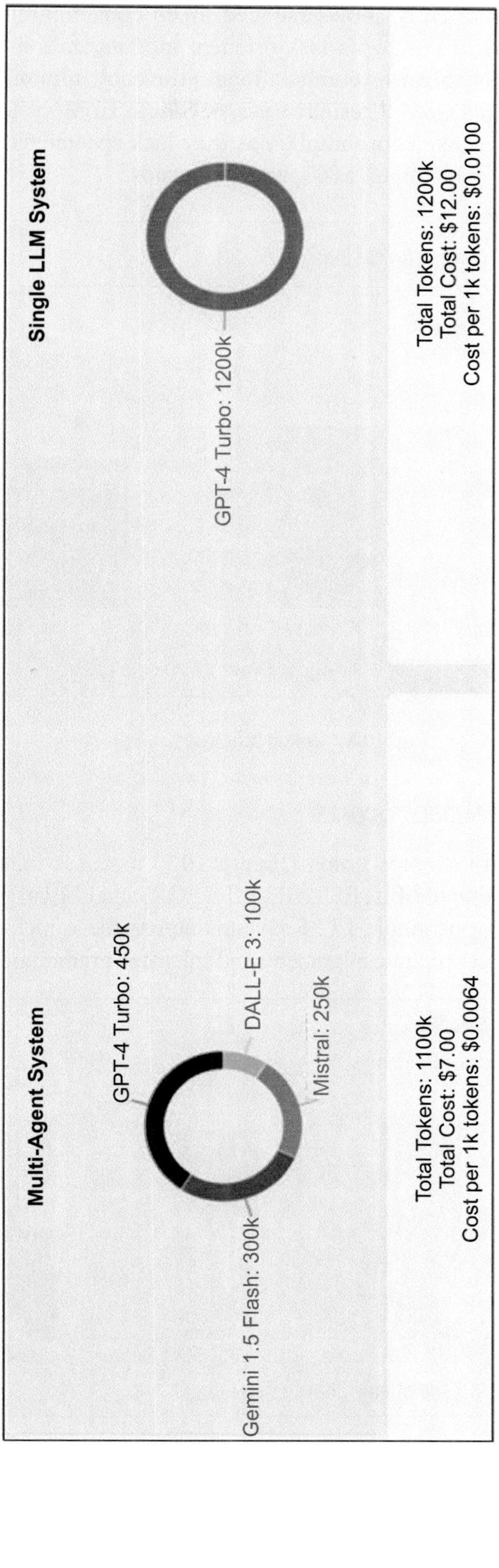

Fig. 10.7 Multi agent vs single agent cost per token.

review. Agentic AI consistently shows reduced token consumption compared to the single LLM, primarily due to its task-oriented modularization. This targeted use of specialized agents enhances computational efficiency, minimizes redundant processing, and optimizes overall resource usage. Single LLM systems, however, show consistently higher token consumption as they lack specialized optimization, reflecting inefficiencies in handling task-specific demands.

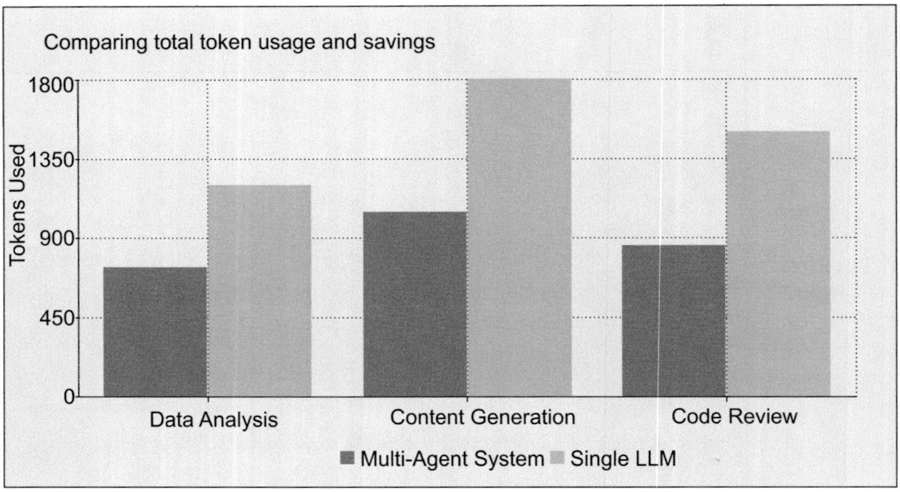

Fig. 10.8 Token efficiency.

24.1.4 Automated Evaluation Scores

The automated evaluation scores graph (Figure 10.9) assesses industry-standard evaluation metrics including BLEU, ROUGE, METEOR, and G-Eval scores. Agentic AI consistently outperforms single LLM systems across these metrics, indicating superior text quality, better semantic alignment, and enhanced grammatical correctness.

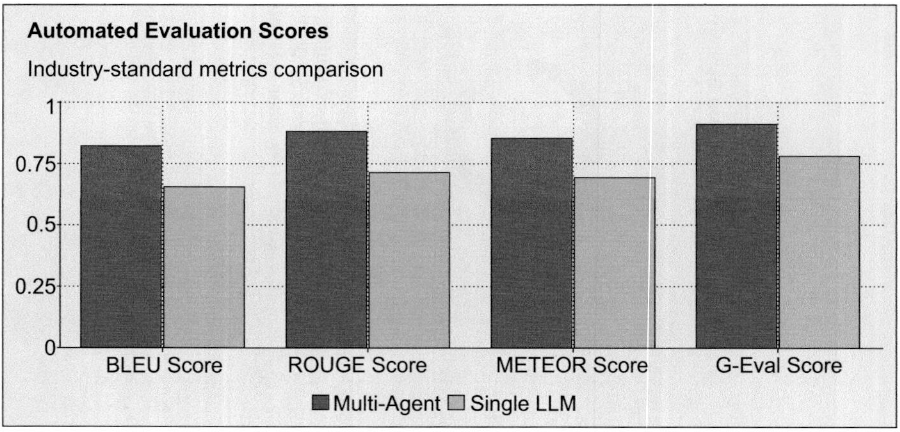

Fig. 10.9 Evaluation metrics for LLMs.

The BLEU score, reflecting precision in generated content, and the ROUGE score, measuring recall or completeness of relevant content, were notably higher for the agentic system. METEOR and G-Eval scores further corroborate these findings, emphasizing the effectiveness of multi-agent collaboration in enhancing linguistic and contextual quality.

24.1.5 Response Time Analysis

The response time analysis graph (Figure 10.10) evaluates performance across different task complexities from simple queries to complex reasoning and multi-step tasks.

This responsiveness is critical for maintaining engagement and ensuring efficient workflow management in educational environments.

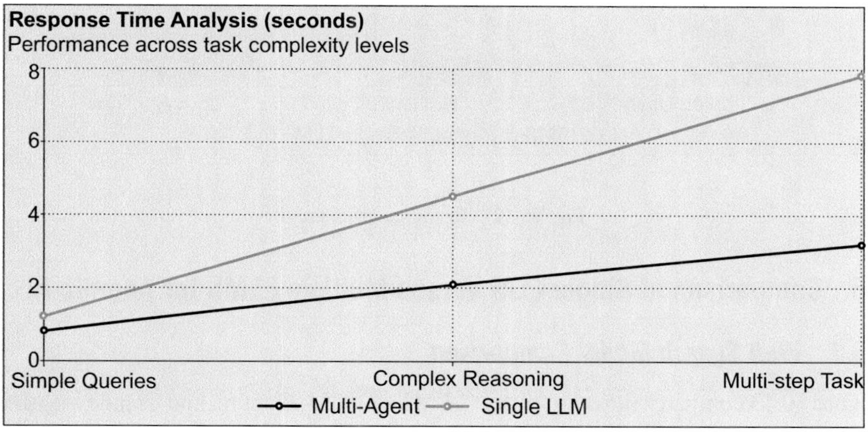

Fig. 10.10 Response time analysis.

24.2 Quality Dimensions

The quality dimensions radar graph (Figure 10.11) presents a comparative analysis of output quality aspects such as coherence, relevance, factuality, completeness, and consistency.

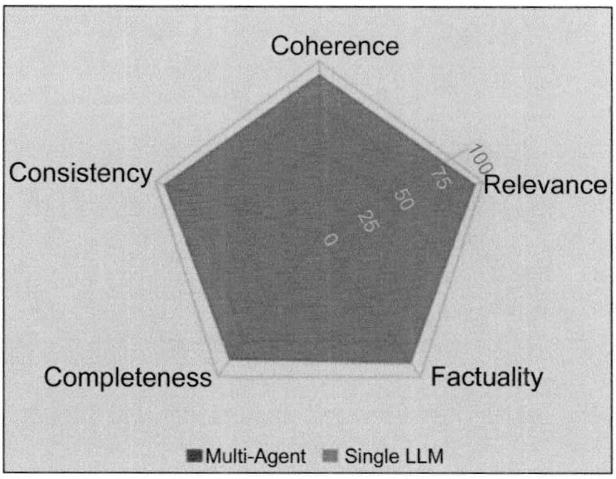

Fig. 10.11 Quality dimensions.

24.3 Resource Efficiency

The resource efficiency graph (Figure 10.12) compares resource utilization between agentic AI and single LLM systems, assessing token usage, API calls, compute time, and memory usage.

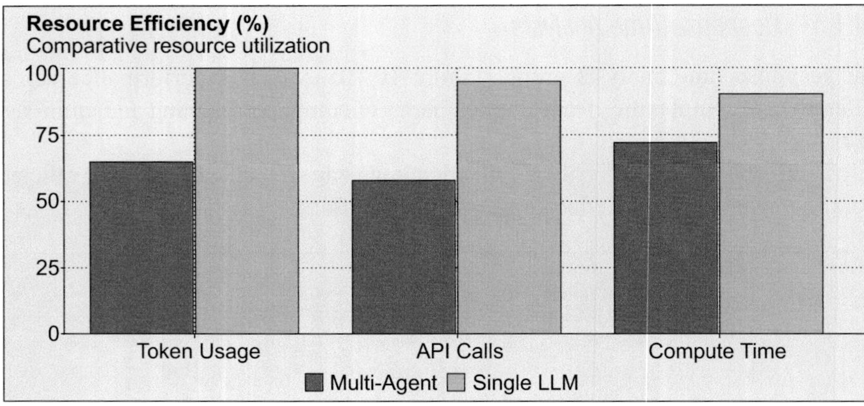

Fig. 10.12 Resource efficiency.

25. Comparison of Single LLM Versus Multiple LLMS for Agentic AI

25.1 Web Search Model Comparison

Figure 10.13 compares three prominent LLMs—GPT-4, Gemini, and Claude—in their effectiveness in handling web search tasks.

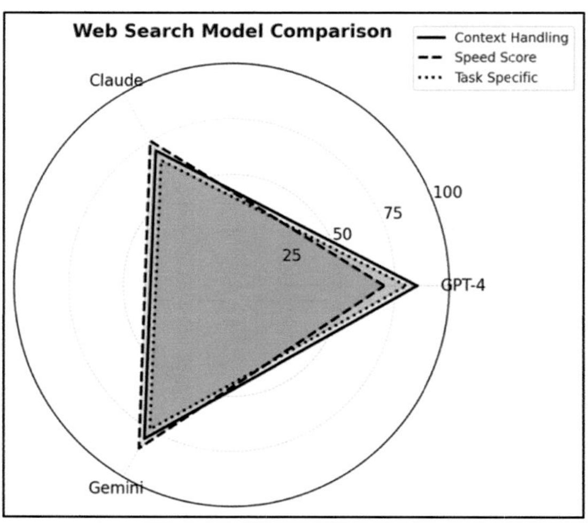

Fig. 10.13 Web search comparison across LLMs.

The metrics evaluated include context handling, speed score, and task specificity. The use of multiple LLMs enables leveraging individual strengths, such as GPT-4's specificity and Claude's context retention, thus optimizing performance for diverse web search tasks compared to a singular generalized model.

25.2 Reasoning Model Comparison

In reasoning tasks, each LLM has distinct strengths and limitations (see Figure 10.14):

- *GPT-4* has a strong capability in handling context and task-specific reasoning, making it highly suitable for complex problem-solving scenarios.
- *Gemini* performs efficiently in terms of speed, advantageous for real-time reasoning applications, yet lags slightly in deep task-specific reasoning.
- *Claude* excels significantly in context handling and task-specific reasoning, positioning it as a highly reliable model for intricate analytical tasks.

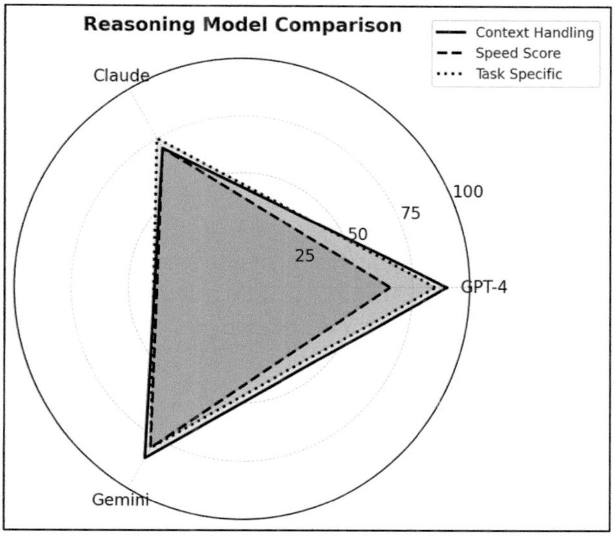

Fig. 10.14 Reasoning model comparison.

Using multiple specialized LLMs allows an agentic AI system to allocate tasks based on specific reasoning demands, enhancing overall reasoning performance compared to relying on a single generalized LLM.

25.3 General Task Handling Comparison

Multiple LLM utilization maximizes task performance by dynamically allocating general tasks based on the individual model strengths, thereby outperforming a single generalized LLM approach (see Figure 10.15).

25.4 LLM-specific Metrics Comparison

Detailed comparisons based on LLM-specific metrics illustrate:

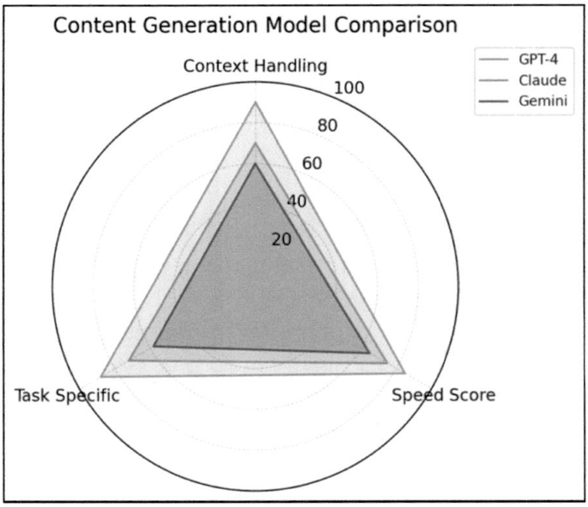

Fig. 10.15 Content generation comparison.

- *Token context length*: GPT-4 and Claude offer extensive context handling capabilities, ideal for tasks needing significant contextual understanding. Gemini has slightly lower context length capability.
- *Response time*: Gemini outperforms GPT-4 and Claude, delivering quicker responses beneficial for real-time interaction scenarios.
- *Instruction following*: GPT-4 and Claude both demonstrate strong adherence to complex instructions, essential for maintaining task accuracy.
- *Cost efficiency*: Gemini provides superior cost efficiency, crucial for scalable deployments.
- *API reliability*: All three models offer robust API reliability, with GPT-4 slightly ahead, ensuring consistent operational stability.

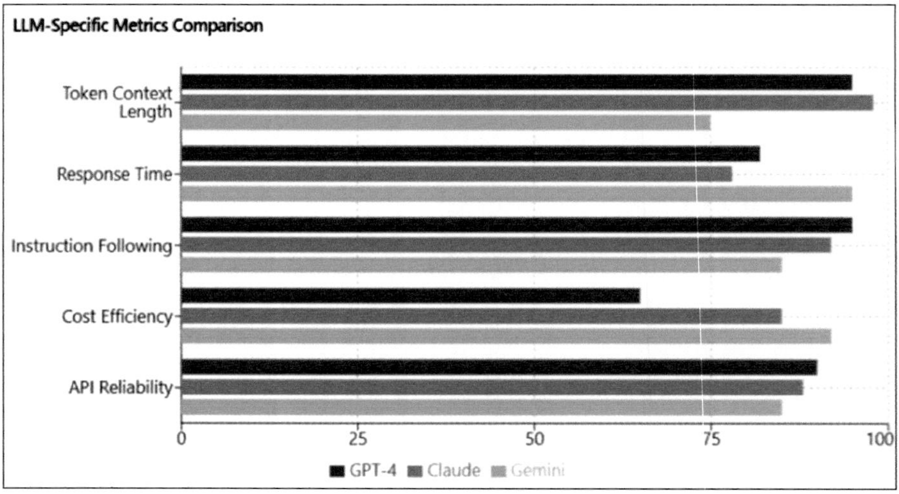

Fig. 10.16 Detailed comparison of LLM providers.

Employing multiple LLMs enables strategic allocation according to these metrics, significantly enhancing overall performance, cost efficiency, and operational reliability compared to a uniform single-LLM system (see Figure 10.16).

26. Conclusion

The comparative analyses underline the strategic advantage of utilizing multiple specialized LLMs within agentic AI systems. Each LLM—whether GPT-4, Gemini, or Claude—possesses unique strengths across various critical metrics. By integrating multiple LLMs strategically, an agentic AI system effectively optimizes its overall performance, resource utilization, and cost efficiency, clearly surpassing the limitations inherent to relying on a single generalized LLM. The continued development of AI in education, especially in the form of multi-agent systems, has generated a completely new class of research opportunities. The systems are now not just about simple automation but are turning into smart, interactive, and context-aware contributors to the learning process. To realize the complete transformation potential of AI, future studies must focus on optimizing system structure, developing agent specialization, implementing the systems across numerous domains, and making them capable of being seamlessly integrated with other education technologies.

References

Ahmed, L., Khan, S. and Taylor, M. (2022). The LearningPath framework: Personalized curriculum planning using AI. *Journal of Intelligent Learning Systems*, *18*(3), 215–231.

American Psychological Association. (2020). *Publication manual of the American Psychological Association* (7th ed.). Available at https://doi.org/10.1037/0000165-000 (Last accessed 10 Feb, 2025)

Anderson, B., Lee, H. and Kumar, P. (2023). The EQUIP framework: Measuring quality in AI-generated educational content. *Journal of Educational Metrics and AI*, *12*(2), 134–150.

Barr, M. and Massa, A. (2023). *Programming embedded systems: With C and GNU development tools* (3rd ed.). O'Reilly Media.

Berger, A.S. (2022). *Embedded systems design: An introduction to processes, tools, and techniques*. CRC Press.

Bloom, B.S., Engelhart, M.D., Furst, E.J., Hill, W.H. and Krathwohl, D.R. (2023). *Taxonomy of educational objectives: The classification of educational goals. Handbook I: Cognitive domain*. Longman.

Catsoulis, J. (2023). *Designing embedded hardware: Create new computers and devices* (3rd ed.). O'Reilly Media.

Chai, S. and Wang, L. (2022). A comprehensive review of curriculum design methodologies for engineering education. *IEEE Transactions on Education*, *65*(2), 145–162.

Chang, A. and Roberts, J. (2023). The MAESTRO system: Hierarchical agent planning for science learning. *International Journal of AI in Education*, *33*(1), 52–70.

Gettinger, M. and Seibert, J.K. (2023). Best practices in setting up adaptive learning progressions in technical education. *Journal of Educational Technology*, *41*(3), 278–295.

Heath, S. (2021). *Embedded systems design* (3rd ed.). Newnes.

Heller, A. (2023). What is agentic AI and why does it matter? Medium. https://medium.com/@alexheller/what-is-agentic-ai-and-why-it-matters-5a3b8f021e4e (Last accessed 12 Feb, 2025)

Jensen, K.L., Smith, J.B. and Thompson, A.R. (2024). Agent-based systems for educational content generation: A comparative analysis. *Educational Technology Research and Development*, *72*(1), 36–52.

Johnson, T.R., Miller, D. and Wong, Y. (2022). The EDUCATOR system: Multi-agent architecture for collaborative content creation. *Educational Software Architecture Review*, *7*(2), 95–112.

Krathwohl, D.R. (2022). A revision of Bloom's taxonomy: An overview. *Theory Into Practice*, *41*(4), 212–218.

Lee, E.A. and Seshia, S.A. (2023). *Introduction to embedded systems: A cyber-physical systems approach* (3rd ed.). MIT Press.

Martinez, R. and Johnson, T. (2023). CurriculAI: AI agents for curriculum sequencing. *International Journal of Educational AI*, *15*(4), 201–220.

Marwedel, P. (2022). *Embedded system design: Embedded systems foundations of cyber-physical systems* (4th ed.). Springer.

Mishra, A.K. and Dubey, A.K. (2024). Integration of automated assessment systems in engineering education. *Journal of Engineering Education*, *113*(2), 289–305.

Nguyen, H. and Patel, K. (2022). Measuring efficiency in educational AI systems. *Journal of Applied Educational Computing*, *19*(2), 167–182.

Noergaard, T. (2023*). Embedded systems architecture: A comprehensive guide for engineers and programmers* (2nd ed.). Newnes.

Patel, R. and Lee, S. (2023). AssessNet: Cognitive-level-balanced assessment generation using agents. *Journal of Intelligent Assessment Systems*, *13*(1), 45–63.

Patel, V., Zhou, Y. and Morris, K. (2022). The EduArch framework: Hierarchical educational content agents. *AI in Education Systems*, *10*(3), 88–106.

Rivera, D. and Chen, J. (2022). Cognitive balance assessment for Bloom's taxonomy. *Educational Evaluation and AI*, *9*(1), 55–74.

Rodriguez, A. and Kim, L. (2023). AdaptEd: A hybrid market-based platform for educational content generation. *Learning Systems Engineering*, *18*(4), 193–212.

Sanchez, M., Li, T. and Ng, J. (2023). MultiEd: A platform for multi-agent adaptive content generation. *Advances in AI and Education*, *7*(3), 120–144.

Schmidt, D.C. and Buschmann, F. (2022). Patterns of software architecture for embedded systems. *IEEE Transactions on Software Engineering*, *48*(5), 456–471.

Singh, H. and Walia, G.S. (2024). Artificial intelligence in educational content development: Current practices and future directions. *AI & Education*, *15*(2), 187–203.

Thompson, A., Gupta, R. and Lin, J. (2022). EvalGen: Multi-agent assessment generation aligned with content. *Journal of Educational Content Automation*, *6*(2), 73–90.

University of Florida Center for Instructional Technology and Training. (n.d.). Bloom's taxonomy—Graphic description. Available at https://citt.ufl.edu/resources/the-learning-process/designing-the-learning-experience/blooms-taxonomy/blooms-taxonomy-graphic-description/

Valvano, J.W. (2023). *Embedded microcomputer systems: Real time interfacing* (4th ed.). Cengage Learning.

Wang, Y. and Ramirez, A.V. (2024). Cognitive load considerations in automated educational content generation. *Educational Psychology Review*, *36*(1), 67–85.

Wei, X., Park, J. and Ahmed, N. (2021). EduMarket: A market-based multi-agent content generator. *Transactions on Learning Technologies*, *13*(4), 304–322.

Williams, J., Ahmed, T. and Rios, M. (2023). Evaluating educational effectiveness in AI-supported content. *Review of Educational Research in AI*, *17*(2), 110–132.

Wilson, C. and Chen, F. (2023). CollabEd: A semantic-node blackboard system for educational AI. *Journal of Collaborative Learning AI*, *5*(3), 84–102.

Wolf, W. (2022). *Computers as components: Principles of embedded computing system design* (5th ed.). Morgan Kaufmann.

Yiu, J. (2023). *The definitive guide to ARM Cortex-M3 and Cortex-M4 processors* (4th ed.). Newnes.

Zhang, L., Zhou, P. and Evans, R. (2022). EdCreator: Multi-agent generation of visual and textual educational content. *Journal of AI Education Platforms*, *9*(2), 153–170.

Chapter 11

An Investigation of AI's Influence on Contemporary Pedagogy via Advanced Deep Learning and Sensor-driven Educational Institutions

M. Ramprasath[1*], *Elangovan G.*[2], *Prakash Duraisamy*[3] and *V. Kavitha*[4]

1. Introduction

Quickly becoming a game-changer across several sectors, artificial intelligence (AI) is already making waves in the academic world. The rise of AI has spurred innovations and improvements in many areas of human existence. Because of its central role in personal and societal growth, education has reaped many rewards from advances in AI. Learners learn, educators impart knowledge, and educational institutions operate differently due to AI's incorporation into educational systems. Considering the significance of incorporating AI into education, it is necessary to contemplate its consequences. Our objective is to examine, via literature review, the possible effects of AI on education. Our primary areas of interest are applications, benefits, and difficulties. One step of the evaluation process is to use the correct terms to search within Scopus's database. We sort, filter, and examine the results to get to the good stuff. We break each theme into subfields to better organize the literature and the problem. Several parts are used to present the results by the main themes and subthemes. We conclude that embracing new technology while implementing safeguards to prevent

[1] Department of Data Science and Business Systems, SRM Institute of Science and Technology, Tamil Nadu, India. http://orcid.org/0000-0002-2667-9184
[2] Department of Data Science and Business Systems, SRM Institute of Science and Technology, Tamil Nadu, India. http://orcid.org/0000-0003-2385-2129
[3] Department of Computer Science and Engineering, University of Wisconsin-Green Bay, USA.
[4] Department of Data Science and Business Systems, SRM Institute of Science and Technology, Tamil Nadu, India. https://orcid.org/0000-0002-2728-5646
* Corresponding author: ramprasm@srmist.edu.in

DOI: 10.1201/9781003639367-11

misuse is the way to go. Natural language processing (NLP) is one of the most essential uses of AI. Machines that can comprehend spoken and written language are the goal of NLP. Specifically, intelligent chatbots are seeing a surge in deployment across a variety of businesses for support along with customer service purposes (Hu & Huang, 2015). Since their inception in 2005, the development of state-of-the-art chatbots has been going strong (Mendell & Heath, 2005). Another area that chatbots have impacted is education (Saini, M.K. & Goel, N., 2019). According to a recent survey, chatbots have been more popular in the classroom (Chiou & Tseng, 2015). According to various research, chatbots can potentially enhance and simplify the educational process for pupils (Stazi et al., 2017). In November 2022, with the release of ChatGPT, the tide turned in society's adoption of AI. Unexpectedly, ChatGPT attracted many users and garnered immense recognition for its exceptional ability to write and comprehend. It was ground-breaking because it brought the promise and urgency of AI to a broader audience than just those involved in machine learning. ChatGPT had the most significant impact on the academic world. There is a lot of back and forth over ChatGPT's potential as a tool for academic dishonesty and its ability to produce intelligent teaching solutions. Secondary and higher education teachers have expressed concern that students could misuse ChatGPT and have demanded that it be restricted. New York City and Seattle educational institutions, as well as the Australian states of Tasmania and Queensland, have all banned ChatGPT from student devices and networks. Similar restrictions are being considered by a few educational institutions (Twumasi et al., 2017). But stopping the children from utilizing AI seems like an insurmountable task. Students in higher education can find solutions to various problems using ChatGPT, as mentioned in Pastor et al. (2022). These problems cover everything from writing essays to creating code. Better student learning outcomes can be achieved by integrating AI into the educational institutions and utilizing its potential. Figure 11.1 shows the impact of AI in education. Because AI can be scaled to accommodate all students, personalized learning is now within reach (Chiou, C.K. & Tseng, J.C., 2015). Applying AI algorithms like reinforcement learning allows for the dynamic adaptation of the learning process to each student's unique demands. In the context of individualized education, intelligent tutoring systems that engage learners and provide insightful comments are within reach. Another noteworthy feature of AI is the computerized assessment process.

Integrating image processing alongside language recognition systems enables the automation of task, test, and refresher grading. Automated grading allows teachers to spend more time interacting with their pupils.

Artificial intelligence's many forms of feedback and analytics can also help educators and students work together more effectively. The use of AI in classrooms exemplifies how intelligent systems can improve teaching and learning. AI has several positive effects, including enhanced educational results, more efficient use of time and money, and more access to high-quality education worldwide. Students, particularly those from underprivileged backgrounds, can benefit from computerized tutoring and customized education. Better educational opportunities will be accessible to students in both wealthy and developing nations, thanks to the worldwide scope and scalability of AI. The educational system stands to save a tonne of money and time with automated grading. The time instructors spend on grading and related tasks is

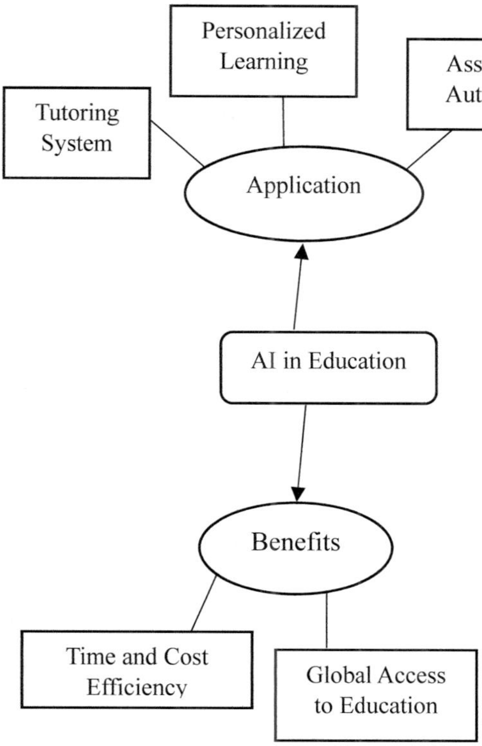

Fig. 11.1 Multifaceted impact of AI in education.

approximately 40% of their total time. If teachers are not weighed down by grading, they will have more time to devote to each student and offer individualized assistance with their coursework. We must be mindful of the possible risks of implementing autonomous machines in education, notwithstanding the attractiveness of AI's educational uses and benefits. AI should only be used for instructional purposes after extensive testing and monitoring, as youngsters are more easily misled than adults.

Possible concerns include issues with data security and confidentiality, prejudice and discrimination, and the connection between teachers and students. Students' private information is necessary for AI uses, like personalized instruction. For example, when given information about a pupil's psychological state or cognitive impairment, AI can tailor its approach and content to meet their specific needs. Although there are many positive uses for individuals' private data, there are also risks associated with privacy and security. Student information can be made more secure by encrypting and anonymizing it. To tackle this problem, though, a thorough plan is necessary. Racism and other forms of prejudice are another critical problem. The internet's inherent biases can affect AI education on public data. The self-learning of bias is another potential issue with AI algorithms.

Due to their high entropy levels, AI algorithms may act unexpectedly. When it comes to using AI in the classroom, one of the biggest obstacles is reducing bias. Throughout its history, new technology has posed the risk of abuse. The destructive nuclear weapons were produced with the discovery of atomic fission. The dark web emerged with the rise of the internet as a means for criminals and other lawbreakers to evade authorities. Nevertheless, with the help of security forces and international cooperation, society has managed to curb the possibility of misusing technology. The advantages of new technology much exceed its disadvantages. It will be better to incorporate new AI technology into the curriculum on equilibrium rather than halt or impede its educational growth.

Investigating smart classroom devices and artificial intelligence in education (AIEd) has recently gained more academic attention. However, studies examining the sensors utilized by smart classrooms are dispersed and poorly organized. Furthermore, there is a lack of studies on integrating AI and sensors into smart classrooms. There is a dearth of research that compiles and reviews the substantial benefits that AI technology offers to sensors.

Through its methodical narration of the investigation, this chapter offers a multifaceted contribution in comparison to the prior survey (Alkhayat et al., 2015)

1. We provide a technological overview of sensor technology and its uses in smart classrooms, discussing its classifications and applications in areas like tracking student attention, taking attendance, and analysing student performance on assessments.
2. We present several uses for smart classrooms that integrate AI with detectors, and we go over recent developments in AI at the level of application technology. We examine and contrast the roles and functions of various AI technologies in sensor technology, and we delve into how AI technology may facilitate innovative learning by integrating with sensors.
3. We discuss where the study might go and investigate the hazards and difficulties of combining sensor and AI technology. By showcasing the most recent developments and research trends in innovative classroom sensor applications and AI technology integration, we may provide future researchers with the essential domain expertise and spark new lines of inquiry.

This chapter emphasizes smart classrooms and thoroughly analyses AI's use in the school. In the second section, we review the many types of sensors used in smart classrooms and how they are evolving. It also draws attention to the ways in which biometric and environmental sensors might improve classroom settings. The third section examines various AI technologies, comparing them to see how they can enhance sensor performance and enable new capabilities in educational contexts. The fourth section discusses the benefits and drawbacks of using AI in the classroom and some good places to look for future studies. The purpose of this part is to connect the dots between current technology and how it might be used in real-life educational settings.

The final section briefly summarizes the chapter's main points and serves as the conclusion. We have thoroughly reviewed and analysed the current educational literature on AI. We have also identified key trends in the uses, benefits, and problems of AI in higher education and its prospective consequences on educational practices. This chapter is an excellent reference for learning about the present and future of AI in the classroom.

2. Use of Smart Sensors and Related Techniques

Using sensors in the classroom involves analysing and improving every facet of the educational process. An innovative classroom sensor system is illustrated in Figure 11.2 as an exemplary example. When it comes to education, smart sensor technologies are all about including all parts of the educational experience. Cameras, thermometers, illumination and temperature sensors, air quality monitors, sound level meters, personnel recognition cameras, and other sensors allow this system to gather data from various sources, including the environment and human behaviours, to provide a valuable learning experience. This section explores the many uses of smart sensors, focusing on the innovative ways in which they might improve teaching and learning. This survey covers various sensor types and their practical uses in smart classrooms. It categorizes environmental and biometric sensors into different kinds of sensors. Emphasizing attributes like referencing information and viewpoints to help create a fulfilling and fruitful curriculum experience.

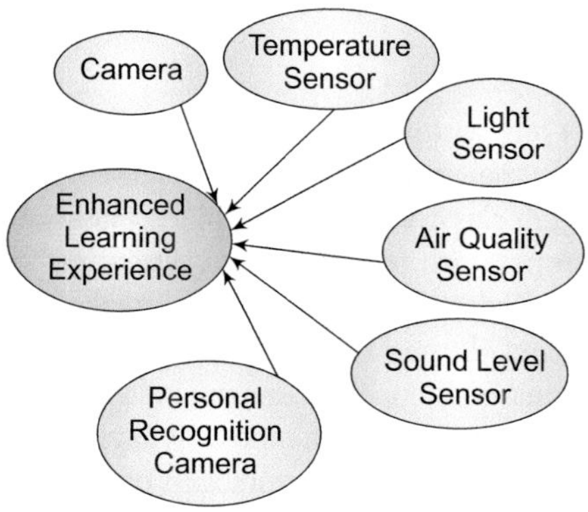

Fig. 11.2 Illustration of an intelligent classroom monitoring system.

2.1 Environmental Monitoring Devices

Sensor technology has allowed for the collection of data on physical surroundings in real time. The usual investigations of smart classroom environmental sensors are summarized in Table 11.1. Typically, these systems have microcontrollers that can control actuators and data displays. This allows us to see, control, and monitor invisible events. The standard for school networks with sensors is the Raspberry Pi–based weather monitoring system (WMS) suggested by Huy and Huang, R. (2015). This system can gather various meteorological data, such as temperature, humidity, and wind direction. Mendell and Heath (2005) noted that to establish pleasant surroundings essential for efficient learning, deploying environmental sensors can help control and regulate the indoor classroom environment.

Everything from airborne dirt to fine particles, carbon monoxide (CO), moisture content, radiation, volatile organic compounds (VOCs), nitrogen dioxide (NO2, from fuel combustion), carbon dioxide (CO2, primarily from humans), acoustics, auditory noise, and electricity was meticulously categorized by Saini and Goel (2019)

Table 11.1 Sensors for the surroundings in smart teaching

Type of Sensor	*Features being Tracked*	*Purpose of Tracking*
Temperature sensor	High temperature	Humidity in the school classroom
Sound detector	Milliseconds detector	Measurement of decibel levels
Wireless Infrared (IR) Detector	Thermal radiation	Observing classroom environments for signs of student engagement
Oxygen detectors	The concentration of carbon dioxide	The tracking of atmospheric concentrations of carbon dioxide
Detector that uses light	Environmental lighting	Keeping an eye on brightness and consistency

as components of a typical classroom physical environment. A smart classroom's use of environmental sensors can be broadly classified into three areas: temperature regulation, illumination management, and noise abatement.

2.1.1 Weather Management

Overall, the air quality—temperature, humidity, and pollution levels—makes up most of the climate element. In their research, Chiou and Tseng (2015) suggested a smart strategy for managing a lab classroom. To control the classroom's exterior conditions (lighting, humidity, temperature, etc.), this investigation built a Wireless sensor Network (WSN) utilizing Zigbee technology. To ensure the suggested system worked, a field trial was carried out. Experiments conducted in real time demonstrated the system's high accuracy and resilience. To enhance classroom air quality and current relaxation, Stazi et al. (2017) suggested an innovative window system that could open and close automatically based on temperature and carbon dioxide readings from a PT100 thermistor. Research comparing two nearby classrooms found that this system met or exceeded expectations for thermal convenience, user happiness, and indoor air quality. A classroom fan was programmed to switch on and off automatically by Twumasi, C., et al. (2017) using wireless sensor that detects infrared light without actively emitting any signal. A Passive Infrared Detector (PIR) sensor detector is activated whenever a pupil approaches the classroom using the infrared radiation that it produces. After that, the data is sent to the microcontroller, which will subsequently "turn on" the fan. Once 10 minutes have passed and no movement is detected, the fan is "turned off" automatically. The fan will not be activated until the temperature in the room rises to 25–30 degrees Celsius. The suggested method enables the simultaneous real-time monitoring of multi-dimensional indicators in common open spaces in higher educational settings, according to Pastor et al. (2022). These indicators include carbon dioxide, temperature, humidity, and particle matter. When activated, this system can automatically regulate the air conditioning or ventilation system in the relevant room.

Keeping the educational setting clean and safe for pupils was paramount during the COVID-19 pandemic. A classroom's temperature can be more easily controlled with infrared cameras, which check kids for fever (Alkhayat et al., 2015). Also, for indoor settings (particularly classrooms), Deepaisarn et al. (2022) suggested a system that records human physical distance using end-to-end cameras. The recording system can monitor students' whereabouts and real-time motion patterns in the classroom. In addition to keeping pupils at a safe distance from one another when inside, it logs their trips to and from their seats, which helps to prevent the spread of contagious diseases.

2.1.2 Power Sources

Classroom lighting is no longer a static fixture for illumination in instructional environments. Researchers have discovered that students are more engaged and have less eye strain when using adaptive lighting systems, which change their intensity depending on the amount of natural light entering the room and the activities taken on within it. Regarding the mental state of acquiring knowledge, a study by Du et al. (2023) suggests that a brightly lit room promotes a greater desire to learn and a higher level of excitement for the process. Following our investigation, we learned that most of the existing literature is devoted to several topics, such as finding ways to automatically change the intensity of interior lighting (Cruz et al., 2021). Each row of classroom lighting equipment could have its intensity adjusted in three zones, thanks to the Amelkina and Duplenkova (2021) system, which included a smooth control function, light sensors, and presence sensors. This setup would significantly reduce energy consumption while maintaining brightness and uniformity under combined lighting. This chapter's findings on the effects of combined lighting on students' physical and mental well-being align with those of Du et al. (2023), who also found a noticeable lack of homogeneity in the calculated illumination on the desktop.

There was a consequent lessening of impact on the environment, waste management expenses, and energy usage. Additionally, a multi-mode lighting management system for classroom sub-areas was suggested by Chen et al. (2017) for use in university settings. This data was then used to program the chip, enabling the automatic light switching and achieving a good energy-saving effect. Environmental noise directly affects productivity and well-being, as pointed out by Marques and Pitarma (2020). High blood pressure and stress are two health problems linked to excessive volume. Conversely, being comfortable listening to music can boost focus, dialogue, and output. The paper also introduced "iSoundIoT", a real-time noise monitoring system that uses Internet of Things (IoT) technology. It is designed for use in classrooms and other indoor environments (see de Valencia et al. (2023).

2.2 Fingerprint Readers

Biosensors are also helpful in the classroom for increasing student engagement and critical thinking skills. Expression, movement, eye gazing, core temperature, electroencephalogram (EEG) data, and heart rate are some physiological signs biosensors frequently track in students and instructors. Students' involvement, emotional reactions, and mental procedures can be better understood with the help of the physiological and behavioural data collected by these sensors. Personalized learning paths based on individual physiological reactions are just one of many potential uses for such devices. Others include tracking levels of attention and stress. In addition, biosensors make accessibility possible for students with disabilities.

Using three categories—engagement analysis, attendance, and accessibility support—this section will show how biosensors are used in smart schools.

2.2.1 Engagement Evaluation

High levels of involvement are associated with better task performance and learning outcomes (Peifer & Engeser, 2021). Various elements about the student, the work at hand, and the educational setting impact the student's attention level (Carroll et al., 2021). Improving the efficacy of online education requires tracking and adjusting students' participation levels during lessons. Most studies surveyed have focused on using the sensors of cameras, eye tracking, or other image-capturing technologies for monitoring student engagement through pictures.

2.2.2 Fingerprint Sensors Based on Images

Much information is included in photos, and numerous methods exist for analysing student engagement using visuals. According to Verner and Dickinson (1967), yawning, fidgeting, sketching, and looking around are the top indicators of students looking distracted in class. Table 11.2 shows that many image-based studies have suggested ways to assess student engagement by analysing indicators such as body movements, emotions displayed by the face, and the most promising methods are devices that detect inattention using cameras and eye movement sensors. Students' body language, expressions, and eye movements can be captured via cameras. The technology behind cameras has advanced dramatically in the past few years. The technology behind cameras is shrinking in size and dropping in price, making it more accessible. Additionally, AI and machine learning advancements have steadily improved software algorithm technologies that can handle visual data such as photographs and videos. Alkabbany et al. (2023) constructed a biometric sensor network (BSN) with a webcam, a wall-mounted camera, and a high-performance computer to evaluate the participation of students in science, technology, engineering, and mathematics (STEM) classes at American colleges.

The purpose of this device was to record the students' facial expressions, body language, and head position. An AI-based model is trained using the collected image data to measure students' emotional and behavioural involvement in a classroom setting. The results showed that the suggested approach better evaluated behavioural and emotional engagement. The studies compared the proposed technique with the state-of-the-art framework, lasting 75 minutes each. Students aged nine to eighteen in elementary and secondary schools can have their unusual seating positions detected by an innovative learning table built by Zhu et al. (2023). At an optimization level above 92%, the system could identify improper sitting postures such as extended leaning time, head tilt, body tilt, and head drooping. Plus, it can stop students from becoming sick from sitting too much and provide them with immediate comments and reminders if they notice a drop in student participation.

Thermal infrared imaging is a tried-and-true method for evaluating emotional states, psychophysiological responses, and vital signs without invasive procedures or physical touch. Using thermal infrared imaging, Kim (2019) assessed the mental health of college students in Korea. To gauge the level of participation across the board, we averaged the temperatures of the students' areas of interest (AOI). The more heated the classroom, the more engaged the students will be. According to Kim's plan, a smartphone app could show instructors how engaged their students were in class using a traffic light system, with green signifying fully engaged, yellow indicating moderate involvement, and red indicating low engagement. Similarly, Hu et al. (2023) devised a strategy for detecting active learning in a virtual reality setting by integrating multimodal features. They used the HTC Vive Facial Tracker to monitor the subject's face and the HTC Vive Pro Eye to monitor the subject's eyes. When linked to a computer, those devices could record eye gazing, facial expressions, and pupil diameter data. Researchers in this study utilized a head-mounted device equipped with a ThinkGear ASIC module chip to record participants' EEGs, and these were then used to gauge participants' level of focus in a virtual reality (VR) setting related to their emotional regulation, behavioural regulation, and cognitive function. Every step of the experiment took place in a university-level geography English lesson.

Compared to the model that only used one dimension or one kind of data, the comprehensive data input (containing all types and dimensions) yielded an F1 score ranging from 0.66 to 0.73, demonstrating substantially better learning concentration recognition. One way to keep tabs on your blink rate and the location of your eyes is with an eye-tracking detector. There has been a recent uptick in interest in using eye-tracking technologies for multimedia learning research. Hu et al. (2023) have transformed conventional pedagogical practices by providing a rapid and ubiquitous way to record and monitor eye movements. Current eye-tracking devices were classified by Wang et al. (2021) into four groups: those that are mounted on a tower, those that are screen-based, those that are head-mounted or wearable, and those that are transportable. Additionally, they note that mobile eye-tracking devices and those worn on the head are better suited to practical uses and everyday learning.

Pupil-Lab's eye-tracking goggles were worn by Shvarts and Abrahamson in a study that included dual-eye-tracking technology (Shvarts & Abrahamson, 2019). Two persons could freely navigate a shared space while this equipment tracked their eye movements in real time. Its placement allowed for the evaluation of the instructor–student dynamic in college psychology classes by monitoring the students' and instructors' eye movements. At the same time, they worked together to concentrate on a single visual object. According to the study, there was strong coordination between instructors' perceptual and students' operational activity. Teachers could build a picture of how their kids learned by keeping track of student actions and determining when verbal assistance was most effective.

Zaletelj (2017) used these features and data on students' eye points, attitude, and other behavioural indicators to estimate how attentive they were in class. The system utilized a Bagged Trees classifier, which, depending on the parameters, achieved an accuracy ranging from 85.0% to 86.9%. Students' attention levels were also significantly connected with specific behaviours, including writing, yawning, head support, and gaze direction. Prieto et al. (2018) also established a measurement system for smartwatches, which included SensoMotoric Instruments (SMI) eye-tracking eyewear and a mobile phone with an altimeter. The purpose of this equipment was to capture audio and data related to the teacher's gaze, including where and what the teacher was looking at in the classroom. It also monitored the instructor's every step in elementary and secondary schools, where pupils aged eleven to twelve were actively engaged, whether seated or on the move. A graphical choreographic that can depict the instructional processes and relationships between individuals throughout a course was created by analysing the data using a machine learning model. This allowed for a thorough understanding of the dynamics in the classroom.

Pupil, an open-source, inexpensive wearable eye-tracking system proposed can replace costly commercial eye-trackers. A low-cost alternative approach has emerged: mobile eye-tracking technology, made possible by the proliferation of personal mobile devices. This technique can recognize faces, eyes, iris or pupil, and gaze angles using the front or back cameras of inexpensive personal smart devices like smartphones and tablets in conjunction with robust software programs (Gunawardena et al., 2022). It is important to note that most modern eye-tracking studies are in a dedicated lab. Using long-term eye tracking in a classroom is an entirely fresh and innovative idea (Jarodzka et al., 2021). To further comprehend the transferability of laboratory study content to real educational settings, we urge future research to be conducted in more authentic educational contexts.

2.2.3 Student Attendance

Taking attendance through a roll call is the standard practice in most classrooms, but it can get tedious quickly, particularly in larger classes. Also, it is easy for this strategy to lead to instances when one individual shows up instead of another. College students' tendency to skip lectures has grown in recent years, impacting both the content and delivery of instruction in these settings. Numerous studies on automatic attendance recording have surfaced to address this issue. At present, the most popular ways to take attendance include using a roll call system, radio frequency technology (RFID) (Dias et al., 2020; Fernández et al., 2013), technologies that are built into smartphones (e.g., Bluetooth, NFC, WIFI [Banepali et al., 2019]), and biometric recognition systems (e.g., fingerprint, face, and voice). Furthermore, several researchers have built recognition apps that use smartphone signal recognition (Zorić et al., 2019). Scan the QR-code to start recording, or use your personal mobile device's Bluetooth, WIFI, or GPS technology to punch in after you have entered a particular range. On the other hand, it is worth noting that some schools have banned the usage of smart gadgets.

Table 11.2 Classifications of Sensory Data and the Goals of Their Interpretation

Category of Sensing	*Information under Observation*	*Objective of Inspection*
Cinematography	Physical characteristics that make up one's appearance	Identifying facial characteristics
Devices Fingerprint Reader	Unique fingerprint designs	Identifying fingerprint characteristics
Sound Recorder	Voice features	Recognition of vocal characteristics
Pressure Gauge	Chair under stress	Find out if the chair is occupied
Infrared Night Vision	An infrared motion picture	Locate motion

Using biometrics for attendance recognition can lessen the need for individual devices. A combination of automatic recognition algorithms and recognition sensors placed throughout the classroom can help monitor attendance.

Table 11.2 shows fingerprint sensors used for attendance. Students must first have their basic biometric features "registered" and stored in an information system as templates to use biometric sensor-based punctuality technology. Such characteristics could include a person's face, iris, voice, or fingerprints. Verification follows, which involves comparing the collected fingerprint characteristics in real time with pre-existing database templates. A successful pairing elevates the attendance record to a recognized landmark. Our research uncovered several studies and methods that might be implemented to raise student attendance in the classroom. Specifically, Ni et al. (2022) proposed a system for taking attendance in college classes that combines the most recent advancements in facial recognition technology with a camera-based roll-call system. The number of attendees and status of pupils can be quickly tallied using this approach. In this configuration, a front-facing camera records the entire classroom and uploads the footage to the school's server, where facial recognition software may analyse it. In Margreat et al. (2021), a chatbot's cameras and the Haar-Cascade facial recognition classifier were used to record attendance. The next step after taking the student's pictures was to identify them and compare them to the databases of known student faces in the system. Alternatively, the robot's internal speaker could be used to get an identifiable result.

Moreover, Gagandeep et al. (2019), and Rahmatya and Wicaksono (2019) use biometrics, fingerprint detection methods for a substantial physical means of keeping track of attendance. Adal et al. (2018) proposed using smart mobile phones for fingerprint recognition attendance since most contemporary mobile telephones have embedded fingerprinting sensors, which are cheaper. Using software that recognizes voices is another way to keep track of attendance.

An attendance system that uses voice biometrics was presented by Amri et al. (2017). Students had to sign up and save their voices before they could do anything else. By comparing the students' real-time speech input with the pre-registered voice data, a success rate of 60% was achieved, indicating that the method could be helpful. Classroom management and attendance can also be automated using the approach described in Sweetlin et al. (2016), which uses the built-in mike of an Humanoid smartphone for voice recognition. Despite their innovative approach, the two researches above did not provide sufficient rigorous empirical proof. Additionally, it should be mentioned that most voice-based attendance systems utilize the mobile phone's internal microphone to capture audio (Hoo & Ibrahim, 2019). This is because smartphones come equipped with user-friendly sound sensors that do not require extra deployment.

2.2.4 Assistance with Accessibility

The implementation of electronic sensors has opened new vistas for individuals with disabilities, helped them make sense of the world around them, and given students with disabilities access to a broader range of educational opportunities. The emergence of sensors has had this profound effect. When interacting with non-disabled people, sign language is the only option for people with trouble hearing or speaking. People who are not impaired should not feel pressured to learn sign language as they already have all the necessary information. One solution that could be implemented in smart classrooms is gesture recognition machinery based on images.

Regarding real-world interactions, static gesture recognition devices and algorithms do not cut it. Leveraging the Kinect from Microsoft gadget, Varshini and Vidhyapathi (2016) suggested an instrument and method for continuous finger identification. The system's speed in processing depth data, detection of dynamic gestures of one or two hands, constant monitoring of finger motions, and production of results from real-world testing make gesture recognition software suitable for real-time field situations, such as schools. Students with trouble hearing or speaking can benefit significantly from this system's ability to assist them in class through improved communication. A more straightforward and more effective smartphone-based motion appreciation system (GazeSpeak) that can understand eye movements in real time was proposed by Zhang et al. (2017). The e-tran board was also the subject of a comparison study in the article (Lathière & Archambault, 2014). According to the achievements, the system provided faster and more reliable transmission than the e-tran board, which led to satisfied users. The study eye-tracking gadget and a computer were the building blocks of EyeDraw, a tool created in 2005 for kids with severe motor difficulties (Hornof & Cavender, 2005). Disabled children could use this technology to create visual art by directing their gaze. The research also showed that including visual elements like pattern, colour, and auditory input into software could enhance the user experience and encourage more in-depth use.

2.3 An Introduction to Sensor Technologies and Their Use in the Classroom

In addition, the Equivital EQ02 is just one example of a research gadget that can track numerous physiological data simultaneously. Classrooms now have smaller, more excellent integration sensors because of the proliferation of smart mobile devices. When students can use their smartphones in class, these devices have also become efficient, portable, and inexpensive sensors. Smartphones equipped with microphones, cameras, accelerometers, and fingerprint identification sensors can be used by teachers and pupils to track who is there and who is not. Naturally, new technology has not replaced the reliance on conventional sensors like microphones, cameras, and outdoor sensors in innovative classroom applications. Image cameras still provide the most abundant and intuitive data. Data fusion and multimodal analysis are becoming increasingly popular, and using sensors in classrooms is a prime example of this trend. The importance of integrating data from multiple sensors, rather than depending on just one, is growing. To better understand and anticipate students' learning status and requirements, it is necessary to conduct a thorough analysis. Developing hardware and implementing programming procedures are crucial for the evolution of sensor technologies in smart classrooms. One way in which AI is improving educational services is by making sensor systems more efficient at processing and analysing data. This, in turn, allows for creating more nuanced and personalized content.

Sensor software has come a long way in the last decade, thanks to improvements in classic algorithms and new forms of AI. Machine learning, rationality and judgement systems, and handling natural languages are only a few areas of AI that have been studied and used in education. With an emphasis on the function of software algorithms, especially the use of AI, this section will outline the developments in the field of sensors in the educational setting. Today, the science and technology behind computer systems' ability to carry out activities typically requiring human intelligence is called AI technology.

Learning (the process of taking in new knowledge and organizing it according to predetermined rules), reasoning (the process of concluding, either approximation or definitive), self-correction, and language comprehension are all part of this category of activities. Baker and Smith Machine Learning offered an Exhaustive description of AI. AI is a machine that can learn and solve problems and other cognitive processes commonly linked with human intelligence. In addition, they stress that AI is an umbrella word for many different approaches and technologies, not just one.

3. Analysing and Making Sense of Data

One of the primary functions of software algorithms is to analyse and interpret the data gathered by sensors. Because of this, the processing system can transform the raw sensor data into usable data signals, make scene interpretation and identification a reality, and lay the groundwork for microprocessor decision-making. Following a literature review, studies focusing on AI-powered data algorithms for recognition are primarily used to comprehend genetic characteristics, such as the behaviour and status of teachers or students. It calls for processing bigger and more complicated data, which is more complex than comprehending the physical environment characteristics of the classroom. Automatic information processing, recognition of large-scale data, generalisability, processing speed and economy, and accuracy in specific scenarios are all areas where deep learning technology shows significant promise. It makes direct training from input to output and end-to-end learning possible. On top of that, it excels at processing music and complicated images. Figure 11.3 shows the challenges

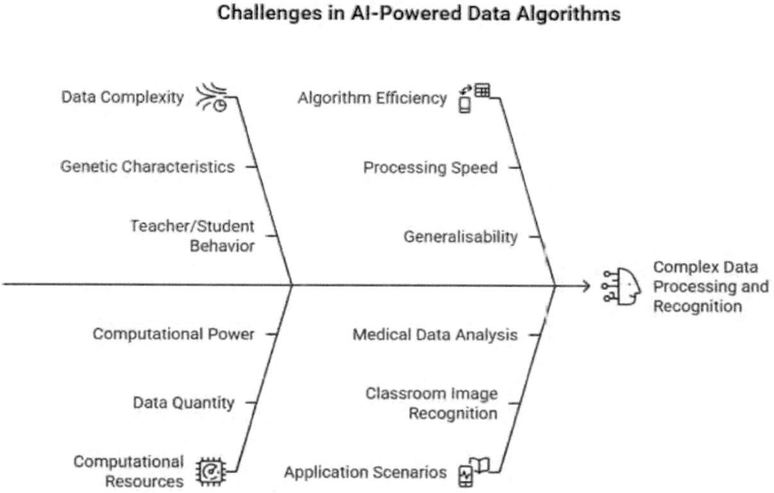

Fig. 11.3 Challenges in AI-powered data algorithm.

in AI-based data algorithms. While there may be constraints on data quantity and computational resources, conventional machine learning methods can handle the data and produce satisfactory results. Given the abundance and development of AI-based learning environments to recognize image examinations. Specifically, it will look at linear problems like electrical brain waves, body temperature, and electrocardiograms (ECGs), and how conventional machine-learning approaches are more straightforward and simpler to debug than deep-learning algorithms.

3.1 Recognizing Things from Images

Though current tools like EEGs, ECGs, and eye-tracking devices can detect emotions, the most promising sensor type is the camera since visual images typically contain the most detailed information and do not require any wearable technology (e Silva et al., 2023). A typical AI system that uses recognition of images. Another reliable technology for taking attendance automatically in schools is facial image recognition pointed out by Tarnowski (2017). Lek and Teo (2023) partitioned the algorithms to detect facial emotions. Deep learning–based face recognition algorithms heavily use deep neural networks (DNNs), a relatively new machine learning technique. The algorithms for deep learning differ primarily from more conventional recognition algorithms in that they are utilized as tools for education. The following procedures make up facial emotion recognition. The first step, as shown in Figure 11.4, is general emotion recognition process which includes preliminary processing (tasks like normalization, dimensionality reduction, detecting faces, and noise and data redundancy reduction). The second step is feature extraction. It encompasses the extraction of properties based on geometry, appearance, or physiological FER characteristics. Third, we can classify emotions using various classifiers to allocate suitable categories to the retrieved feature expressions, making recognizing feelings possible.

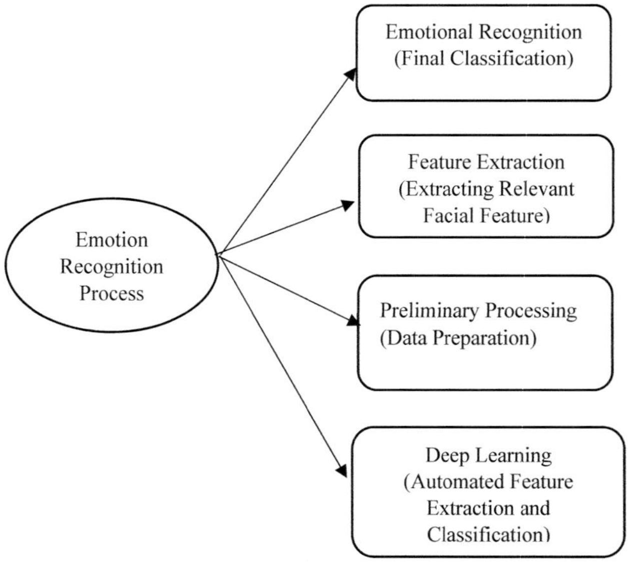

Fig. 11.4 Emotion recognition process.

Emotion recognition in conventional FER algorithms consists of characteristics of expression that are retrieved by hand and categorized by means other than non-deep neural machine learning techniques for networks. On the other hand, deep learning-based FER immediately does feature extraction and classification automatically, doing away with the requirement for human feature extraction. Feature extraction and classification are tasks carried out by DNNs.

Table 11.3 Artificial intelligence systems that use recognition of images

Classification of Detection Devices	*Identified Characteristics*	*Objective of Recognition*	*Classification Algorithm*	*Correctness*
System Camera	Photographs of pupil faces	Classifying emotions	CNN	96%
Security camera for the classroom	Photographs of pupil faces	The ability to assess student progress in many ways	CNN	72%
Security camera for the classroom	Photographs of pupil faces	Participation tracking system	MTCNN	99%
Security camera for the classroom	Photographs of pupil faces	Pupil proof of identity, instructional sign-in, recognition of feelings, and facial identification of features	KNN	97%
Security camera for the classroom	Photographs of pupil faces	Recognition of faces in hazy environments	YOLOv5	81%

CNN–Convolutional neural networks; MTCNN – Multi-Task, You Only Look Once version 5 Cascaded, KNN – K-Nearest Neighbors Convolutional Neural Network

Table 11.3 shows the uses of AI in recognition of images. In some instances, both approaches can achieve high recognition efficiencies. Currently, one common approach to emotion classification in traditional machine learning is support vector machines (SVMs). When it comes to accurately analyzing facial movements and recognizing emotions among people, FER systems often employ neural network models. Additionally, convolutional neural networks (CNNs) rank among the most often employed deep learning classifiers.

Some researchers, like Lek and Teo (2023), have noted that most literature reviews use CNNs for feature extraction. The software classified students' emotions based on static frontal facial pictures. Just before applying face observation techniques, these procedures are taken: After the initial step of converting the photos to black and white and stretching their brightness, the following steps are taken: noise reduction by filtering, detection of regions of interest, and normalisation of the images to make sure they are all the same size and orientation. For human facial monitoring algorithms to extract features and perform recognition, the photos must undergo certain pre-processing procedures. Findings from the research suggest implementing an innovative classroom learning status management system (Lek & Teo, 2023). Various sensor technologies are utilized by it, such as cameras, thermometers, heart rate monitors, and facial recognition software, to identify and gather the following:

features are extracted using the Grey-Level Co-Occurrence Matrix (GLCM), images are segmented using the skin-colour technique, and the localization of the eyes and mouth is done using the face model technique. After the preceding feature extraction and image processing processes, the SVM regression classifier is used for emotion categorization.

This data is further analysed using a Bayesian clustering network to deduce the learning state of students. In addition, the system has a feedback mechanism that not only notifies students who are deemed unfocused in class but also provides instructors with the results of real-time analysis of their learning status.

3.2 Analysis of Data and Creation of Content

We discussed the enormous promise of AI in the preceding part, wherein sensors may be assisted in comprehending data about the classroom setting. But it is also quite intriguing to see how AI may improve data analysis and the production of instructional content. Instead of immediately processing the data acquired by the sensors, data analysis and content development methods equip the sensors with more robust capabilities to adapt to diverse learning situations. Personalized learning, efficient teaching content support, intelligent homework correction, and intelligent feedback are all possible thanks to analytics and content generation algorithms for students (Zawacki-Richter et al., 2019). Educators, meanwhile, can benefit from these algorithms for educational decision-making, student performance prediction, automatic assessment of teaching quality, and management of stress reduction.

4. Feedback from Students and Evaluations of Classroom Instruction

Historically, assessments, surveys, or grades were the main tools used to assess the effectiveness of a teacher's lessons. Due to the subjective nature of the procedure, the system for evaluating performance may be inaccurate (Lin, 2021). AI technology can improve sensor data processing, leading to a more thorough and unbiased evaluation of instructional methods. Furthermore, assessment processes can be mechanized using AI technology. A hybrid Gaussian process (Luckin & Holmes, 2016) and an enhanced correlation vector machine model were suggested by Wenming (2021) as an active learning algorithm. Through careful sample selection and labelling, this method enhanced the reliability and validity of English Language Teaching (ELT) evaluation. Educators at all levels have the critical obligation of evaluating the learning outcomes of their students. In terms of written assignments, it is among the most difficult, tedious, and time-consuming jobs (Del Gobbo et al., 2023). A method for autonomously evaluating software applications using algorithms for learning was initially suggested in Srikant and Aggarwal (2014).

There are typically two types of questions on student assignments: closed and open. Due to the fixed number of possible correct responses in closed-question correction, grading generally is as simple as comparing student responses to reference answers; furthermore, practical methods for correcting such problems already exist. However, there are frequently no standard responses for open-ended, subjective, short-answer questions; these questions demand a lot of focus from teachers and are susceptible to grading subjectivity. The development of deep learning algorithms is anticipated to stimulate study in this area, and automated machine grading will significantly alleviate this issue. A computerized approach for scoring semi-open-ended brief answers was created by Zhang et al. (2022).

4.1 Classroom Algorithms for AI-Integrated Sensors

Here, we look at how AI and sensor software algorithms work together. It is seen via the lenses of algorithms for recognition, analysis, and prediction, as well as for content generation, highlighting the breadth of its potential uses. AI technology has made data processing and analysis far more efficient and accurate. For instance, recognition algorithms have significantly benefited from advancements in DNN technology, which offers tremendous room for improvement. CNNs, a type of deep learning algorithm, excel at handling massive volumes of video and picture data.

Advancements in AI widened the range of sensors. Aside from enhancing identification efficiency, machine learning technology may also analyse data on students' learning behaviours to forecast their academic performance or suggest individualized learning routes by learning the underlying patterns within the data. With the use of AI, sensors can double as "teaching supervisors" and "grading assistants", collecting data on students' progress in class, assessing assignments thereafter (auto-grading), and providing instructors with objective feedback on their pedagogical performance. NLP technologies performed exceptionally well in voice recognition and sentiment analysis.

5. Conclusion

Smart classrooms combining sensor technology with AI are reshaping traditional teaching practices, improving the "perception" capability and bolstering intelligent, interactive, and personalized instruction. This chapter provides an overview of how smart classrooms utilize sensor technologies and how these technologies are deeply integrated with AI algorithms. It also summarizes the key trends and issues related to these applications. A smart classroom's sensor system can multitask. Its goal is to provide the best possible learning environment by constantly tracking the classroom's physical parameters. As a teaching supervisor, the sensor system can record students' and teachers' nonverbal cues, such as body language, vocalizations, and even brain waves, to objectively assess their performance.

Moreover, this study uncovered several tendencies in the smart classroom's use of sensors. There has been a meteoric rise in using personal smart mobile devices and wearable sensor devices in smart classrooms. Significant progress has been made in sensor device miniaturization, integration, and wearability. To further accommodate the "virtualization" of classrooms, new sensing devices for distant virtual learning environments will develop alongside virtual reality and remote teaching technologies. These tendencies cannot be considered independent of the concomitant advancements in AI. Deep learning and massive language model technologies have recently emerged, and they hold tremendous promise for enhancing the software algorithmic application of sensors. Data privacy and security, implementation expenses, availability, and novel application contexts are significant obstacles to adopting innovative classroom technology. Addressing these concerns and creating algorithms and technologies that are safer, more efficient, and easier to use should be the focus of future research.

References

Adal, H., Promy, N., Srabanti, S. and Rahman, M. (2018). Android based advanced attendance vigilance system using wireless network with fusion of bio-metric fingerprint authentication. In *Proceedings of the 20th International Conference on Advanced Communication Technology (ICACT)* (pp. 217–222). Institute of Electrical and Electronics Engineers, Chuncheon, Republic of Korea.

Alemdag, E. and Cagiltay, K. (2018). A systematic review of eye tracking research on multimedia learning. *Computers & Education*, *125*, 413–428.

Alkabbany, I., Ali, A. M., Foreman, C., Tretter, T., Hindy, N. and Farag, A. (2023). An experimental platform for real-time students engagement measurements from video in STEM classrooms. *Sensors*, *23*(3), 1614.

Alkhayat, A.H., Bagheri, N., Ayub, M.N. and Noor, N.F.M. (2015). Fever detection & classroom temperature adjustment: Using infrared cameras. In *2015 IEEE International Conference on Consumer Electronics-Taiwan* (pp. 240–241). IEEE.

Amelkina, S.A. and Duplenkova, K.A. (2021). Justification lighting control system using in the classes on the basis of lighting scene simulation. *Light & Engineering*, *29*(3), 26–36.

Amri, U.F., Hashim, N.N.W.N. and Hanif, N.H.H.M. (2017). Speech-based class attendance. In *IOP Conference Series: Materials Science and Engineering*, Vol. 260, No. 1, p. 012008. IOP Publishing.

Banepali, A., Kadel, R., Guruge, D.B. and Halder, S.J. (2019). Design and implementation of wi-fi based attendance system using Raspberry Pi. In *2019 29th International Telecommunication Networks and Applications Conference (ITNAC)* (pp. 1–6). IEEE.

Carroll, M., Lindsey, S., Chaparro, M. and Winslow, B. (2021). An applied model of learner engagement and strategies for increasing learner engagement in the modern educational environment. *Interactive Learning Environments*, *29*(5), 757–771.

Chen, J.B., Xue, R., Liu, P.X. and Zhang, X. (2017). Design of sub-regional lighting control system for classroom. In *2017 7th International Conference on Applied Science, Engineering and Technology (ICASET 2017)* (pp. 238–242). Atlantis Press.

Chiou, C.K. and Tseng, J.C. (2015). An intelligent classroom management system based on wireless sensor networks. In *2015 8th International Conference on Ubi-Media Computing (UMEDIA)* (pp. 44–48). IEEE.

Chiu, C.K. and Tseng, J.C. (2021). A Bayesian classification network-based learning status management system in an intelligent classroom. *Educational Technology & Society*, *24*(3), 256–267.

Cruz, I.Z., Bernabe, J.G., Benítez, R.A.G. and Toxqui, J.M. (2021). Design of a prototype for automatic lighting control in classrooms at Faculty of Accounting and Administration Xalapa based on the IoT paradigm. In *2021 IEEE International Conference on Engineering Veracruz (ICEV)* (pp. 1–9). IEEE.

de Valencia, A.C., Caballero, L., Cilli, Á., González, J., Rojas, I. and Torres, B. (2023). Development of a noise monitoring and control sensor network system for the enclosed spaces within a university environment. In *2023 IEEE Latin-American Conference on Communications (LATINCOM)* (pp. 1–5). IEEE.

Deepaisarn, S., Angkoonsawaengsuk, A., Arunkit, C., Srisumarnk, C., Nimmanwatthana, K., Linphrachaya, N. and Sornlertlamvanich, V. (2022). Camera-based log system for human physical distance tracking in classroom. In *2022 Asia-Pacific Signal and Information Processing Association Annual Summit and Conference (APSIPA ASC)* (pp. 1977–1983). IEEE.

Del Gobbo, E., Guarino, A., Cafarelli, B., Grilli, L. and Limone, P. (2023). Automatic evaluation of open-ended questions for online learning: A systematic mapping. *Studies in Educational Evaluation*, *77*, 101258.

Dias, B., Mohammad, A., Xu, H. and Tan, P. (2020). Intelligent student attendance management system based on RFID technology. In *Complex, Intelligent, and Software Intensive Systems: Proceedings of the 13th International Conference on Complex, Intelligent, and Software Intensive Systems (CISIS-2019)* (pp. 578–586). Springer International Publishing.

Du, X., Zhao, S., Zhang, D. and Yu, Y. (2023). Comparative analysis of light environment perception, eye movement and physiology in university professional classroom based on virtual reality experiment. *Indoor and Built Environment*, *32*(6), 1152–1169.

e Silva, L.C., Sobrinho, Á.A.D.C.C., Cordeiro, T.D., Melo, R.F., Bittencourt, I.I., Marques, L.B. and Isotani, S. (2023). Applications of convolutional neural networks in education: A systematic literature review. *Expert Systems with Applications*, *231*, 120621.

Fernández, M.J.L., Fernández, J.G., Aguilar, S.R., Selvi, B.S. and Crespo, R.G. (2013). Control of attendance applied in higher education through mobile NFC technologies. *Expert Systems with Applications*, *40*(11), 4478–4489.

Gagandeep, Arora, J. and Kumar, R. (2019). Biometric fingerprint attendance system: An internet of things application. In *Innovations in Computer Science and Engineering: Proceedings of the Fifth ICICSE 2017* (pp. 523–530). Springer Singapore.

Gunawardena, N., Ginige, J.A. and Javadi, B. (2022). Eye-tracking technologies in mobile devices using edge computing: A systematic review. *ACM Computing Surveys*, *55*(8), 1–33.

Hoo, S.C. and Ibrahim, H. (2019). Biometric-based attendance tracking system for education sectors: A literature survey on hardware requirements. *Journal of Sensors*, *2019*(1), 7410478.

Hornof, A.J. and Cavender, A. (2005). EyeDraw: Enabling children with severe motor impairments to draw with their eyes. In *Proceedings of the SIGCHI Conference on Human Factors in Computing Systems* (pp. 161–170). Association for Computing Machinery.

Hu, R., Hui, Z., Li, Y. and Guan, J. (2023). Research on learning concentration recognition with multi-modal features in virtual reality environments. *Sustainability*, 15(15), 11606.

Hu, Y. and Huang, R. (2015). Development of weather monitoring system based on Raspberry Pi for technology rich classroom. In *Emerging Issues in Smart Learning* (pp. 123–129). Springer Berlin Heidelberg.

Jarodzka, H., Skuballa, I. and Gruber, H. (2021). Eye-tracking in educational practice: Investigating visual perception underlying teaching and learning in the classroom. *Educational Psychology Review*, *33*(1), 1–10.

Kim, P.W. (2019). Ambient intelligence in a smart classroom for assessing students' engagement levels. *Journal of Ambient Intelligence and Humanized Computing*, *10*(10), 3847–3852.

Lathière, B. and Archambault, D. (2014). Improving accessibility of lectures for deaf and hard-of-hearing students using a speech recognition system and a real-time collaborative editor. In *Computers Helping People with Special Needs: 14th International Conference*, ICCHP 2014, Paris, France, July 9–11, Proceedings, Part II 14 (pp. 490–497). Springer International Publishing.

Lek, J.X.Y. and Teo, J. (2023). Academic emotion classification using FER: A systematic review. *Human Behavior and Emerging Technologies*, *2023*(1), 9790005.

Lin, L. (2021). Smart teaching evaluation model using weighted Naive Bayes algorithm. *Journal of Intelligent & Fuzzy Systems*, *40*(2), 2791–2801.

Luckin, R., Holmes, W., Griffiths, M. and Forcier, L.B. (2016). Intelligence Unleashed: An Argument for AI in Education. London: Pearson Education. ISBN: 978-0992424886

Margreat, L., Paul, J.J. and Mary, T.B. (2021). Chatbot-attendance and location guidance system (ALGs). In *2021 3rd International Conference on Signal Processing and Communication (ICPSC)* (pp. 718–722). IEEE.

Marques, G. and Pitarma, R. (2020). A real-time noise monitoring system based on internet of things for enhanced acoustic comfort and occupational health. *IEEE Access*, *8*, 139741–139755.

Mendell, M.J. and Heath, G.A. (2005). Do indoor pollutants and thermal conditions in schools influence student performance? A critical review of the literature. *Indoor Air*, *15*(1), 27–52.

Ni, L., Shi, J., Han, B., Zhang, N., Lan, Q. and Su, Z. (2022). Classroom roll call system based on face detection technology. In *2022 10th International Conference on Information and Education Technology (ICIET)* (pp. 42–46). IEEE.

Pastor, F.J.F., Horcajadas, M.P., Davo, J.A.L. and Iglesias, V.G. (2022). Developing environmental adaptative comfort using internet of things and business process management: Application in a university building. In *International Conference on Ubiquitous Computing and Ambient Intelligence* (pp. 535–546). Springer International Publishing.

Peifer, C. and Engeser, S. (2021). *Advances in Flow Research*. Cham, Switzerland: Springer.

Prieto, L.P., Sharma, K., Kidzinski, Ł., Rodríguez-Triana, M.J. and Dillenbourg, P. (2018). Multimodal teaching analytics: Automated extraction of orchestration graphs from wearable sensor data. *Journal of Computer Assisted Learning*, *34*(2), 193–203.

Rahmatya, M.D. and Wicaksono, M.F. (2019). Design of student attendance information system with fingerprints. In *IOP Conference Series: Materials Science and Engineering*, *662*(2), p. 022039. IOP Publishing.

Saini, M.K. and Goel, N. (2019). How smart are smart classrooms? A review of smart classroom technologies. *ACM Computing Surveys (CSUR)*, *52*(6), 1–28.

Shvarts, A. and Abrahamson, D. (2019). Dual-eye-tracking Vygotsky: A microgenetic account of a teaching/learning collaboration in an embodied-interaction technological tutorial for mathematics. *Learning, Culture and Social Interaction*, *22*, 100316.

Srikant, S. and Aggarwal, V. (2014). A system to grade computer programming skills using machine learning. In *Proceedings of the 20th ACM SIGKDD International Conference on Knowledge Discovery and Data Mining* (pp. 1887–1896).

Stazi, F., Naspi, F., Ulpiani, G. and Di Perna, C. (2017). Indoor air quality and thermal comfort optimization in classrooms developing an automatic system for windows opening and closing. *Energy and Buildings*, *139*, 732–746.

Sweetlin, J.D., Aswini, V. and Dhanusha, R. (2016). Speech based attendance application register. In *2016 International Conference on Recent Trends in Information Technology (ICRTIT)* (pp. 1–5). IEEE.

Tarnowski, P., Kołodziej, M., Majkowski, A. and Rak, R. J. (2017). Emotion recognition using facial expressions. *Procedia Computer Science*, *108*, 1175–1184.

Twumasi, C., Dotche, K.A., Banuenumah, W. and Sekyere, F. (2017). Energy saving system using a PIR sensor for classroom monitoring. In *2017 IEEE PES PowerAfrica* (pp. 347–351). IEEE.

Varshini, M.L. and Vidhyapathi, C.M. (2016). Dynamic fingure gesture recognition using KINECT. In *2016 International Conference on Advanced Communication Control and Computing Technologies (ICACCCT)* (pp. 212–216). IEEE.

Verner, C. and Dickinson, G. (1967). The lecture, an analysis and review of research. *Adult Education*, *17*(2), 85–100.

Wang, Y., Lu, S. and Harter, D. (2021). Multi-sensor eye-tracking systems and tools for capturing student attention and understanding engagement in learning: A review. *IEEE Sensors Journal*, *21*(20), 22402–22413.

Wenming, H. (2021). Simulation of English teaching quality evaluation model based on Gaussian process machine learning. *Journal of Intelligent & Fuzzy Systems*, *40*(2), 2373–2383.

Zaletelj, J. (2017). Estimation of students' attention in the classroom from kinect features. In *Proceedings of the 10th International Symposium on Image and Signal Processing and Analysis* (pp. 220–224). IEEE.

Zawacki-Richter, O., Marín, V.I., Bond, M. and Gouverneur, F. (2019). Systematic review of research on artificial intelligence applications in higher education–Where are the educators? *International Journal of Educational Technology in Higher Education*, *16*(1), 1–27.

Zhang, L.S., Huang, Y.W., Yang, X., Yu, S.Q. and Zhuang, F.Z. (2022). An automatic short-answer grading model for semi-open-ended questions. *Interactive Learning Environments*, *30*, 177–190.

Zhang, X.Y., Kulkarni, H. and Morris, M.R. (2017). Smartphone-based gaze gesture communication for people with motor disabilities. In *Proceedings of the ACM SIGCHI Conference on Human Factors in Computing Systems (CHI)* (pp. 2878–2889). Association for Computing Machinery, Denver, USA.

Zhu, Y., Wang, S., Peng, Y., Chen, Y., Chen, R., Ke, G. and Li, J. (2023). Research and implementation of intelligent learning desk based on visio sensor in AI IoT environments for smart education. *Sensors & Materials*, 35.

Zorić, B., Dudjak, M., Bajer, D. and Martinović, G. (2019). Design and development of a smart attendance management system with Bluetooth low energy beacons. In *2019 Zooming Innovation in Consumer Technologies Conference (ZINC)* (pp. 86–91). IEEE.

Chapter 12

Roadmap of Artificial Intelligence in Education and Teaching for 5G-enabled Ecosystems: Challenges, Opportunities, and Solutions

R. Felista Sugirtha Lizy[1*] and I. Gusti Bagus Yosia Wiryakusuma[2]

1. Introduction

Digital innovations have revolutionized the way we learn and teach. Among all upcoming technologies, artificial intelligence (AI) sits at the centre of reshaping how teaching and learning are offered, managed, and experienced in education. The rollout of 5G can facilitate real-time intelligent educational services due to ultra-low latency, enhanced mobile broadband, and massive machine-type communications. Figure 12.1 illustrates AI and 5G-enabled smart classrooms.

Fig. 12.1 AI and 5G-enabled smart classroom.

1 Department of Computer Science, A.P.C. Mahalaxmi College for Women, Tamil Nadu, India. https://orcid.org/0000-0003-3648-0205

2 Department of International Business Management, Universitas Ciputra, Indonesia. https://orcid.org/0000-0001-5845-8661

* Corresponding author: 21felistaa@gmail.com

DOI: 10.1201/9781003639367-12

The purpose of this chapter is to offer a complete guide for each of the three dimensions of the use of AI in teaching and learning in 5G-enabled systems: the challenges, the opportunities, and the solutions.

2. Literature Review

Artificial intelligence and 5G are emerging as key focus areas in education.

2.1 *Smart Education: The Role of AI*

As noted by Luckin et al. (2016), AI can enhance learning experiences through adaptive feedback, personalized content recommendations, and real-time tutoring simulations. Intelligent tutoring systems (ITS) (VanLehn, 2011) are claimed to be as effective as student-effectiveness for some learning tasks when students have access to support from, say, natural language processing and learner modelling.

Holmes et al. (2019) categorized the use of AI in education into four broad categories: learner support (e.g., chatbots), formative and summative assessment, administrative automation, and content generation. The authors went on to highlight biases, ethics, and equitable access as potential problems in the use of AI in education.

2.2 *5G in Smart Learning Environments*

The 5G technology, which can provide ultra-low latency, high reliability, and massive levels of connectivity, will provide new ways to learn, inclusive of immersive augmented reality (AR), virtual reality (VR), real-time analytics, and the Internet of Things (IoT) on a large scale. According to both ITU (2020) and Zhou et al. (2021), 5G networks enable data-rich contexts for applications around high-resolution video streaming, synchronous online work, and real-time adaptive learning systems, irrespective of whether those ecosystems are in a remote or mobile situation.

Gupta et al. (2020) examined the learning benefits of 5G-connected classrooms and discovered greater engagement with simulations and remote labs.

2.3 *Convergence of AI and 5G in Education*

Artificial intelligence combined with 5G reflects a significant change in the manner that education systems can conducted. Tang et al. (2020) state that AI tools can operate inside of the infrastructure to provide real-time personalization, proactive intervention, and a seamless delivery of content in smart campuses which can unlock a new AI-enhanced learning ecosystem. Mady et al. (2022) suggest, similarly, that AI + 5G ecosystems create an "education-as-a-service" opportunity for layered learning that is customized, mobile, and available any time.

A report from the United Nations Educational, Scientific and Cultural Organization (UNESCO, 2023) emphasizes that pairing AI and 5G creates new learning opportunities but it also lays ground for consideration about how we can combine these technology innovations in ways that support equity and inclusion. They identify that necessary digital policies should be in place for ethical governance, especially as they scale in unprepared low-resource realities.

2.4 Challenges and Concerns

While these developments show potential, several studies highlight a number of challenges, significant challenges, that still face the integration of AI and 5G into education. Selwyn (2019) argues there is "datafication of education", suggesting that students may be offered automated systems and similar dependencies on data, and automated systems represent a move away from what many see as the humanistic aspects of education. Williamson and Hogan (2020) also outlined risks involved with surveillance, data privacy, and commercialization of education that are also a part of the landscape of new and emerging AI education–enabled platforms for learning.

Infrastructure challenges, particularly in developing nation contexts, are also significant limitations to equitable options for a rollout of 5G and AI systems for education (Pedro et al., 2019). Further results from Chen et al. (2020) also showed that a lack of teacher preparedness and inadequate institutional support hinder the adoption of AI in schools and classrooms.

2.5 Gaps in Current Research

Despite a significant amount of work focused on AI in education and growing literature focused on 5G-enabled learning, large-scale studies, across multiple sites, that examine the convergence potential of both AI and 5G-enabled pedagogy are almost non-existent. Existing studies focus primarily on the technological feasibility aspect and isolated case studies. More integrative frameworks and roadmaps like the research we have presented in this chapter will be necessary to address system and technology integration, but to also consider and address ethical-pedagogical implications.

3. Roadmap for AI Adoption in 5G-Enabled Education Ecosystems

The roadmap for AI adoption in 5G-enabled education ecosystems provides a model framework for the integration of AI technologies in next-generation digital learning environments connected through 5G, as large-scale adoption of technologies by education systems are underway. The roadmap includes far-reaching objectives about digitally transforming educational practices by anchoring pedagogies into intelligent automation, sophisticated data analytics, and personalized learning, all made more impactful thanks to the capabilities of 5G networks. It describes progressive phases that include milestones for action based on the advancement of educational adoption as defined by elements of digital transformation. It delineates in foundational stages the adoption of AI technology within the education system, such as the upgrade of digital infrastructure, building digital literacy for educators, and ethical governance of data; not to mention equity. It describes mid-stage developments that allow for the use of AI technologies in tools such as intelligent tutoring systems, learning analytics dashboards, and virtual labs that must all be connected to take advantage of the speed of 5G to ensure responsiveness in real time. Advanced goals begin to incorporate intelligent environments, situated learning, and AI fusion with AR/VR, and strong integrated networks of edge computing utility and cross-device interoperability. However, the model has meaningful parts for the AI-related collaboration across sectors, addressing the access policies of inclusion, and continued monitoring to understand the implications of AI on pedagogical practice, equity, and privacy. Clearly this roadmap guide is catered to education systems in an attempt to futureproof "teaching and learning" in complex intelligent digital ecosystems that are hyperconnected with other socio-technical sectors.

- The initial phase, *Foundational Integration*, is by far the most important phase in the 5G-enabled education ecosystem AI-enabled model. During this phase, the focus is on establishing the necessary foundation to harness advanced technologies.
- Infrastructure readiness focuses on converting traditional classrooms into smart classrooms, leveraging interactive digital boards, sensors, connected devices, and high-speed internet capabilities needed to support institutional technologies. Schools and universities rolling out 5G networks will also gain access to ultra-low latency and high-bandwidth connectivity, required for AI-powered applications such as real-time analytics, immersive AR/VR applications in education, and adaptive learning systems.
- Policy and governance models for AI use are important to ensure AI implementation in education is ethical, equitable, transparent, and compliant with local and global data protection laws. Institutions need to create actionable plans that clearly define the parameters of how AI tools are selected, how data will be collected and used, and how student privacy and digital rights will be protected. Governance structures should also clearly outline accountability for decisions made by AI, and how and when these implementation decisions will be reviewed.
- Pedagogical and support staff training around AI tools is vital in the effectiveness of this stage. We need teachers to be ready and equipped to use AI-based platforms in terms of personalized learning, authentic assessment, and student engagement. Professional development for educators needs to include teaching with AI-based tools, knowledge about AI and its impact on education, boundaries and limitations of the tools, and the position of AI when arranging aligned units of study.
- In short, Phase 1 establishes the technical, governance, and human foundation for the potential and sustainability of further academic integration of AI and 5G-related technologies in education.
- *Phase 2: Intermediate Implementation* signifies the step where educational institutions begin using AI technologies in earnest, within the 5G-enabled environment built in Phase 1. The implementation of AI technologies moves beyond planning and preparation and into the practical implementation and testing of intelligent systems in a way that directly or indirectly influences teaching and learning.
- The AI-based learning management system (LMS) is one of the key milestones of Phase 2. Next-generation LMS platforms support AI components that amplify functionality beyond delivery of content. For example, AI features could leverage automated grading, wisely recommended content, or predictive analytics to help identify students who may be at risk of falling behind. The baseline 5G is effectively geared to performance, meeting the demands of resource-laden data or multimedia and at scale practices from devices.

- Adaptive learning is another important implementation within Phase 2 that offers tailored learning experiences based on individual student needs. AI-based learning algorithms consider all of the learners' pace, behaviour, and preferences while continuing to evolve in adaptive ways (Figure 12.2). With the supporting environment and performance of 5G, the adaptive platforms within a learning context merge meaningfully into immediate feedback interfaces or interaction across multiple devices in seamless, higher-engagement learning experiences, supporting positive outcomes for students.
- In order to assess the actual implications of AI tools in practice, it is important to initiate pilot programmes in blended and remote learning. Schools and universities are beginning controlled trial programmes where AI-based systems serve the needs of in-person and remote learners. These pilot programmes provide an opportunity to assess the effectiveness of the tools, gather user perspectives, and adjust before scaled implementation.

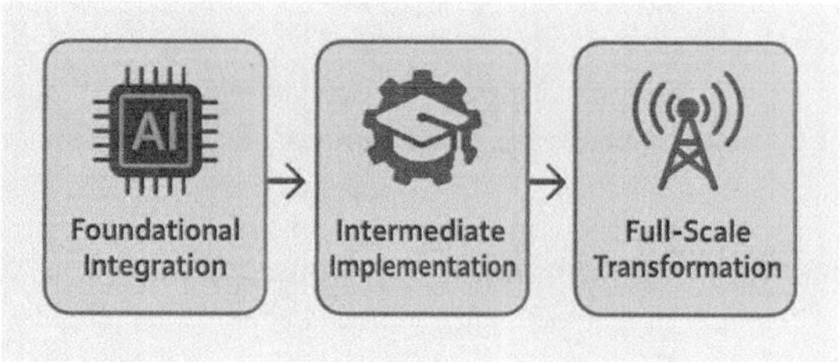

Fig. 12.2 The three phases: Foundational integration, intermediate implementation, and full-scale transformation.

- In summary, Phase 2 will entail the embedding of AI into ongoing, everyday education practices to take advantage of 5G capabilities and contexts to create responsive, scalable, and personalized learning opportunities. Phase 2 serves as a space for experimentation and refinement of tools and strategies which will lay the groundwork for transitions into Phase 3.
- *Phase 3: Full-Scale Transformation* is the final and most complex stage of AI exemplified for 5G-enabled education ecosystems. Phase 3 demonstrates the seamless implementation of AI in every layer of a fully intelligent, inclusive, and an immersive educational ecosystem that integrates AI throughout the teaching and learning process.
- Partial innovations using AI include developing curriculum co-design and dynamic course delivery that allow teachers and AI systems to jointly create responsive learning paths. AI is able to draw upon student performance data, real-time interests, and career goals to suggest or create curriculum content that can be tailored for individual learners rather than static, pre-packaged online learning paths or syllabi. Rather than treating course content as static, one-size-fits all curriculum guidelines, AI provides a dynamic and evolving course material with curriculum components that support some level of oversight, while relying on the heavy lifting from AI in terms of curating content and designing instructional approaches.

- Using AR/VR + AI for immersive learning experiences, along with 5G, allows for augmented and virtual reality to be integrated into classrooms with ultra-low latency and ultra-high speed materials. While AR and VR put students in immersive learning experiences, AI enhances those environments by altering simulations in real time along the way based on learner responses and progress updates. For example, students can tour a human body virtually in biology or past events in history through an AR experience, while AI is monitoring for levels of comprehension and adjusting the experiences so the education continues on an appropriate trajectory that supports engagement and comprehension. Figure 12.3 is a schematic diagram of the infrastructure, adaptive instructional platforms, and immersive AR/VR environments.

Fig. 12.3 Infrastructure, adaptive learning platforms, and immersive AR/VR learning.

- Real time multilingual support and emotion-aware systems make education more inclusive and emotionally intelligent. AI-enabled language translations provide learners with instant multilingual content delivery and aid for students who may have experience learning English as a second language or facing other linguistic challenges. Emotion-aware systems use computer vision and sentiment analysis to detect stress, confusion, or disinterest amongst students in real time so the system or educator can recognize the need for intervention and establish the appropriate response to support student learning—whether that be through support, encouraging the student, or adjusting the content being presented for further engagement and learning (see Table 12.1).

Table 12.1 Phases and its outcomes

Phase	*Key Focus Areas*	*Examples/Tools*	*Outcomes*
Foundational Integration	Infrastructure readiness, policy and governance, teacher training	Smart classrooms, 5G deployment, AI ethics policies	Prepared digital ecosystem, trained staff
Intermediate Implementation	AI-powered LMS, adaptive platforms, pilot blended learning	Adaptive learning systems, intelligent tutoring	Personalized learning, pilot validations
Full-scale Transformation	AI-driven curriculum co-creation, immersive AR/VR, emotion-aware systems	AR/VR labs, emotion analytics, real-time multilingual support	Fully adaptive, immersive education

In conclusion, Phase 3 is the transition to a smart education ecosystem where learning is personalized and data-informed, emotionally adaptive, highly interactive, and accessible for all. This phase is the full realization of AI and 5G integration in education to focus education on the learner and allow education to be a lifelong endeavour. Figure 12.4 illustrates the process by which AI algorithms customize learning content for individual students.

Fig. 12.4 AI algorithms customizing learning content for individual students.

4. Opportunities and Benefits

The combination of AI and 5G in education presents numerous opportunities for transformational change, enhancing learning outcomes, improving accessibility, and driving institutional efficiencies. Table 12.2 presents the opportunities for and benefits of 5G in education.

Table 12.2 Opportunities and benefits

Opportunities	*Description*
Enhanced personalization	AI tailors materials and feedback based on learner data
Inclusive education	Accessibility via speech-to-text, translation, emotion AI
Smart learning environments	Real-time engagement tracking with AI and 5G
Data-driven decision-making	Predictive analytics for retention, enrolment, curriculum

4.1 Improved Personalization

The AI algorithms analyse how students engage with course content and assessments as well as their learning speed and academic goals to present course material, assessments, and feedback that are more tailored to each learner. This ensures that each learner is receiving the support they require based on their needs, therefore assisting with their motivation to engage and ultimately with their performance.

4.2 More Inclusive Education

The AI tools such as speech-to-text, real-time translation, and emotional detection help to provide more accessibility for students with disabilities, students speaking a language other than English, and students with special needs. This provides the framework for all learners to engage in learning and education in an inclusive way.

Figure 12.5 illustrates the AI-powered accessibility features—speech-to-text, real-time translation, and emotion detection—helping diverse learners.

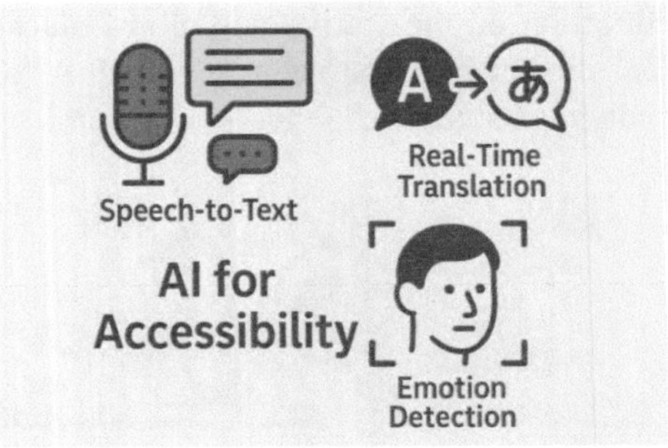

Fig. 12.5 AI-powered accessibility features.

4.3 Intelligent Learning Environments

Integrating AI with 5G provides the ability to assess student engagement, attentiveness, and performance using real-time data, and allows for the opportunity to skilfully adapt course content, style of delivery, or pace of learning to fit the needs of a learner, allowing for more engaging and effective learning.

4.4 Data-informed Decision-making

Education systems can leverage the analytics powered by AI, and when educated on diversified learning analytics, can begin to make informed decisions on student admissions, risk of student attrition, course offerings, and how to allocate resources across institutions. Predictive scenarios offer administrators and educators opportunities to address challenges proactively, ultimately leading to sustained academic success. Figure 12.6 illustrates the use of predictive analytics for enrolment, retention, and curriculum improvement with AI dashboards.

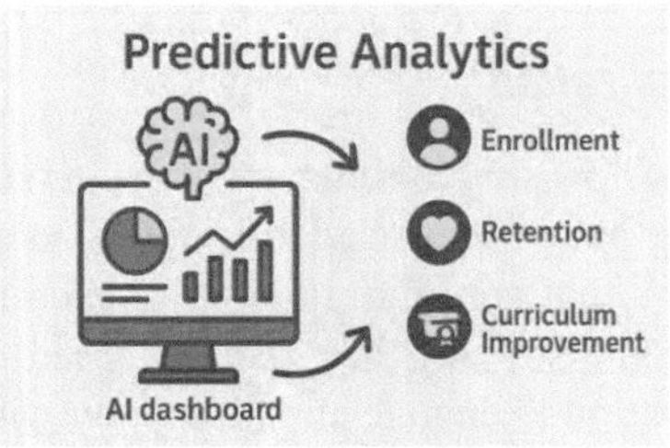

Fig. 12.6 Use of predictive analytics.

5. Key Challenges

Artificial intelligence and 5G offer exciting possibilities for education, but many challenges must be identified and addressed to develop equitable and meaningful applications and adoption.

Fig. 12.7 Gaps, privacy concerns, skill gaps among teachers, and lack of standardization.

Figure 12.7 illustrates a collaborative discussion, which may reflect broader challenges such as gaps, privacy concerns, and skill development.

5.1 Infrastructure and Digital Divide

Many rural areas and economically disadvantaged populations do not have access to 5G networks or current digital infrastructure or services. The digital divide impacts the potential penetration of AI-driven educational tools into rural populations, further exacerbating educational inequity.

5.2 Ethical and Privacy Issues

Artificial intelligence uses vast amounts of personal data to train and produce predictions. A focus on using this information raises significant issues about privacy, data security, bias in algorithms, informed consent, and other concerns. With no universally applied policies, it opens up some misuse and behaviour patterns that could lead to surveillance.

5.3 Professional Development and Skills Gap among Educators

Many educators may not have access to the skills or confidence to use AI tools effectively. Inadequate training, perceived barriers, and/or resistance to change may stall successful technology adoption in the classroom.

5.4 Standardization and Interoperability

A lack of common standards for AI tools within education system leads to siloed systems that may not work well altogether. This begins to raise challenges to integration between the approaches and to sharing data and ultimately scale across entities and platforms over time (see Table 12.3).

Table 12.3 Challenges

Challenges	*Description*
Infrastructure and digital divide	Limited 5G and digital access in underserved regions
Ethical and privacy concerns	Data collection risks, bias, surveillance
Skill gaps among educators	Lack of AI training, resistance to adoption
Standardization and interoperability	Fragmented AI platforms, lack of unified standards

6. Strategic Solutions and Recommendations

To address risks and fully reap the rewards of AI in 5G-enabled forms of education, we need to implement specific strategies and collaborate with others.

6.1 Policy and Regulation

Governments and education authorities need to develop regulations that will define and regulate how AI is utilized. Regulations to ensure ethical data use, data privacy, and equitable access to all learners are needed. Regulations ensure the setting up of a trust that enables responsible AI to be used in education.

6.2 Building on Capacity

Professional development programmes to enable continuous educator and administrator and student learning are essential to improving their AI literacy. Importantly, the professional development programmes will also need to include the use of AI tools, an assessment of their pedagogical impact, and how learning processes can be embedded in real life and curriculum connections.

6.3 Collaborative Ecosystems

Educational institutions must develop partnerships with technology companies, internet service providers, and telecommunications organizations in order to build ecosystems. Partnerships help to provide opportunities to innovate and share resources. Collaborative ecosystems are essential to the co-creation of educational AI solutions that are identified as addressing the diversity of educational needs of learners, educators, support staff, and the community.

6.4 Monitoring and Evaluating

As we introduce and integrate AI into the teaching and learning process, we must implement good monitoring and evaluation mechanisms that include measures about the responses AI will elicit in teaching and learning. The implementation of data-led feedback loops would allow for ongoing dialogue asking if AI was effective, equitable, and whether it could be improved, or if it aligns with educational purposes (Table 12.4).

Table 12.4 Monitoring and evaluating

Strategic Solutions	*Actions*
Policy and regulation	Develop ethical AI policies emphasizing inclusivity
Capacity building	Train educators and students on AI tools
Collaborative ecosystems	Foster partnerships among ed-tech, telecoms, and academia
Monitoring and evaluation	Implement systems to assess AI impact on learning

7. Case Studies and Global Models

This section provides a first look at the cutting-edge ideas informing how countries all around the world are using the convergence of AI and 5G to innovate in how education is delivered to students, providing different models and innovations around the world.

- *China's AI-enabled 5G Classrooms*: China offers many advanced smart classrooms that combine AI with 5G technology in education. Smart classrooms in China, for example, employ facial recognition technology that conduct automated attendance, mitigating the administrative burden with respect to attendance. Emotion analytics using AI to process identifying facial and emotional information enable educators to monitor facial expressions in students' rooms in real time, allowing them to assess engagement, understanding, and emotional state, and alter their instruction delivery accordingly.
- *EdTech Initiatives in Finland*: Finland is utilizing its existing nation-wide municipal 5G footprint to provide schools with AI-powered adaptive learning platforms. The delivery is designed to allow AI to provide responsive adaptive learning for schools, focused to tailor content based on student progression and past preferences to ultimately provide more engaged and student-centred learning. However, it is noteworthy that Finland's approach collects the responsibilities of multiple stakeholders and industries and looks at how they can work collaboratively, government and educational institutions alike, and provide equitable access to push forward innovative learning in new technologies to continuously innovate in education.
- *India's National Digital Education Architecture (NDEAR)*: NDEAR is a government-led initiative that seeks to establish a consolidated digital ecosystem for K-12 education, leveraging AI, cloud, and broadband connectivity (including 5G). Its aims include offering personalized learning experiences, providing teachers with real-time analytical information, and providing inclusive access to quality education for both urban and rural areas. NDEAR also aims to support teachers by providing them with AI tools at the school level, while developing a data-led education system through to the national level.

These case studies illustrate the potential for and different approaches to using AI and 5G to support the creation of more responsive, inclusive, and effective education systems around the world.

8. Future Directions

Going forward, the combination of AI, 5G, and emerging technologies such as quantum computing, blockchain, and AR will be a catalyst for changes in education. Should quantum computing become mainstream, the speed of AI's data processing to create productive AI will increase significantly with the Object Size Function (OSF). As well as providing additional opportunities for personalized learning and the ability to create and adapt learning in the moment, blockchain will allow for the secure, transparent management of academic records, credentials, and learning paths. This will build trust and ensure credentials can be transferred across institutions. Similarly,

AR applications combined with appropriate computers and the public measure of speed (5G) will enhance the possibilities of enriched and immersive learning which will abstract the boundaries between the physical world and the digital world.

The key areas for future research include explainable AI, which aims to make the decision-making processes of AI in education transparent and comprehensible by the educator and learner to facilitate trust and accountability; AI for lifelong learning, which ensures that learning experiences continue beyond formal education, which adapts to lifelong developments in careers and personal lives, with tools that can evolve as the needs of society change; and importantly, the development of frameworks for governance of AI, as ethical, privacy, and equity issues will arise as AI becomes more and more embedded in the educational ecosystem.

Combined together, these trajectories can help facilitate an education future based on inclusion, innovation, and trust in the combined use of AI, 5G, and other innovative technologies.

9. Conclusion

The AI and 5G technologies are on track to create a revolution in education, from traditional, standardized approaches towards dynamic, personalized, and responsive learning ecosystems. This roadmap illustrates an approach to leverage these innovations, while giving meaning to thoughtful planning, policy development, and inclusive approaches. There are important challenges to address proactively so that stakeholders can ensure that all learners and educators have access to the benefits of AI and 5G technologies to create an effective, equitable, and future-ready education system.

References

Chen, L., Chen, P. and Lin, Z. (2020). Artificial intelligence in education: A review. *IEEE Access*, *8*, 75264–75278.

Gupta, B., Nair, D.S. and Dandekar, S. (2020). 5G technology and education: A perspective. In *2020 IEEE International Conference on Communication Technology*, Institute of Electrical and Electronics Engineers, October 28-31, 2020, Nanning, China (pp. 123–128). IEEE.

Holmes, W., Bialik, M. and Fadel, C. (2019). *Artificial Intelligence in Education: Promises and Implications for Teaching and Learning*. Center for Curriculum Redesign.

International Telecommunication Union (ITU). (2020). *The Impact of 5G on the Future of Education*. Available at https://www.itu.int (last accessed September 9, 2025)

Luckin, R., Holmes, W., Griffiths, M. and Forcier, L.B. (2016). *Intelligence Unleashed: An Argument for AI in Education*. Pearson Education.

Mady, A., Elnagar, A. and Hassanein, A. (2022). Education-as-a-service in 5G-enabled environments: The role of AI and cloud computing. *Education and Information Technologies*, *27*(5), 6453–6474.

Pedro, F., Subosa, M., Rivas, A. and Valverde, P. (2019). *Artificial Intelligence in Education: Challenges and Opportunities for Sustainable Development*. UNESCO.

Selwyn, N. (2019). *Should Robots Replace Teachers? AI and the Future of Education*. Polity Press.

Tang, T., Wang, Y. and Liu, S. (2020). Smart education in the era of 5G and AI: Trends, opportunities, and challenges. *International Journal of Emerging Technologies in Learning (iJET)*, *15*(20), 4–15.

UNESCO. (2023). *AI and Education: Guidance for Policy-makers*. Paris: UNESCO.

VanLehn, K. (2011). The relative effectiveness of human tutoring, intelligent tutoring systems, and other tutoring systems. *Educational Psychologist*, *46*(4), 197–221.

Williamson, B. and Hogan, A. (2020). Commercialisation and the governance of education data: Proposals for policy in the age of analytics. *Learning, Media and Technology*, *45*(3), 270–282.
Zhou, Y., Li, J. and Zhang, X. (2021). Application of 5G in smart education: Opportunities and challenges. *Journal of Educational Technology Development and Exchange*, *14*(1), 1–12.

Chapter 13

AI in Education Technologies: New Development and Innovative Practices

Usman-Hamza Fatimah Enehezei[1*], *Oluwakemi Christiana Abikoye*[2], *Muyideen Abdulraheem*[3], *Joseph Bamidele Awotunde*[4], *Paul Olujide Adebayo*[5], *Idowu Dauda Oladipo*[6] and *Evelyn Damilola Temitayo*[7]

1. Introduction

One of the most important scientific and technological endeavours of the twenty-first century is artificial intelligence (AI), which is radically changing economies, societies, and industries. Fundamentally, AI is an interdisciplinary field that aims to comprehend and develop intelligent entities or beings with the capacity for independent observation, learning, reasoning, and action. AI's beginnings can be found in the middle of the twentieth century, when novel ideas started to emerge in a variety of disciplines, such as computer science, mathematics, economics, philosophy, and neurology. When John McCarthy coined the term "AI" at the Dartmouth Conference in 1956, the field was formally born.

Since AI encompasses a broad range of goals and approaches, defining the field can be difficult. In their work, Russell and Norvig (2010) categorized the definitions of AI along two main lines: whether they focus on behaviour or cognitive processes and reasoning, and whether they evaluate success in comparison to human performance or an ideal standard of rationality. This categorization leads to four primary methods:

1. *The cognitive modelling approach to thinking humanly*: This approach focuses on creating computer algorithms that simulate human mind. To understand

[1-3, 5-7] Department of Computer Science, University of Ilorin, Nigeria.

[4] Department of Computer Science, Faculty of Communication and Information Sciences, University of Ilorin, Ilorin, 240003, Nigeria

https://orcid.org/0009-0004-1115-4555[1]; https://orcid.org/0000-0002-8912-6333[2]; https://orcid.org/0000-0003-4568-7566[3]; https://orcid.org/0000-0002-1020-4432[4]; https://orcid.org/0000-0002-4088-0634[5]; https://orcid.org/0000-0002-1472-2237[6]; https://orcid.org/0009-0006-5483-2891[7]

* Corresponding author: usman-hamza.fe@unilorin.edu.ng

DOI: 10.1201/9781003639367-13

how the human mind functions and then mimic those cognitive processes in machines, disciplines like cognitive science and psychology are involved.

2. *Acting humanly (Turing test approach)*: This approach, which was first presented by Alan Turing in 1950, defines intelligence as the capacity of a computer to exhibit intelligent conduct that is either identical to or indistinguishable from that of a human. Whether a human interrogator can distinguish between a computer and a human throughout their contact is the last criterion.
3. *Thinking rationally (Laws of thought approach)*: This viewpoint emphasizes the importance of logical thinking and seeks to develop intelligent systems that can follow established laws of thought and arrive at correct conclusions through logical deduction.
4. *Behaving rationally (Rational agent approach)*: This contemporary viewpoint focuses on developing rational agents, which are systems built to seek the greatest possible outcome or, in the face of uncertainty, the best expected outcome. According to this perspective, solely logical cognitive processes take primacy as long as reasonable actions achieve the intended outcomes. This rational agent paradigm is used by the majority of contemporary AI systems, especially in real-world applications, to produce agents that are aware of their surroundings and take action to improve their effectiveness (Russell & Norvig, 2010).

The emergence of machine learning (ML), a branch of AI that allows systems to learn from data without complex programming, is one of the main drivers of AI's progress. Deep learning, a subset of ML, has further transformed the discipline and led to significant advancements in areas like speech recognition, image recognition, and natural language processing (NLP) by employing multi-layered neural networks to analyse enormous datasets. Because AI can learn, adapt, and make predictions or judgements based on complex data, it has created incredible prospects in almost every industry, from finance and healthcare to transportation and, most importantly, education. Figure 13.1 displays approaches to AI.

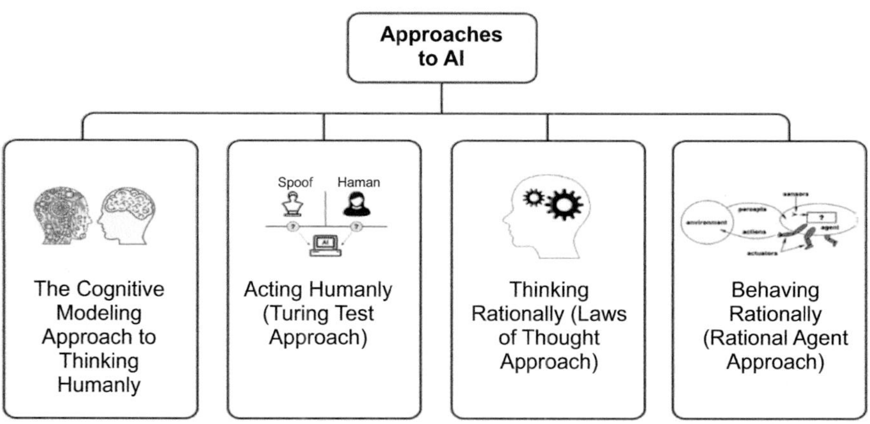

Fig. 13.1 Approaches to AI.

1.1 Role of AI in Education Technologies

A significant change is represented by the incorporation of AI into education technology (EdTech), which goes beyond conventional digital resources to offer more dynamic, personalized, and adaptable learning environments. AI's ability to process data, recognize patterns, and alter answers is proving to be crucial for resolving a number of persistent issues in education, including managing vast volumes of educational data, accommodating various learning styles, and offering individualized feedback. The following significant categories can be used to generally classify the impact of AI in EdTech:

- First, adaptive tutoring systems and personalized learning are important aspects of AI's impact. AI-enhanced platforms can assess a learner's performance, learning preferences, and cognitive speed to adjust content, difficulty levels, and teaching strategies in real time. Intelligent tutoring systems (ITS), which are made to offer individualized instruction and real-time feedback in a style akin to that of a human tutor (Holstein, McLaren, & Aleven, 2017), are a significant use of AI in the education sector. Because of these systems' capacity to recognize knowledge gaps, provide focused exercises, and modify the learning path at any time, every student will receive an educational experience that is precisely suited to their unique needs. This shift moves away from a "one-size-fits-all" approach in an attempt to boost efficacy and engagement (Popenici & Dotson, 2017).
- Second, AI is changing methods for evaluation and feedback; conventional evaluation techniques frequently take a long time and offer feedback after a wait. AI has the ability to autonomously evaluate a variety of assignments, such as essays and multiple-choice questions, using NLP algorithms, providing students with immediate feedback on their work. In addition to basic grading, AI-driven systems can offer qualitative feedback, highlighting common mistakes, recommending areas for improvement, and assisting students in better understanding the material. Providing students with immediate feedback is essential for enhancing their learning and enabling them to quickly fix errors, which speeds up the learning process (Chen, Chen, & Lin, 2020).
- Finally, AI plays a major role in educational data mining (EDM) and learning analytics. By analysing massive datasets generated by student interactions with online learning platforms, AI algorithms may find hidden trends, forecast student performance, identify at-risk learners, and even assess the effectiveness of various teaching styles (Baker & Siemens, 2014). With the use of these insights, educators and educational institutions can make data-driven decisions that enhance curriculum design, enhance teaching methods, and provide early support to challenging students. This ability to turn raw data into insightful knowledge promotes ongoing gains in educational outcomes.

1.2 Significance of AI-driven Innovation in Learning

Beyond simple personalization or automation, AI-driven advancements in education mark a profound change in the ways that knowledge is gathered, shared, and used. Rather than merely enhancing current educational systems, these advancements are actively generating completely new possibilities for both teachers and students. The transformative influence can be examined using a number of crucial factors.

To begin with, advancements in AI are greatly improving learning results and student engagement. Through the provision of highly customized and adaptive learning experiences, AI guarantees that the content is presented to each student at the best possible pace and in the most efficient format. Meta-analyses demonstrating the positive impacts of AI-based learning on student achievement and enthusiasm indicate that this results in increased skill mastery, better knowledge retention, and deeper comprehension (Hwang & Tu, 2021). The interactive and reactive aspects of AI-powered products further boost student engagement, turning learning from a passive to a dynamic process.

Artificial intelligence is also crucial for promoting more accessibility and equality in education. AI-powered resources can accommodate a variety of learning needs, including those of kids with disabilities. Features like speech-to-text conversion, real-time translation, adaptive interfaces, and material distribution that are tailored to each learner's unique learning issues might help break down traditional educational boundaries. This makes high-quality learning experiences more accessible, ensuring that education is equitable and reaches a wider global audience.

Finally, AI enhances educators' abilities by enhancing their knowledge and freeing up critical time. By automating repetitive tasks like grading, administrative work, and even basic lesson planning, AI frees up teachers to focus on more complicated teaching responsibilities. This entails attending to students' socio-emotional needs, promoting critical thinking, facilitating group projects, and providing one-on-one guidance. AI enables more significant and meaningful interactions, transforming the role of the instructor from one of a mere knowledge source to one of mentor and facilitator (Luckin & Holmes, 2017).

2. Evolution of AI in Education

2.1 Early Foundations: Mechanical Teaching Machines and CAI (1920s–1970s)

Long before the digital era, AI in education was being developed. In an effort to automate the evaluation process, psychologist Sidney Pressey created the first mechanical teaching machine in 1924. It displayed multiple-choice questions and offered instant response (Pressey, 1926). Computer-aided instruction (CAI) systems began to appear during the 1960s. One such system was IBM's 1500 Instructional System, which could accommodate up to thirty-two student workstations with audio-visual capabilities. Dr Patrick Suppes of Stanford University oversaw the prototype of this method at Brentwood Elementary School in East Palo Alto, California (Watters, 2021).

2.2 Emergence of Intelligent Tutoring Systems (ITS) (1980s–1990s)

Intelligent tutoring systems (ITS), which used increasingly complex AI algorithms to mimic human tutors, emerged in the 1980s and 1990s. Figure 13.2 shows the evolution

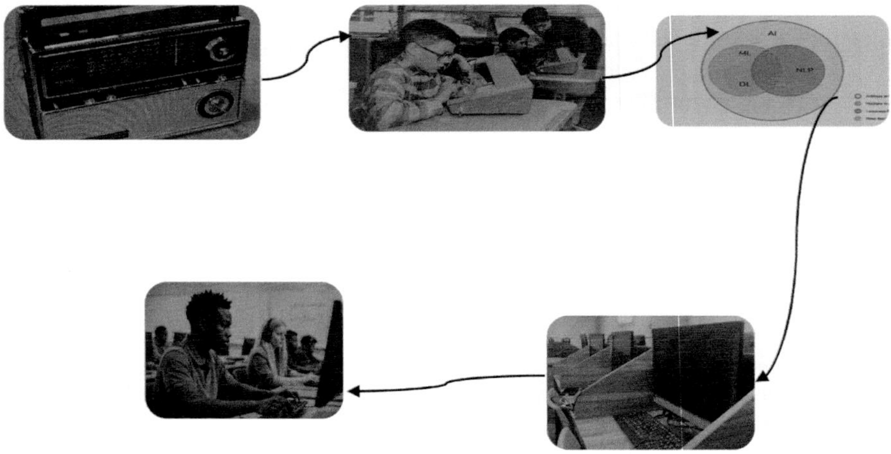

Fig. 13.2 Evolution of AI in education.

of AI in education. Innovative programs like SHERLOCK and SOPHIE showed how well ITS can raise student engagement and performance (Anderson et al., 1995). These systems provided real-time feedback, adaptive learning routes, and data-driven insights to guide instructional tactics by utilizing cognitive models to adjust to the needs of individual learners.

2.3 Advancements in NLP and Dialogue Systems (1990s–2000s)

The 1990s and 2000s saw developments in NLP, which created new opportunities for AI in education. NLP made it possible to create conversation systems, which let students use natural language to communicate with AI-powered virtual tutors. The potential of NLP-based dialogue systems to improve learning outcomes and encourage student involvement was demonstrated by programs like AutoTutor and ITSPOKE (Graesser et al., 2005).

2.4 Adaptive Learning and Personalized Education (2000s–2010s)

Artificial intelligence entered adaptive learning systems in the 2000s as digital learning platforms proliferated. These systems analysed student data using ML algorithms to deliver recommendations and tailored information. Knewton and Smart Sparrow, two adaptive learning platforms, showed how effective AI is at resolving knowledge gaps, fostering mastery, and customizing learning experiences for each learner (Tan et. al., 2025).

2.5 Integration of AI in Educational Technologies (2010s–Present)

The way AI was incorporated into educational technologies saw a dramatic change in the 2010s. Platforms driven by AI started to provide real-time performance data, automated grading, and intelligent content production. For example, AI algorithms are used by automated essay grading systems, such as ETS's e-rater, to evaluate and score student essays according to pre-set standards, expediting the grading process and offering instant response. Personalized dynamic content creation and AI-driven student profile have also given teachers a deeper understanding of each student's learning preferences, learning deficits, and strengths, allowing them to make tailored recommendations for the best teaching methods (SACE Board of South Australia 1968).

Table 13.1 captures the key stages in the evolution of AI in education, highlighting technological milestones, their characteristics, and representative tools or systems from each era.

2.6 AI-powered Virtual Assistants and Chatbots

The use of chatbots and virtual assistants driven by AI has revolutionized the field of education. These resources offer students individualized support by responding to inquiries, providing clarifications, and helping them work through issues. For instance, South Australia has led the way in integrating AI into student assessments by using chatbots to test migrant students' English competence in just 52 seconds, compared to 30 minutes for teachers in the past. In a similar vein, Andrew Ng launched the EdTech company Kira Learning with the goal of transforming education by using AI-powered teaching assistants to lighten the workload of teachers by managing lesson planning, grading, and student performance analysis (Kira Learning, 2023).

2.7 AI in Higher Education and MOOCs

Artificial intelligence has significantly impacted higher education as well. AI is included in all the programmes offered by universities such as Universal AI University in India, which offers undergraduate, graduate, and doctoral courses that incorporate AI into its architecture (Universal AI University, India, 2018). Furthermore, the growing significance of AI literacy is reflected in massive open online courses (MOOCs), such as the "Elements of AI" course, which was introduced in 2018 by the University of Helsinki and MinnaLearn with the goal of teaching the fundamentals of AI to a wide audience (Finnish Government & University of Helsinki, 2018).

2.8 Ethical Considerations and Challenges

Although the development of AI in education offers many advantages, there are drawbacks as well as moral dilemmas. Given that AI systems need access to enormous volumes of private student data, worries around data security and privacy are critical. Furthermore, existing educational disparities may be reinforced by the possible bias in AI systems. For AI to be successfully incorporated into education, these issues must be resolved, and transparent and moral AI deployment must be guaranteed (Gaisie et al., 2025).

For example, Greenwich Public Schools in the US are thinking about putting in place a formal policy to direct how staff and students use AI, with a focus on ethical and responsible use that stays within the bounds of professional standards and academic integrity.

Table 13.1 Key stages in the evolution of AI in education

Period	*Sub-title/Phase*	*Description*	*Examples of Tools/Systems*
1920s–1970s	Early foundations: Mechanical teaching machines and CAI	Sidney Pressey developed the first mechanical teaching machine (1924) that presented multiple-choice questions and provided instant feedback. In the 1960s, computer-aided instruction (CAI) systems like IBM's 1500 Instructional System emerged, supporting multiple audio-visual student workstations.	Pressey's mechanical teaching machine (1924), IBM 1500 Instructional System, CAI prototypes at Stanford
1980s–1990s	Emergence of intelligent tutoring systems (ITS)	ITS uses AI algorithms to simulate human tutors, offering adaptive learning paths, real-time feedback, and cognitive modelling to personalize instruction.	SHERLOCK, SOPHIE, ITS programs
1990s–2000s	Advancements in natural language processing (NLP) and dialogue systems	NLP-enabled AI tutors to engage students through natural language conversations, enhancing interaction and learning outcomes.	AutoTutor, ITSPOKE
2000s–2010s	Adaptive learning and personalized education	Machine learning algorithms analysed student data to tailor content and recommendations, addressing knowledge gaps and fostering mastery.	Knewton, Smart Sparrow
2010s–Present	Integration of AI in educational technologies	AI-driven platforms began offering real-time performance analytics, automated grading, and intelligent content creation, providing deeper insights into student learning styles and needs.	ETS e-rater automated essay grading, AI-powered dynamic content platforms

3. New Developments in AI for Education

Artificial intelligence is revolutionizing education by enhancing instruction worldwide. AI tools like NLP and ML are utilized in education to personalize and increase accessibility to learning. Students' content is personalized via adaptive learning algorithms according on their performance. AI chatbots and virtual tutors offer ongoing academic support, and automated grading systems reduce the effort for educators. By providing resources like text-to-speech and real-time translation, AI is improving education and helping students who struggle with language or disability. It aids in curriculum design improvement, cheating detection, and student outcome prediction in higher education. However, considerations like bias, ethics, and data privacy are significant challenges. Though its effectiveness depends on careful deployment and ongoing monitoring, AI offers a substantial shift in education by offering prospects for improved learning experiences.

3.1 Connect with New AI Development

In the rapidly evolving field of education today, keeping abreast of new advancements in AI is essential. AI developments have enormous potential to enhance administration, instruction, and learning. Educators who stay up to date can use these tools to address different student needs, manage classrooms more effectively, and deliver better individualized learning. Recent developments like GPT-4 have revolutionized the way that training is carried out by enabling the creation of dynamic content, automatic feedback, and real-time linguistic assistance. Educators and legislators can better address ethical concerns like bias, privacy, and access by keeping abreast of AI developments. It supports the digital skills required for future employment and equips institutions to use new tools responsibly. Frequent engagement with AI advancements encourages continuous innovation and development, guaranteeing that students gain from new technologies while maintaining teachers' efficacy and flexibility in a setting that is changing quickly.

3.2 AI-powered Personalized Learning

Artificial intelligence–powered personalized learning adapts the pace and content of instruction to each learner using intelligent algorithms. Through real-time performance analysis and adaptive feedback, this approach enhances engagement, comprehension, and outcomes. AI is used by adaptive learning systems to analyse student performance and modify course contents accordingly. It identifies problem areas, provides targeted assistance, and develops individualized learning experiences. As a result, students are more interested, stay involved, and perform better in a variety of learning environments. These AI-powered content customization platforms include DreamBox, Knewton, and Khan Academy.

The AI-powered solutions highlight students' areas of strength and improvement while offering continuous feedback on their performance. Teachers may adapt their lessons and help students better manage their learning with the help of this real-time information. These systems make learning more individualized which increases engagement and accelerates success. AI tutors and grading schemes that adapt to students' performance are two examples.

The AI algorithms are used in learning path optimization to analyse student data, including behaviour, preferences, and performance. It suggests individualized learning plans based on the requirements and development of every learner. This strategy seeks to increase productivity and motivation while modifying curriculum to promote improved learning results. For example, students can use predictive analytics to recommend courses and study schedules. AI is used in intelligent content production to create and edit learning resources such as multimedia, tests, and textbooks. These tools reduce teachers' workloads, personalize instruction, and facilitate the creation of content. More effective and interesting learning experiences can result from AI systems' ability to adapt information to suit various learner demands.

3.3 AI-generated Learning Materials

According to Aggarwal et al. (2023), AI-generated learning materials produce personalized textbooks, tests, and multimedia materials that adjust to the demands of each learner. It saves teachers' time when creating content and allows for personalized learning by updating according to student progress. This invention increases student engagement and boosts educational effectiveness. Leading tools in this field include ScribeSense and Content Technologies. ScribeSense reduces teachers' workload by automatically grading handwritten work using optical character recognition. Deep learning is used by Content Technologies to transform curricula into unique textbooks that satisfy curriculum requirements. These resources demonstrate how AI may improve educational opportunities.

One tool that employs AI to create exams and quizzes from learning materials is automated question generation. It saves teachers' time, offers different question types, and adjusts difficulty to fit students' level. This helps with ongoing learning and finding areas where students need to improve. Multimodal learning resources combine text, audio, video, and interactive features to fit different learning styles and boost engagement. Tools like AI-made simulations and games help users understand better by providing various sensory experiences. This approach improves accessibility and memory, increasing the effectiveness and inclusivity of learning. One of these AI-generated tools is OpenAI's Sora.

3.4 NLP in Education

By assisting robots in comprehending human language, NLP enhances educational communication and learning (Shaik et al., 2022). It contains language learning resources, automatic essay grading, and AI chatbots for tutoring. NLP offers personalized feedback and aids diverse learners by interpreting text and speech. Its increasing use boosts engagement and efficiency in education.

3.5 Conversational AI Tutors

Conversational AI tutors use NLP to create human-like interactions that provide personalized help, answer questions, and give instant feedback (Liu et al., 2024). They are available 24/7, and thereby enhance learner engagement and accessibility. The tutors adjust to each learner's needs, promoting active learning and skill development.

Chatbots and virtual assistants, such as Duolingo's GPT-based tutor, use AI to offer personalized language practice and conversational learning. They simulate real-world dialogues to boost motivation and fluency while making sure every learner has easy access to support.

The AI essay grading systems evaluate writing creativity, grammar, and quality using technology. It speeds up the review process by offering individualized advice and regular feedback. Students' writing abilities are enhanced and teachers' workload is reduced with the use of these resources. However, complex judgements still require human review. Furthermore, AI-powered tools improve language acquisition through the use of NLP, speech recognition, and tailored feedback. Adaptive exercises and immediate corrections are provided by programs such as Rosetta Stone and Duolingo, which increase user engagement and retention. By supporting immersive practice and accommodating different learning styles, these technologies improve the effectiveness and accessibility of language acquisition.

3.6 AI for Administrative and Teacher Support

Grading, scheduling, attendance monitoring, and report generation are all handled by AI-driven automated administrative chores, which simplify education administration. Teachers can concentrate more on instruction and student support since these systems lessen their administrative workload. Automation facilitates rapid decision-making and resource allocation by increasing institutional efficiency, accuracy, and data management (Kumar, 2024; Adepoju et al., 2022). This change improves operational efficacy and overall educational quality.

The AI teacher assistants support educators by managing repetitive duties including lesson planning, grading, and monitoring student progress. It employs data to identify pupils who might require further support, enabling prompt interventions. These aides reduce teachers' burdens so they may concentrate on individualized instruction and student engagement. IBM Watson Education, Civitas Learning, Squirrel AI, and Gradescope are a few AI systems that help with lesson design and detecting students who are at risk. To help teachers in their work, each of these technologies has special capabilities including automatic grading insights, predictive analytics for student achievement, and personalized content recommendations. By concentrating on certain abilities and pedagogical approaches, AI-driven professional development offers instructors individualized training. It encourages continuous learning and improved teaching strategies by using data analytics to recommend helpful materials and courses. In the end, this strategy improves student results by assisting teachers in developing their abilities and adjusting to new educational technologies.

3.7 Accessibility and Inclusivity through AI

With features like text-to-speech, speech-to-text, and real-time translation, AI enhances accessibility and inclusion in education (Fitas, 2025). Students with impairments and those who are experiencing linguistic problems benefit from these resources. In order to promote equity and participation, they also design customized learning experiences that cater to different needs. AI-powered platforms give marginalized populations access to additional educational opportunities while fostering a friendly environment that benefits all students. By providing individualized learning for everyone, regardless of location or financial status, AI contributes to more equitable education. AI is used by programs like Khan Academy to reach students worldwide, including those in underserved areas. Additionally, AI adjusts to various learning requirements, guaranteeing that every learner receives the support they require. Even if AI reduces prices and makes access possible from any location, issues with technological access and equitable AI use still need to be resolved for true educational equity.

By offering individualized help, AI technologies increase access to education for students with impairments. Visually challenged students can navigate their environment with the aid of tools like Microsoft's Seeing AI. While AI communication devices help students with speech difficulties, text-to-speech and speech-to-text applications help people who struggle with reading or writing. By encouraging involvement and self-reliance, these resources broaden the scope of education. AI assists impaired students in overcoming educational challenges by customizing resources and providing real-time support.

Additionally, by eliminating language barriers, AI-powered localization and translation tools enhance education. They make it possible for students from various backgrounds to access native-language content. For a variety of learning scenarios, real-time text, speech, and video translation is provided by programs like Google Translate and Microsoft Translator. By adapting content to particular cultural situations, localization improves comprehension. By encouraging inclusion and international collaboration, these technologies increase access to education on a global scale. In contexts of multilingual education, they also help with comprehension and communication.

3.8 Challenges and Ethical Considerations

The digital gap, algorithmic prejudice, and data privacy are some of the issues facing AI education. To avoid perpetuating inequality, it is critical to guarantee equitable and transparent AI use. Human contact in learning can also be impacted by an over-reliance on technology (Zhai et al., 2024). For AI to be used in education in a responsible and equitable manner, these problems must be resolved.

The goal of data privacy and security in AI education is to protect student data from misuse and illegal access. Building confidence requires adherence to laws like General Data Protection Regulation (GDPR). Personal information may be safeguarded and detrimental breaches can be avoided for both students and institutions by implementing robust encryption, anonymization, and secure data handling procedures. Algorithmic bias, which occurs when AI systems inadvertently favour some groups over others due to skewed data or subpar design, is another difficulty. This can lead to unjust treatment or inaccurate assessments of students' abilities in the classroom, which can exacerbate already existing disparities.

Utilizing a variety of data, being open, and routinely assessing AI technologies to make sure they promote justice and equity are all crucial in combating bias. Over-reliance on technology in the classroom can impede creativity, critical thinking, and interpersonal communication. Over-reliance on AI could reduce the importance of teachers and students' ability to solve problems. To encourage full learning and ensure that AI complements human judgement and social interaction rather than replaces it, it is critical to strike a balance between technology and conventional approaches. Similarly, teacher and student readiness for using AI in education requires building digital skills and adapting to new technologies. Teachers need training to use AI tools effectively, and students must learn to think critically about AI in their learning. Preparing both groups helps to reduce resistance and increase the advantages of technology in education, leading to a smooth adoption of new tools.

4. Innovative Practices in AI-Powered Education

4.1 Insights into AI-powered Education

When "artificial intelligence" or AI is mentioned, one easily pictures robots that resemble humans or a brain connected to wires and computer chips. These are popular images associated with AI, but they do not accurately represent what AI is all about. In everyday life, especially in education, AI can be likened to intelligent software that learns from data and helps personalize learning, rather than a robot walking around. Having this understanding helps us see how AI can truly change the way we learn and teach. AI technologies are revolutionizing education by minimizing administrative burdens, sharing valuable intelligence, and facilitating professional growth. These advancements empower educators and learners to focus on creative and strategic aspects of teaching and learning, respectively, thereby improving the quality of education.

Artificial intelligence–powered education refers to the application of intelligent computer programs and technologies designed to enhance teaching effectiveness and improve student learning outcomes (Renz & Vladova, 2021). It utilizes data and AI-driven tools to create a more effective and accessible learning experience for everyone. AI-powered tools, such as gamification, virtual reality (VR), and augmented reality (AR), can supplement traditional techniques (Arya & Verma, 2024). Agatova and Latipova (2025) investigated the implementation and advancement of personalized learning in universities with the support of AI tools, analysing the current landscape of AI-driven educational technologies in two developing countries and highlighting the key challenges and future opportunities in this area. But, challenges remain, such as the lack of personal communication and difficulties in recognizing emotions.

By improving learning experiences, assisting teachers, encouraging creative behaviours, and opening the door for more study in the area, Jabeen et al. (2024) advanced our knowledge of how AI and ML may revolutionize education. A number of international campaigns have been started to make AI education available to everyone. To provide students, developers, and professionals with the requisite AI capabilities, OpenAI's Academy in India, for example, seeks to increase access to AI education in the nation (*The Economic Times*, 2025). Similarly, in order to foster innovation and maintain economic growth, China's educational reform incorporates AI into every level of its educational system (Reuters, 2025). In order to adjust education to the demands of a quickly evolving technological context, these contributions are essential.

Innovative approaches to education that fully utilize AI are revolutionizing institutional procedures and learning environments globally. These methods consist of:

1. Personalized and adaptive learning systems/emotional technologies
2. Intelligent tutoring systems
3. Administrative and assessment tasks automation
4. Immersive, interactive learning platforms
5. Ethical considerations and future directions in AI-powered education

4.2 Personalized and Adaptive Learning Systems/Emotional Technologies

By adjusting the pace and content of instruction to each learner's particular interests, shortcomings, and abilities, AI-driven personalized learning platforms are revolutionizing classrooms. Real-time analysis of student performance data by these systems allows them to modify the difficulty of lessons and provide resources that support each learner's unique learning objectives. In order to guarantee that everyone advances at their own ideal rate, adaptable platforms, for instance, can skip unnecessary content for expert students or offer focused assistance for those who struggle. This method has been particularly successful for a variety of groups, such as students with disabilities or different educational backgrounds, and it has made high-quality, individualized instruction widely available. By having meaningful discussions with kids and providing coping mechanisms, AI-powered chatbots offer ongoing mental health support. Learning becomes less daunting and more emotionally supportive as a result of the reduction of worry, stress, and loneliness (Cao et al., 2023; Shahzad et al., 2024). By offering individualized learning environments, adaptive learning systems and emotionally intelligent e-learning systems (EIES) are revolutionizing the way that people learn (He et al., 2022; Dehbozorgi & Kunuku, 2023). Figure 13.3 displays the innovative practices in AI-powered education.

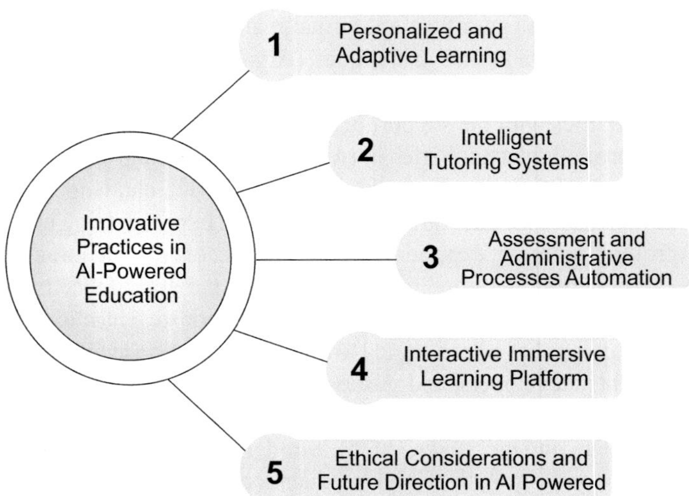

Fig. 13.3 Innovative practices in AI-powered education.

4.3 Intelligent Tutoring Systems

In many classrooms, smart tutoring systems—which use intelligent technology to tailor learning materials for specific students—have become commonplace (Molenaar, 2021). AI is used by intelligent tutoring systems (ITS) to simulate the function of a human tutor, offering prompts, guided help, and instant feedback. In contrast to traditional standard teaching approaches, ITS are able to pinpoint specific misconceptions and modify explanations or practice tasks accordingly. Because these technologies provide individualized assistance precisely when it is needed, they not only improve academic achievement but also inspire students to acquire autonomous problem-solving skills.

Intelligent tutoring systems are a useful tool for filling educational gaps in primary, secondary, and post-secondary education because of their ability to provide scaled, tailored support. Students now have access to real-time guidance and feedback, thanks to the introduction of virtual assistants and ITS driven by AI. These resources mimic one-on-one tutoring sessions by providing clarifications, responding to questions, and helping students understand difficult ideas. Through constant accessibility, the gap between classroom instruction and personalized study is bridged, promoting continuous learning (Maity & Deroy, 2024).

In terms of emotional technology, AI systems are becoming more and more outfitted with tools like sentiment analysis, speech tone identification, and facial expression analysis to identify and decipher students' emotional states as they are engaged in educational activities. By recognizing signs of frustration, boredom, or engagement, AI can adapt content delivery, provide timely encouragement, or suggest breaks, thereby creating a more responsive and supportive learning environment. EIES utilize Bayesian models to anticipate students' emotions and adapt feedback in real time, enhancing motivation and engagement (Arya & Verma, 2024).

4.4 Administrative and Assessment Tasks Automation

The AI-powered tools can automate administrative tasks, allowing educators to allocate more time to meaningful interactions with students (Eden et al., 2024). AI is streamlining administrative and assessment processes, freeing educators from repetitive tasks such as grading, scheduling, and data entry (Siddiqui, 2024).

Automated grading tools can efficiently evaluate multiple-choice, short-answer, and even some essay responses, delivering prompt feedback to students and allowing teachers to focus on more complex instructional duties. Additionally, administrators are able to make better decisions about curriculum design and budget allocation, thanks to AI-powered analytics that assist them in identifying trends in student performance. In addition to improving productivity, this automation helps create more timely and knowledgeable educational interventions.

4.5 Interactive Immersive Learning Platforms

Through gamified experiences, VR, and interactive simulations, AI-powered immersive learning platforms are revolutionizing engagement. By traversing historical events in a simulated world or doing virtual science experiments, these settings enable students to practically investigate difficult ideas. By modifying scenarios in response to user input, AI personalizes these encounters and keeps every learner engaged and challenged. The end effect is a dynamic, interactive learning process that improves comprehension and memory much more than passive approaches can.

4.6 AI-driven Virtual Classrooms

The AI-driven virtual classrooms combine VR and AI to improve online learning. These classrooms offer individualized guidance and realistic simulations that improve learning results and student engagement. NLP improves student–teacher communication in online learning environments. For example, Bhashi et al. (2025) created a VR classroom that incorporates AI virtual assistants to provide personalized feedback, interactive tests, and real-time assistance. Their research showed that by catering to a variety of learning styles, these settings greatly boost student motivation and engagement. Furthermore, in order to create immersive and captivating learning

Table 13.2 Shown the key AI technologies transforming education

S. No.	*AI-powered Education Technologies*	*Description*	*Case Studies/Examples*
1.	AI-driven Personalized Learning Systems	Platforms that analyse learner data to tailor educational content and pace to individual needs.	Squirrel AI and Microsoft Reading Coach customize learning paths to student strengths and styles.
2.	Adaptive Learning Technologies	Real-time adjustment of difficulty and content based on student performance to keep learners challenged and motivated.	Duolingo adapts language exercises dynamically based on user accuracy and speed.
3.	Automated Grading and Feedback	AI algorithms automatically grade assignments and provide instant, detailed feedback to students.	Grammarly and Turnitin offer automated writing feedback; Recursive's AI grades handwritten math answers rapidly.
4.	Intelligent Tutoring Systems (ITS)	AI systems that provide personalized instruction, real-time feedback, and adapt content to learner needs.	ITS are used in K-12, higher education, and corporate training to enhance engagement and automate grading.
5.	Virtual Reality (VR) and Augmented Reality (AR) Learning	Immersive technologies that enable experiential learning through virtual environments and simulations.	VR/AR is used to explore historical sites, conduct virtual experiments, and visualize complex subjects.
6.	Gamification and Immersive Learning	Utilization of game elements and immersive technology to engage students and boost motivation.	Kahoot! and Minecraft: Education Edition use AI for adaptive quizzes and simulations. ClassDojo incorporates gamified activities, such as scavenger hunts and quizzes, to boost participation.

environments, smart classrooms are being introduced with interactive dashboards, AR, and virtual assistants driven by AI. For instance, AR apps that show human anatomy and intricate structures in three dimensions help engineering and medical students grasp concepts better than they could with conventional approaches (Ejeh & Igbokwe, 2025). In Nigerian classrooms, where individualized attention is frequently limited by big class sizes, these advances have a significant impact.

4.7 Gamification and AI

Gamification refers to the incorporation of fun, game-inspired elements into everyday activities to make learning more engaging and motivating. Thanks to AI, this approach has become even smarter. AI can tailor the learning experience to each student, making lessons feel more personal and interactive. This not only keeps students interested but also helps them learn more effectively. Gamification makes learning more engaging, motivating, and memorable for students (Renz & Vladova, 2021). By introducing elements of play, it transforms lessons into engaging and meaningful learning experiences. This approach also encourages healthy competition, teamwork, and the development of problem-solving skills. Some leading examples of AI-powered gamification platforms are Kahoot!, 360learning, Thirst, Classcraft, ClassDojo, and Duolingo. For example, Kahoot!+ is an all-in-one teaching toolkit for engaging lessons, homework and assignments, review, and assessment. These tools use game design features and personalized learning paths to create richer and more interactive educational experiences.

4.8 AI for Special Education

Artificial intelligence-driven technologies have shown great promise in supporting special education by providing personalized learning experiences tailored to the unique needs of each student. These AI tools can adapt content delivery, offer real-time feedback, and assist educators in tracking progress, thereby enhancing both engagement and learning outcomes for students with diverse abilities (Renz & Vladova, 2021; US Department of Education, 2024). Students with disabilities get motivated with these assistive technologies such as text-to-speech and speech-to-text recognition. AI-powered tools like Microsoft's Immersive Reader assist students with dyslexia or reading difficulties by reading text aloud or segmenting it into digestible pieces, making content more accessible.

4.9 Ethical Considerations and Future Directions in AI-powered Education

Some scientists and engineers argue that AI is still far from achieving human-level consciousness. The investigation of consciousness does not precipitate an urgent impetus to address ethical questions it brings about (Siau & Wang, 2020).

The study of AI ethics includes the wide-ranging set of moral principles and ethical considerations, as well as guidelines, policies, rules, and laws that shape AI. The use of AI in schools, however, is challenged by a number of hurdles (Renz & Vladova, 2021). The implementation of AI education opens up incredible opportunities for progress and innovation, but poses complex challenges that require careful attention.

Educational institutions tackle these challenges by deploying strong policies and improving systems. The complete realization of AI benefits alongside fairness protection demands critical involvement from technology experts and educators for the establishment of trust and inclusion within educational environments. The establishment of ethical frameworks and guidelines by stakeholders is essential to secure the transparent and fair deployment of AI technologies in institutional settings (Porayska-Pomsta et al., 2024). The equilibrium between technological advancement and the essential nature of responsible implementation emerges as a critical factor to unlock AI's potential benefits for future development in education (Arya & Verma, 2024).

In conclusion, AI ought to be viewed as an auxiliary partner in education, which is expected to be used thoughtfully and with a strong adherence to ethical standards. These innovative practices in AI-powered education present lots of potential opportunities and are greatly influencing education positively, but are not without challenges and future directions.

5. Case Studies and Real-World Applications (Ongoing)

5.1 Personalized Learning at Carnegie Learning

One of the first companies to use AI in individualized education is Carnegie Learning (Bhutoria, 2022; Maghsudi et al., 2021). Its systems for adaptive math instruction helped achieve this. The system analyses student performance in real time using AI algorithms. In a similar vein, it may modify educational materials appropriately, allowing for customized learning paths. This method offers tailored feedback that fits each learner's pace and learning preferences in addition to assisting in the identification of knowledge gaps. Although this kind of customization could lead to better academic results and greater learner engagement, it also raises concerns about data privacy and the pedagogical effects of less instructor control. However, a wider trend towards utilizing AI to scaffold individualized learning in K-12 and higher education environments is suggested by Carnegie Learning's paradigm (Chang et al., 2022).

5.2 Intelligent Tutoring Systems

Computer-based platforms known as ITS mimic some characteristics of human tutoring in an effort to provide instantaneous, individualized teaching. Drawing on cognitive science, these systems use adaptive algorithms to modify teaching strategies in real time according to students' mental processes (Cai & Hu, 2018). AutoTutor, for example, uses natural language conversations to promote deeper learning. Though they struggle with sophisticated student responses and emotional cues, ITS are excellent in math and programming. Accordingly, their growing use points to a move towards more scalable types of tailored educational assistance (Aleven et al., 2023).

5.3 AI-driven Assessment Tools

The AI-powered evaluation instruments are a major advancement in the assessment of student learning. In addition, these tools provide data-driven and automated insights that enhance conventional approaches. Large amounts of student data can be processed by these programs to provide instant feedback. Additionally, they can recognize learning trends and even forecast future performance (Yadav, 2020; Minn, 2022). This trend is best illustrated by technologies like real-time analytics dashboards and automated essay scoring. Although they have the potential to simplify evaluation and lessen the workload for teachers, questions remain over their capacity to capture intricate, context-dependent skills like creativity or moral reasoning. Additionally, in order to reduce algorithmic bias and guarantee equitable results, their use requires thorough validation. As such, AI-enhanced assessment is best viewed as a supplement to rather than a replacement for human evaluative judgement.

5.4 Language Learning Applications

The AI-powered language learning applications, such as Duolingo and Babbel, demonstrate how ML and NLP can support personalized language acquisition (Hashim et al., 2022; Rebolledo Font de la Vall & González Araya, 2023). These platforms adapt exercises based on learner performance. They also offer immediate feedback and scaffolded practice across listening, speaking, reading, and writing tasks. By utilizing speech recognition and adaptive algorithms, they provide immersive and interactive opportunities that extend beyond traditional classroom settings. However, these programs might not be able to provide the cultural insights and nuanced feedback that come from human instructors. Therefore, it is likely that they perform best when incorporated into all-encompassing, communicative language learning frameworks.

5.5 Predictive Analytics in Student Retention

In order to forecast dropout risks and guide prompt responses, predictive analytics in education uses AI to evaluate both historical and current student data. By examining patterns in academic achievement, attendance, study involvement, and backgrounds, schools can determine whether kids are at danger of falling behind academically, disengaging from learning, or being at risk of dropping out (Hlosta et al., 2022; Toktarova & Pashkova, 2022). Systems such as Blackboard Analytics and Civitas Learning, for example, offer dashboards that let advisers make data-driven choices. These techniques raise ethical questions about student profiling and the possibility of perpetuating already existing imbalances, even as they hold out the promise of increased retention and institutional efficiency. Therefore, openness, stakeholder participation, and continual assessment of its effects are necessary for the proper application of predictive analytics.

5.6 AI in Special Education

Because AI technologies can tailor support to each student's unique needs, they are essential in special education. This potential is demonstrated by tools like emotion recognition software for individuals on the autistic spectrum and speech-to-text apps for dyslexic children (Almgren Bäck et al., 2024; Bhola, 2022; Dantas & Do Nascimento, 2022). Even while these technologies have the potential to improve academic results, communication, and engagement, their application must be ethically sound and inclusive. Importantly, AI should be used as a component of a larger framework of adaptive, responsive instruction, not as a substitute for human expertise.

5.7 Virtual Learning Assistants

Virtual learning assistants utilize conversational AI to offer on-demand support to students. They help learners with course content, manage tasks, and answer academic questions. Examples such as IBM Watson Tutor and AI chatbots integrated into learning management systems help reduce the workload for instructors (Baha et al., 2021; Shilowaras & Jusoh, 2022). They provide scalable support, particularly in online courses with large student enrolments. Their effectiveness, however, hinges on advanced NLP and contextual adaptability. These tools can help make learning faster and improve learner satisfaction. But, they should support, not replace, the important roles that human teachers play in building relationships and teaching.

5.8 Adaptive Learning Platforms

Adaptive learning platforms utilize AI to dynamically modify content presentation based on continuous assessment of a learner's performance. These platforms, such as Smart Sparrow and Knewton, construct personalized pathways that respond to students' progress in real time, thereby fostering mastery learning. They meet the varied demands of diverse classrooms by providing individualized instruction at scale. The effectiveness of the underlying pedagogical models and the openness of adaptive algorithms, however, are critical to their success. Furthermore, an over-reliance on automation could unintentionally diminish the interpretive role of the teacher. Therefore, rather than taking the role of educator-guided learning, adaptive systems should be seen as cooperative instruments that supplement it.

6. Future Trends and Challenges of AI in Education

Examining the intersecting paths and latent tensions that accompany AI's rapid dissemination is crucial as technology becomes more and more integrated into modern educational settings. It is possible that generative AI advancements, the drive for morally sound algorithms, and initiatives to make intelligent systems more accessible to all will change educational standards in ways that are not entirely clear now (Perkins et al., 2024; Valdivieso & González, 2025; Mittal et al., 2024). However, these opportunities are accompanied by persistent worries about data ethics, surveillance, the loss of teacher agency, and the coherence of curriculum goals in AI-mediated settings. The following conversation seeks to critically analyse these advancements, addressing both the revolutionary possibilities and the open questions that AI presents for educational theory and practice.

6.1 Integration of AI into Curriculum Design

The integration of AI into curriculum design represents a significant shift in educational philosophy and practice, presenting both exciting opportunities and challenges. A move towards more responsive pedagogical models is suggested by emerging trends, such as adaptive learning systems that adjust instructional material, pacing, and evaluation in response to unique student profiles (Gligorea et al., 2023; Wang et al., 2024). However, these developments must be viewed in light of the limitations of current curriculum frameworks, which frequently thwart quick change. One major worry is whether AI technologies could automate current systems instead of facilitating truly creative and inclusive curriculum reform (Akavova et al., 2023). It will take collaborative frameworks that combine technological innovation with pedagogical discernment to design courses that critically interact with AI's epistemological assumptions, particularly its preference for quantitative outcomes.

6.2 Ethical Considerations in AI Implementation

Ethical considerations will be crucial in discussions about implementing AI in education. This is important as algorithms developed exert an increasing influence over learner trajectories and assessments. Issues such as algorithmic bias, opacity, and the dehumanization of pedagogical processes pose serious ethical dilemmas (Hartong & Piattoeva, 2021). The future challenge lies not only in designing ethically aligned AI systems but also in ensuring that ethical principles are embedded across the entire lifecycle of educational AI—from data collection to deployment and evaluation. Given the dynamic and context-sensitive nature of ethics in education, static guidelines are insufficient; instead, iterative and participatory ethical frameworks may be required to keep pace with rapid technological developments. Also, students and educators must be equipped with critical AI literacy to engage meaningfully with these technologies and challenge unjust outcomes when necessary. Ethical implementation, therefore, is not a technical add-on but a core concern that demands interdisciplinary and stakeholder-informed solutions.

6.3 Data Privacy and Security

The rapid expansion of AI technologies in education raises pressing concerns regarding data privacy and security, particularly as learning platforms increasingly collect extensive and sensitive information about student behaviour, performance, and even emotional states. Although predictive analytics and learner modelling present significant pedagogical advantages, they simultaneously expose students to potential risks of data misuse and surveillance (Pretorius, 2023). A critical challenge moving forward involves establishing a robust data governance framework. As educational data becomes an essential asset for AI training, institutions face the delicate task of maximizing its utility while minimizing intrusion. Consequently, protecting learner data will require not only advanced technical measures but also comprehensive policy reforms, the cultivation of stakeholder trust, and potentially new legal frameworks explicitly tailored to the educational sector.

6.4 Access and Equity in Education

Artificial intelligence has the potential to make access to high-quality education more equitable for everyone; however, disparities in infrastructure, digital literacy, and localized content development may exacerbate existing inequalities. The gap in digital access is a significant concern in low- and middle-income nations, where dependable internet connectivity and devices for AI technology are frequently insufficient (UNESCO, 2023). Plus, AI systems often rely on datasets that underrepresent marginalized communities, resulting in solutions that fail to generalize across diverse educational contexts. Looking ahead, equitable AI in education will require inclusive design practices, participatory data collection, and policies that prioritize infrastructural development in underserved areas. On top of that, ongoing efforts must focus on ensuring that AI tools support, rather than supplant, local pedagogical practices. Without intentional and context-aware interventions, the risk remains that AI will reinforce educational privilege rather than mitigate it.

6.5 Teacher Training and Professional Development

Artificial intelligence is changing not just teaching methods but also instructors' professional identities as it continues to infiltrate educational settings. It is arguable that this changing environment necessitates a rethinking of teacher training, one that goes beyond superficial digital skills to develop sophisticated pedagogical strategies for interacting with AI as a collaborator as well as a tool (Li & Zou, 2021). However, institutional inertia, varying degrees of technical proficiency among teachers, and wider social conflicts around automation in education hamper efforts to create such training (Santos Guimarães Carvalho, 2020). Therefore, professional growth must be continuous, context-sensitive, and responsive to teachers' interpretative agency. Importantly, educators should be regarded as co-creators of morally and pedagogically sound AI applications, rather than passive adopters.

6.6 Support for Diverse Learning Needs

By addressing students whose needs frequently lie outside the purview of traditional pedagogical models—those managing disabilities, linguistic variety, or unique cognitive pathways—AI holds the potential to broaden the scope of inclusive education. Learning environments that are sensitive to individual trajectories might be imagined thanks to tailored distribution methods and adaptive technology (Fernandes et al., 2023). However, structural and epistemological constraints temper such promise. Under the pretence of personalization, current AI algorithms run the risk of re-inscribing exclusionary norms because they are often developed on homogenous datasets. As Dubova (2022) suggests, the nuanced interplay between AI systems and culturally embedded cognitive practices remains underexplored. Advancing inclusive AI will require not only technical refinement but also ethically grounded, participatory design grounded in pluralist educational imaginaries.

6.7 Blended Learning Environment

The rise of blended learning, where face-to-face instruction intersects with AI-driven technologies, marks a notable shift in how educational experiences are conceived and delivered (Nawale, 2022; Perach & Alexandron, 2022). While AI offers tools that may streamline instruction, personalize learning pathways, and monitor engagement (Alam, 2023), its integration into pedagogical settings remains far from seamless. The challenge lies not only in aligning technological capacities with educational values but also in ensuring that transitions between algorithmic systems and human instruction do not fracture the learning process. Plus, disparities in infrastructure and educator readiness suggest that equitable implementation is far from guaranteed. Meaningful innovation will likely depend on flexible, context-aware designs that preserve the human dimensions of teaching.

6.8 Assessment and Feedback Mechanisms

Artificial intelligence may well reconfigure the landscape of educational assessment by enabling more personalized and adaptive forms of feedback through technologies such as ITS, NLP, and learning analytics (Hooda et al., 2022; Shaik et al., 2022). While these innovations hold promise for enhancing both formative and summative practices, their adoption invites critical scrutiny. Automated assessments often struggle to apprehend the complexities of human cognition—especially in evaluating creativity, critical engagement, or culturally situated reasoning. Plus, such systems risk perpetuating algorithmic bias and diminishing the relational, interpretive dimensions that are central to meaningful pedagogy. It is perhaps more constructive, then, to envision a future in which AI complements rather than supplants human judgement—anchoring assessment in ethical transparency, contextual sensitivity, and the preservation of learner autonomy.

6.9 Collaboration among Stakeholders

The advancement of AI in education appears to depend not merely on technological innovation but also on sustained, dialogical collaboration among educators, developers, policymakers, students, and researchers (Eguchi, 2022; Ramasamy, 2024). While such coalitions may foster ethically grounded and pedagogically responsive systems, the task is complicated by differing institutional logics and epistemic priorities. Developers may emphasize scalability and efficiency, whereas educators often advocate for contextual sensitivity, relational care, and equity. Without careful negotiation, these tensions risk producing AI applications that are misaligned with classroom realities. Hence, it is possible that the future of educational AI lies in participatory design practices that cultivate mutual intelligibility and honour the complex, human-centred nature of teaching and learning.

6.10 Sustainability and Scalability of AI Solutions

Concerns about AI's long-term viability and egalitarian scaling have become more pressing as technology gets more integrated into educational environments. Although early adoptions frequently produce positive results, spreading these innovations across diverse educational systems reveals enduring conflicts, from differences in institutional capacity and technological literacy to financial precarity and infrastructure fragility (Zawacki-Richter et al., 2019). Without adaptable governance structures, open-source substitutes, and consistent investment in local knowledge, AI-enhanced education might end up escalating rather than resolving current disparities. Therefore, future initiatives need to stay aware of the changing pedagogical ethos influencing educational futures as well as contextual specificity.

6.11 Evolving Student–Teacher Dynamics

Artificial intelligence's subtle recalibrating of teacher–student relations in educational contexts offers both generative opportunities and disconcerting difficulties. Such technologies run the risk of redefining pedagogical authority in ways that could marginalize the human presence at the centre of education, even as they may relieve educators of administrative demands and reveal patterns in student learning (Selwyn, 2019). AI has the potential to change the classroom from a place of relational co-construction to one of automated instruction if critical involvement is not maintained. Teachers who are not only proficient with technology but also aware of the ethical and affective aspects of education are needed to navigate this terrain.

6.12 Research and Evidence-based Practices

Even as interest in using AI in education is growing, there is still a big gap between the rate of technology advancement and the breadth of empirical knowledge. Table 13.3 shows the current trends/challenges of AI in education with future directions. All too

Table 13.3 Shows the current trends/challenges of AI in education with future directions

Trend/Challenge	*Current Issues*	*Future Directions*
1. Integration of AI into Curriculum	Adaptive learning systems; risk of automating existing frameworks rather than enabling creative reform; slow institutional change.	Encourage teachers and technology experts to work together. Update school lessons so students think about how AI works and what it means. Make sure new lessons are creative, fair, and fit the needs of each school and community.
2. Ethical Considerations	Algorithmic bias, lack of transparency, risk of dehumanizing education; static ethical guidelines insufficient.	Develop iterative, participatory ethical frameworks; embed ethics throughout AI lifecycle; enhance critical AI literacy among all stakeholders; ensure interdisciplinary and stakeholder-informed solutions.
3. Data Privacy and Security	Sensitive student data at risk; potential for misuse and surveillance; lack of robust governance.	Establish comprehensive data governance and legal frameworks; invest in advanced technical and policy safeguards; build stakeholder trust; develop sector-specific privacy standards.
4. Access and Equity	Digital divide in infrastructure and literacy; underrepresentation in datasets; risk of exacerbating inequalities.	Prioritize inclusive design and participatory data collection; invest in infrastructure for underserved regions; ensure AI tools complement local pedagogies; develop policies to promote equitable access.
5. Teacher Training/ Professional Development	Teachers need more than technical skills; unwillingness to try new things; variable proficiency; risk of teachers as passive adopters.	Create continuous, context-sensitive professional development; position teachers as co-creators of AI solutions; support interpretative agency; integrate AI pedagogy into teacher education.

frequently, AI tools are used without thorough, context-sensitive evaluation, leading to interventions whose educational impact may be speculated at best (Luckin, 2017). A research culture that values methodological variety, critical reflexivity, and ongoing communication between academics and practitioners appears to be desperately needed. AI's educational potential can only be reframed as a responsive, morally sound addition to human learning through such integrative efforts, particularly those that embrace longitudinal and qualitative discoveries.

7. Conclusion

Artificial intelligence is not merely a technological advancement—it is a transformative force reshaping the educational landscape. From personalized learning pathways and ITS to administrative automation and inclusive tools for learners with diverse needs, AI is redefining how education is delivered, experienced, and evaluated. Its integration into education has enabled more adaptive, efficient, and equitable learning environments, empowering both students and educators. AI's ability to analyse vast amounts of data in real time allows for highly individualized instruction, ensuring that learners receive content tailored to their pace, preferences, and performance. This personalization enhances engagement, fosters deeper understanding, and supports mastery-based learning.

Moreover, AI-driven tools such as chatbots, virtual tutors, and predictive analytics are providing timely support and interventions, reducing dropout rates, and improving academic outcomes. For educators, AI offers significant relief from routine tasks such as grading, scheduling, and lesson planning. This shift allows teachers to focus on higher-order instructional strategies, emotional support, and creative pedagogy. AI also supports professional development by offering personalized training and insights into student learning patterns, enabling more informed and responsive teaching. However, the integration of AI into education is not without challenges. Ethical concerns around data privacy, algorithmic bias, and the potential dehumanization of learning must be addressed with urgency and care. Ensuring transparency, inclusivity, and fairness in AI systems is essential to prevent the reinforcement of existing inequalities. Furthermore, over-reliance on AI may risk diminishing the critical human elements of education—empathy, mentorship, and social interaction.

Looking ahead, the future of AI in education lies in thoughtful, collaborative, and ethically grounded implementation. Policymakers, educators, technologists, and learners must work together to co-create AI systems that are not only innovative but also just and human-centred. Continuous research, inclusive design, and robust governance frameworks will be vital in navigating the evolving educational ecosystem. In essence, AI should be embraced as a powerful ally in education—one that augments human capabilities rather than replaces them. When applied responsibly, AI holds the promise of democratizing access to quality education, fostering lifelong learning, and preparing learners for the complexities of a rapidly changing world.

Artificial intelligence is revolutionizing education by making learning more personalized, engaging, and effective. It supports teachers by automating administrative tasks and providing insights into student performance. AI tools are not just enhancing the learning experience but also ensuring that education is accessible to all students, regardless of their individual needs. As AI continues to evolve, its role in education will only grow, offering new opportunities for innovation and improvement in teaching and learning processes.

References

Adepoju, A.H., Austin-Gabriel, B., Eweje, A. and Collins, A. (2022). Framework for automating multi-team workflows to maximize operational efficiency and minimize redundant data handling. *IRE Journals*, *5*(9), 663–664.

Agatova, O. and Latipova, L. (2025). Exploring AI-powered personalized learning in universities. *International Journal of Evaluation and Research in Education*, *14*(2), 837–848.

Aggarwal, D., Sharma, D. and Saxena, A.B. (2023). Exploring the role of artificial intelligence for augmentation of adaptable sustainable education. *Asian Journal of Advanced Research and Reports*, *17*(11), 179–184.

Akavova, A., Temirkhanova, Z. and Lorsanova, Z. (2023). Adaptive learning and artificial intelligence in the educational space. *E3S Web of Conferences*, *451*, 06011.

Alam, A. (2023). Harnessing the power of AI to create intelligent tutoring systems for enhanced classroom experience and improved learning outcomes. *Lecture Notes on Data Engineering and Communications Technologies*, *171*, 571–591.

Aleven, V., Baraniuk, R., Brunskill, E., Crossley, S., Demszky, D., Fancsali, S., Gupta, S., Koedinger, K., Piech, C., Ritter, S., Thomas, D.R., Woodhead, S. and Xing, W. (2023). Towards the future of AI-augmented human tutoring in math learning. *Communications in Computer and Information Science*, *1831* CCIS, 26–31.

Almgren Bäck, G., Lindeblad, E., Elmqvist, C. and Svensson, I. (2024). Dyslexic students' experiences in using assistive technology to support written language skills: A five-year follow-up. *Disability and Rehabilitation: Assistive Technology*, *19*(4), 1217–1227.

Anderson, J.R., Corbett, A.T., Koedinger, K.R. and Pelletier, R. (1995). Cognitive tutors: Lessons learned. *Journal of the Learning Sciences*, *4*(2), 167–207.

Arya, R. and Verma, A. (2024). Role of AI in education. *International Journal of Advanced Research in Science, Communication and Technology*, *2*(3), 589–594.

Baha, T.A.I.T., Hajji, M.E.L., Es-Saady, Y. and Fadili, H. (2021). Towards highly adaptive Edu-Chatbot. *Procedia Computer Science*, *198*, 397–403.

Baker and Siemens. (2014). Educational Data Mining and Learning Analytics. In Handbook of Learning Technologies (pp. 253–264). Springer. (Likely related to https://onlinelibrary.wiley.com/doi/abs/10.1002/9781118956588.ch16.

Bhashi, V., Krishna, V. and Srikanth, S. (2025). Virtual reality classrooms with AI generated scenarios. *International Journal of Future Modern Research*, *7*(1).

Bhola, N. (2022). Effect of text-to-speech software on academic achievement of students with dyslexia. *Integrated Journal for Research in Arts and Humanities*, *2*(4), 51–55.

Bhutoria, A. (2022). Personalized education and AI in the United States, China, and India: A systematic review using a Human-In-The-Loop model. *Computers and Education: Artificial Intelligence*, *3*, 100068.

Cai, Z. and Hu, X. (2018). Autotutor: An intelligent tutoring system and its authoring tools. In *Deep Comprehension: Multi-Disciplinary Approaches to Understanding, Enhancing, and Measuring Comprehension* (pp. 140–153), eds K. K. Millis, D. Long, J. Magliano and K. Wiemer. Routledge.

Chang, Q., Pan, X., Manikandan, N. and Ramesh, S. (2022). Artificial intelligence technologies for teaching and learning in higher education. *International Journal of Reliability, Quality and Safety Engineering*, *29*(05), 2240006.

Chen, Chen and Lin. (2020). Artificial intelligence in education: A review. IEEE Access, 8, 75264–75278.

Dantas, A.C. and Do Nascimento, M.Z. (2022). Recognition of emotions for people with autism: An approach to improve skills. *International Journal of Computer Games Technology*, 2022.

Dehbozorgi, N., and Kunuku, M. (2023). Exploring the influence of emotional states in peer interactions on students' academic performance. *IEEE Transactions on Education*, *67*, 405–412.

Deroncele-acosta, A., Luis, J., Ayala, M. and Barrasa, A. (2024). Integrating artificial intelligence to assess emotions in learning environments: A systematic literature review.

Dubova, M. (2022). Building human-like communicative intelligence: A grounded perspective. *Cognitive Systems Research*, *72*, 63–79.

Eden, C.A., Chisom, O.N. and Adeniyi, I. S. (2024). Integrating AI in education: Opportunities, challenges, and ethical considerations. *Magna Scientia Advanced Research and Reviews*, *10*(2), 6–13.

Eguchi, A. (2022). AI-powered educational robotics as a learning tool to promote AI and computer science education. In *Robotics in Education. RiE 2021. Advances in Intelligent Systems and Computing*, vol 1359 (pp. 279–287), eds M. Merdan, W. Lepuschitz, G. Koppensteiner, R. Balogh and D. Obdržálek. Springer, Cham.

Ejeh, P.C. and Igbokwe, A.C. (2025). Artificial intelligence and the future of Nigerian university students. *OCHENDO: An African Journal of Innovative Studies*, *6*(1), 45–56.

Fernandes, C.W., Rafatirad, S. and Sayadi, H. (2023). Advancing personalized and adaptive learning experience in education with AI. EAEEIE 2023—Proceedings of the 2023 32nd Annual Conference of the European Association for Education in Electrical and Information Engineering.

Finnish Government & University of Helsinki. (2018). Elements of AI—Massive open online course. Available at https://www.elementsofai.com. (last accessed 29th July, 2025).

Fitas, R. (2025). Inclusive education with AI: Supporting special needs and tackling language barriers. arXiv preprint arXiv:2504.14120.

Forbes-Riley, K. and Litman, D. (2011). Designing and evaluating a wizarded uncertainty-adaptive spoken dialogue tutoring system. *Computer Speech & Language*, *25*(1), 105–126.

Gaisie, E., Owusu-Boateng, O., Yidana, I. and Ghansah, B. (2025). Ethical considerations and data privacy in artificial intelligence. In *Unlocking the Potential: Artificial Intelligence in Education* (pp. 58–81), ed. O. Owusu-Boateng. Deep Science Publishing.

Gligorea, I., Cioca, M., Oancea, R., Gorski, A.T., Gorski, H. and Tudorache, P. (2023). Adaptive learning using artificial intelligence in e-learning: A literature review. *Education Sciences*, *13*(12), 1216.

Graesser, A.C., Chipman, P., Haynes, B.C. and Olney, A. (2005). AutoTutor: An intelligent tutoring system with mixed-initiative dialogue. *IEEE Transactions on Education*, *48*(4), 612–618.

Hartong, S. and Piattoeva, N. (2021). Contextualizing the datafication of schooling–a comparative discussion of Germany and Russia. Critical Studies in Education, *62*(2), 227–242.

Hashim, S., Omar, M.K., Ab Jalil, H. and Mohd Sharef, N. (2022). Trends on technologies and AI in education for personalized learning: Systematic literature review. *International Journal of Academic Research in Progressive Education and Development*, *11*(1), 884–903.

He, T., Li, C., Wang, J., Wang, M., Wang, Z., and Jiao, C. (2022). An emotion analysis in a learning environment based on theme-specified drawing by a convolutional neural network. *Frontiers in Public Health*, *10*, 958870.

Hlosta, M., Herodotou, C., Papathoma, T., Gillespie, A. and Bergamin, P. (2022). Predictive learning analytics in online education: A deeper understanding through explaining algorithmic errors. *Computers and Education: Artificial Intelligence*, *3*, 100108.

Holstein, McLaren and Aleven. (2017). Intelligent Tutoring Systems. In The SAGE Encyclopedia of Educational Technology (pp. 378–382). SAGE Publications.

Hooda, M., Rana, C., Dahiya, O., Rizwan, A. and Hossain, M.S. (2022). AI for assessment and feedback to enhance student success in higher education. *Mathematical Problems in Engineering*, 2022, 5215722.

Jabeen, Z., Mishra, K., Dayal, R. and Mishra, B. (2024). Transforming Education in the World of AI. 2, 113.

Kira Learning (2023). AI Teaching Assistant. Available at https://support.kira-learning.com/ai-teaching-assistant, 3rd August, 2025.

Kumar, D. (2024). AI-driven automation in administrative processes: Enhancing efficiency and accuracy. *International Journal of Engineering Science and Humanities*, *14*(Special Issue 1), 256–265.

Li, Y. and Zou, L. (2022). The application of AI teachers in facilitating game-based literacy learning: An introduction to theories and evidence-based tools. In Handbook of research on acquiring 21st century literacy skills through game-based learning (pp. 381–395). IGI Global.

Liu, D., Huang, R., Chen, Y., Adarkwah, M.A., Zhang, X., Li, X., ... and Da, T. (2024). Personalized tutoring through conversational agents. In: Dejian Liu, Ronghuai Huang, Ying Chen, Michael Agyemang Adarkwah, Xiangling Zhang, Xin Li, Junjie Zhang, Ting Da. *Using Educational Robots to Enhance Learning: An Analysis of 100 Academic Articles* (pp. 59–85). Singapore: Springer Nature Singapore.

Luckin and Holmes. (2017). Intelligence Unleashed: An Argument for AI in Education. UCL Institute of Education Press.

Luckin, R. (2018). Machine Learning and Human Intelligence. The future of education for the 21st century. UCL institute of education press.

Maity, S. and Deroy, A. (2024). Generative AI and its impact on personalized intelligent tutoring systems. arXiv preprint arXiv:2410.10650.

Maghsudi, S., Lan, A., Xu, J. and Van Der Schaar, M. (2021). Personalized education in the AI era: What to expect next. *IEEE Signal Processing Magazine*, *38*(3), 37–50.

Minn, S. (2022). AI-assisted knowledge assessment techniques for adaptive learning environments. Computers and Education: *Artificial Intelligence*, 3, 100050.

Mittal, U., Sai, S., Chamola, V. and Sangwan, D. (2024). A comprehensive review on generative AI for education. *IEEE Access*.

Molenaar, I. (2021). Towards hybrid human-AI learning technologies (Conference keynote). Sixteenth European Conference on Technology Enhanced Learning, Bozen-Bolzano, Italy.

Nawale, A.M. (2022). Blending learning: Towards transforming Indian higher education. *Shanlax International Journal of English*, *10*(4), 30–36.

Perach, S. and Alexandron, G. (2022). A blended-learning program for implementing a rigorous machine-learning curriculum in high-schools. L@S 2022 – Proceedings of the 9th ACM Conference on Learning @ Scale (pp. 267–270).

Perkins, M., Furze, L., Roe, J. and Macvaugh, J. (2024). The AI assessment scale (AIAS): A framework for ethical integration of generative AI in educational assessment. *Journal of University Teaching and Learning Practice*, *21*(6), 49–66.

Popenici and Dotson. (2017). The future implications of artificial intelligence on teaching and learning in higher education. *Research and Practice in Technology Enhanced Learning*, 12(1), 1–17. (From https://link.springer.com/article/10.1186/s.

Porayska-Pomsta, K., Holmes, W. and Nemorin, S. (2024). The ethics of AI in education. arXiv preprint arXiv:2406.11842.

Pressey, S.L. (1926). A simple apparatus which gives tests and scores—and teaches. *School and Society*, *23*(586), 373–376.

Pretorius, A. (2023). A pedagogy to support learning analytics: a case for a specific higher education institution. In Handbook of research on AI and machine learning applications in customer support and analytics (pp. 56–79). IGI Global.

Ramasamy, T. (2024). AI in higher education. *International Journal of Scientific Research in Engineering and Management*, *8*(6), 1–5.

Rebolledo Font de la Vall, R. and González Araya, F. (2023). Exploring the benefits and challenges of AI-language learning tools. *International Journal of Social Sciences and Humanities Invention*, *10*(1), 7569–7576.

Renz, A. and Vladova, G. (2021). Reinvigorating the discourse on human-centered artificial intelligence in educational technologies. *Technology Innovation Management Review*, *11*(5), 5–16.

Reuters. (2025). China to rely on artificial intelligence in education reform bid. 17 April. Available at https://www.reuters.com/world/asia-pacific/china-rely-artificial-intelligence-education-reform-bid-2025-04-17/ (last accessed 5th August, 2025).

Russell and Norvig. (2010). Artificial intelligence: A modern approach. Prentice Hall. (Based on https://dai.fmph.uniba.sk/courses/ICI/References/rn.chap1.pdf) (last accessed 5th August, 2025).

SACE Board of South Australia (1968). AI guidelines for educators. Available at https://www.sace.sa.edu.au/teaching/assessment/assessment-and-academic-integrity/guidelines-for-using-ai

Santos Guimarães Carvalho, I. (2020). The construction of digital teaching skills: A challenge for teacher training. *ACM International Conference Proceeding Series*, 892–895.

Selwyn, N. (2019). What's the problem with learning analytics?. *Journal of Learning Analytics*, *6*(3), 11–19.

Shaik, T., Tao, X., Li, Y., Dann, C., McDonald, J., Redmond, P. and Galligan, L. (2022). A review of the trends and challenges in adopting natural language processing methods for education feedback analysis. *IEEE Access*, *10*, 56720–56739.

Shilowaras, M. and Jusoh, N.A. (2022). Implementing AI chatbot in Moodle learning management system. *Engineering, Agriculture, Science and Technology Journal*, *1*(1), 70–75.

Siau, K. and Wang, W. (2020). AI ethics: Ethics of AI and ethical AI. *Journal of Database Management*, *31*(2), 74–87.

Siddiqui, M.T. (2024). Innovative Pedagogical Practices: Transforming Education with Artificial Intelligence. Available at SSRN 5059335.

South for Education—EdChat AI Pilot Program. https://www.education.sa.gov.au/parents-and-families/curriculum-and-learning/ai/edchat, (last accessed 5th August 2025).

Tan, L.Y., Hu, S., Yeo, D.J. and Cheong, K.H. (2025). Artificial intelligence-enabled adaptive learning platforms: A review. *Computers and Education: Artificial Intelligence*, *9*, 100429.

The Economic Times. (2025). OpenAI launches Academy in India; to expand access to AI education. (last accessed 5th August 2025)

Toktarova, V. I. and Pashkova, Y. A. (2022). Predictive analytics in digital education: Analysis and evaluation of students' learning success. *Siberian Pedagogical Journal*, *1*, 97–106.

U. Mittal, S. Sai, V. Chamola and D. Sangwan. (2024). A Comprehensive Review on Generative AI for Education. IEEE Access, vol. 12, pp. 142733–142759

UNESCO. (2023). *AI in Education: Current Status and Policy Directions*. Paris: UNESCO.

Universal AI University, India. (2018). Academic programs. Available at https://www.universalai.in. (last accessed 19 August 2025).

US Department of Education. (2024). AI and the Future of Teaching and Learning.

Valdivieso, T. and González, O. (2025). Generative AI tools in Salvadoran higher education: Balancing equity, ethics, and knowledge management in the Global South. *Education Sciences*, *15*(2), 214.

Wang, X., Huang, R.T., Sommer, M., Pei, B., Shidfar, P., Rehman, M.S., ... and Martin, F. (2024). The efficacy of artificial intelligence-enabled adaptive learning systems from 2010 to 2022 on learner outcomes: A meta-analysis. Journal of Educational Computing Research, *62*(6), 1348–1383.

Watters, A. (2021). *Teaching Machines: The History of Personalized Learning*. MIT Press.

Yadav, A. (2020). Assessment of AI virtual learning applications & conventional classroom learning methods. April. Available at https://www.researchgate.net/profile/Sameer-Mohammad-4/publication/354876299_Assessment_of_Artificial_Intelligence_virtual_Learning_Applications_Conventional_Classroom_Learning_Methods/links/615279aa154b3227a8b3e357/Assessment-of-Artificial-Intelligence-v. (last accessed 19th August 2025).

Zawacki-Richter, O., Marín, V.I., Bond, M. and Gouverneur, F. (2019). Systematic review of research on artificial intelligence applications in higher education–where are the educators?. International journal of educational technology in higher education, *16*(1), 1–27.

Zhai, C., Wibowo, S. and Li, L.D. (2024). The effects of over-reliance on AI dialogue systems on students' cognitive abilities: A systematic review. *Smart Learning Environments*, *11*(1), 28.

CHAPTER 14

Transforming Learning Environments: The Synergistic Roles of AI and 5G Technologies in Education

Mohan Singh[1*], *Seema Singh*[2], *Kapil Dev Tyagi*[3] and *Vaibhav Bhushan Tyagi*[4]

1. Introduction

Artificial intelligence (AI) and Fifth-generation (5G) technology applications have huge potential to transform education and teaching methods, providing unparalleled opportunities to build dynamic, personalized, and accessible learning ecosystems. Education will be one of the first aspects that will gradually begin to integrate AI into daily life so seamlessly that people will never know the difference. 5G networks are also to leapfrog these revolutions by delivering the high-speed, low-latency connectivity that enables AI-powered applications to work, and by enabling richer immersive experiences (through enhanced virtual reality and augmented reality experiences) and real-time data processing. This convergence forms ecosystems that personalize individual student experiences, provide immersive educational encounters, and empower educators through intelligent support (Kohnke & Zaugg, 2025). Educational institutions are introducing AI to provide innovative solutions for meeting diverse student needs and also enhancing teaching outcomes beyond conventional approaches (Dey, 2024).

[1] Department of ECE, G.L. Bajaj Institute of Technology and Management, Uttar Pradesh, India. https://orcid.org/0000-0002-2973-3676

[2] Department of Education, Lloyd Institute of Management & Technology, Uttar Pradesh, India. https://orcid.org/0009-0009-4776-2241

[3] Department of ECE, Jaypee Institute of Information Technology, Uttar Pradesh, India https://orcid.org/0000-0002-4596-2321

[4] Department of ECE, ISBAT University, Uganda https://orcid.org/0000-0001-8153-3607

* Corresponding author: mohansingh571@gmail.com

DOI: 10.1201/9781003639367-14

Current education blends have been emphasizing the importance of personalized learning to tailor educational content and utilization of AI to meet the demands of each learner depending on their specific needs, interests, and pacing of learning (Jian, 2023). This method is especially important when considering different learning styles and speeds in classrooms, ensuring that all students are included and improving the overall academic potential. Even AI-powered tools that automate administrative tasks can be used to free up educators to maximize their ability to engage students and develop their curricula. While a rollout of AI compatible with 5G is conceivable, it is also a possible source of a myriad of new issues ranging from implementation to equity. Such challenges range from data privacy issues to the digital divide, and all-encompassing teacher training. A strategic plan that outlines how to take the next steps of embracing these technologies in the academic world and a dive into a culture of innovation and improvement are the only methods able to combat these challenges.

As a backdrop, this chapter aims to provide a systematic guide to the integration of AI and 5G technologies in the areas of education, the challenges related to these technologies, their potential opportunities, and possible solutions, as shown in Figure 14.1. The increased data throughput that 5G networks provide is critical to supporting the needs of next-generation educational technologies (i.e., virtual reality) to offer an immersive learning experience (Ahmad et al., 2023). This chapter thus aims to instruct potentially involved educators, policies, and technology developers on the specific opportunities offered by 5G and AI and their joint implications on education for the future, by breaking down what exactly constitutes a future-proof imagined education. It is expected to have a flexible and long-term synergistic role that can grow with the changing needs of the education industry and utilize new technologies.

Furthermore, this chapter works to fill the existing gap in the literature exploring the strategic use of AI in education, understanding both the vast opportunities and challenges, and proposing a model for practical implementation (Baker, 2021). AI and other technologies can be powerful allies in enriching the learning process for students, but we must not sacrifice the expertise we have in human teachers or the interactions they provide for students. They could be game changers for education and the transformative potential of education, as AI and 5G have the power to personalize learning, automate administrative work, and deliver immersive educational experiences, improving student outcomes and preparing them for future workforce needs.

1.1 Background and Motivation

The intersection of AI with 5G is a transformative development within the sphere of education, allowing us to construct customized learning environments like never before while making education more accessible and efficient. Owing to the diverse learning capabilities of students, the traditional educational models find it very difficult to address the learning differences among students. Bearing in mind the struggle behind the fourth industrial revolution, a few AI-driven solutions were fostered as promising tools to combat such challenges through adaptive learning that accounts for the unique needs and learning styles of individual students. This chapter stemmed from the perspective that although AI and 5G technologies have enormous potential to revolutionize education, effective integration of such technologies

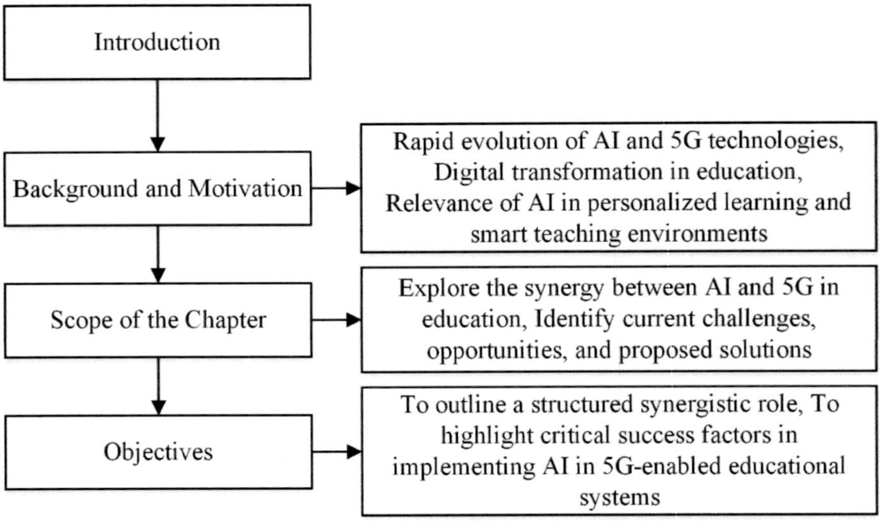

Fig. 14.1 Introduction to AI in emerging education.

into educational contexts requires thoughtful enterprise planning and strategic implementation, as well as an awareness of the issues and opportunities they can present (Owoc et al., 2019).

- *Rapid evolution of AI and 5G technologies*: New developments in AI and 5G technologies emerge frequently, challenging educational methods to adapt and incorporate the most innovative features.
- *The growing significance of tailored learning experiences*: The need for personalized learning experiences has never been more crucial, stemming from the growing recognition of the diverse needs of students.
- *The need for fast and accessible education*: In the context of the demand for fast and accessible education, there is a pressing need for innovative approaches that can deliver quality education to individuals in remote locations and underserved populations. These include the challenges of traditional educational models—which rarely pragmatically address the learning needs of students, and lead to discrepancies in academic results and engagement levels.
- *Digital transformation in education*: With the global education mechanism changing rapidly, education institutions actively pursue digital transformation by taking advantage of digital tools and platforms to facilitate the teaching process and improve the learning experience. Just in the past few years, we have seen a rise in AI-enabled learning systems that can deliver personalized learning material, adapt to the needs of individual students, etc. However, the integration of AI in education is not without its drawbacks. The major risks of using AI in education concern algorithmic bias, data privacy, and the importance of human teachers. AI's academic literature includes those on its applications to personalized and adaptive learning systems, as well as intelligent tutoring and learning analytics, which represent major steps forward in attempting to both

personalize and socialize the learning experience (Mahmoud & Sørensen, 2024). There is no limit to how much AI can help improve education outcomes.

- *Area of relevance–AI personalized learning and smart teaching environments:* AI algorithms can process several terabytes of data to tailor learning material and experiences around students, in addition to helping educators with predictive analytics to address these challenges through optimizing administrative tasks and providing timely interventions. The use of educational nurturing by AI in the form of customized feedback along with administrative automation and improvement of educational quality is bound to be transformative in higher education where AI can assume human responsibilities through ability prediction and personalized interventions. Learning environments and teachers are increasingly recognizing the potential for improving learning outcomes associated with new AI tools in education (Adewale et al., 2024).

1.2 Scope of the Chapter

This study initiates by critically exploring the intersection of AI and 5G communication technology and how they relate to educational innovation and digital pedagogy. Next, we will turn our attention to AI applications for the delivery of personalized learning experiences. It seeks to explore ways in which AI-enabled tools and platforms can help tailor the curriculum for every student and facilitate better learning results for them. With the promise of ultra-fast and low-latency connectivity, 5G power can facilitate bandwidth-heavy applications for virtual reality (VR), augmented reality (AR), and real-time collaboration in education. To be honest, it will encompass many of the issues regarding the compliance of education via AI and 5G technology which include data privacy, efficacy, ethical aspects, and the digital divide (Jiang et al., 2021).

- *AI and 5G in education*: AI technology has the potential to be better used in educational applications through providing speed and low latency, and through supporting such immersive learning processes as VR and AR. The integration of Artificial Intelligence (AI) and 5G technologies is transforming the education landscape offering intelligent, on-demand, and ultra-personalized services. Learning becomes interactive as AI enables adaptive learning experience, automated grading and multilingual support; and with 5G, it enables ultra low-latency and ultra fast connections, delivering experience of accessing content in a frictionless environment with technologies. What this chapter will provide are recommendations to policymakers, educators, and technology developers on how to capitalize on AI and 5G technologies to facilitate more engaging, equitable, and effective learning environments.
- *Identify current challenges, opportunities, and proposed solutions*: There are challenges around deploying AI and 5G in education like data privacy, safety/security, ethical issues, and the digital divide. AI and 5G applications in education The combination of AI and 5G in education creates new potential for coding, for instance through 1-1 tutoring, personalizing learning curriculums, and even "immerse digital classrooms" where (virtual) reality and the real word interact in real-time. But there are still barriers; i.e. disparate infrastructure, data privacy issues, and limited readiness among educators. To respond to these, smart

solutions include increasing 5 G access through public-private partnerships, developing ethical AI frameworks and investing in teacher training for tech-enabled pedagogy. The chapter intends to help practitioners in both identifying the opportunities as well as ethical dilemmas of K-12 classroom uses of AI, and provides an overview of instructional resources that teachers may use to engage students with AI (Akgun & Greenhow, 2022).

- *Structure of the chapter*: The chapter will begin with a general description of how the education system can be revolutionized using AI and 5G with methodology and intelligent education hardware—AI-assisted personalized learning, intelligent tutoring systems, and more intelligent educational hardware to provide an improved experience and education output. We will talk more about the use of these technologies in personalized learning platforms, intelligent tutoring systems, and automated assessment tools, and how 5G is creating a convergence world for smoother, faster, stable connectivity and enabling learners to run AR/VR interactive environments for richer experiences. It will give a big picture of the challenges of introducing AI and 5G to the education space concerning data privacy and security issues, ethical concerns, and the digital divide. Next, we will dive into potential answers to these dilemmas; like the need for ethical guidelines in AI creation, the carbon footprint of the technology used in schools, addressing data privacy and security as concerns in the classroom, and finally how we can bridge the digital divide so that all students have equal access to technology. It will end with a brief list of findings, as well as recommendations for policymakers, educators, and technology developers on how to harness the power of AI and 5G to build more engaging, equitable, and effective learning contexts.

1.3 Objectives

The primary focus of this chapter is to explain a comprehensive model for the integration of AI components used within educational-service-oriented teaching across connected 5G cloud-based models. The goal is to explore how AI-powered tools and platforms can tailor educational content to the needs of each student and the impact this has on learning outcomes. There are many such technologies; AI and 5G in combination could completely change how education and guidance take place, bringing a new era of more personalized, engaging, and effective features of education for students of all ages (Perez-Ortiz et al., 2021).

- *A systematic synergistic role*: This chapter aspires to offer teachers, policymakers, and technology builders a working framework for exploring how to integrate AI into the classroom, whether in curriculum design, learning delivery, student assessment, or support services. It emphasizes the need to eliminate the digital divide and ensure that all children have equal access to technology, as there is no reason any child should be left behind due to their socio-economic status or geographical location. In this chapter, it unfolds a strategic summary for educators that navigates the transformative path of Artificial Intelligence (AI) in educational contexts. It outlines possible courses of action to take to help guide procurement decisions and strategy for the deployment of AI technologies for educational benefit and efficiency.

- *To identify key success factors in the application of AI in 5G-powered educational systems*: This is about a healthy confluence of educational best practices, learning theories, and principles of instructional design, grounded in educational tools and platforms driven by learning and AI. In this chapter, we highlight the importance of professional development programmes for educators that provide them with the training and knowledge needed to effectively leverage AI tools within the classroom and foster progressive pedagogical approaches.

2. Fundamentals of AI and 5G in Education

2.1 Introduction of General AI in Education

Artificial intelligence is revolutionizing many sectors, and education is no exception. Artificial intelligence in education (AIED), then, strives to utilize AI techniques to provide personalized instruction, automate administrative functions, and offer real-time and data-driven insights for instructors. But AI also has the power to transform teaching and learning, as shown in Figure 14.2. AI solutions could also offer feedback on student performance, identify areas of concern, and ensure that students with special needs or language barriers overcome any related challenges. As an emerging field, AIED promises to address many of the key challenges facing contemporary education systems, including personalization, teacher burnout, and the need for lifelong learning opportunities. Here exists an opportunity for a tailor-made solution of an effective, scalable AI system that can not only personalize learning as per student pace, demand, and choice but also automate the manual processes of any school system such as exam schedulings, grading, and delivery of exam content. AI-driven solutions can analyse students' data and identify learning gaps and sources for standardized accommodation, support students with disabilities using assistive technologies, and provide language learning through real-time translation and feedback (Lampou, 2023).

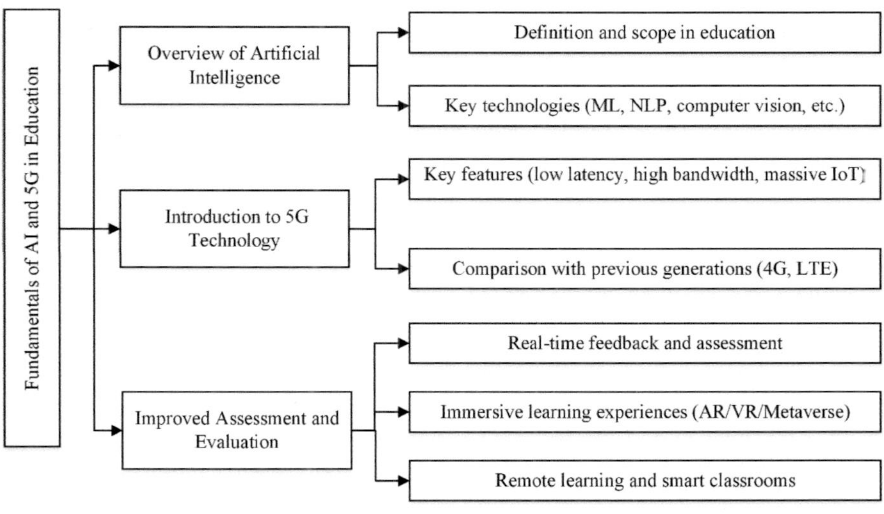

Fig. 14.2 Fundamentals of AI and 5G in education.

- *Definition and scope in education*: AI has applications in education at many levels—adaptive learning platforms, intelligent tutoring systems, automated assessment and feedback, administrative automation, and more. AIED itself is a broader domain, encompassing learning in both formal and informal contexts across the lifespan, including online learning. The software adopts AI techniques to create tailored learning solutions as well as streamline administrative processes and offer data-driven suggestions to enhance the quality of learning. The field now situated within AIED comprises diverse applications from personalized learning environments and intelligent tutoring systems, to assessment and feedback systems based on machine learning. All of these—intelligent tutoring systems, chatbots, and automated grading and assessment—help create efficiencies that enable teachers to spend more time teaching and produce more accurate and consistent feedback. Using machine learning, AI can process its understanding over time, creating learning environments that cater to the pace and learning preference of the student and providing instant feedback and aids, enabling individualized, self-paced learning (Dimitriadou & Lanitis, 2023).
- *Some of the main technologies powering AIED may include the following*: The key technologies of AIED are machine learning (ML), natural language processing (NLP) and computer vision. Although these are major application domains of ML, in AIED it targets a wider array of technologies that is not limited to ML based systems but spans a variety of mechanisms to improve student learning. Generation systems like those in machine learning, where powerful algorithms train in small batches of data, have synergies with methods designed for personalization or adaptive learning, as data-based system features can be adapted to data, and can thus improve over time. NLP is one of the applications in this domain so that the systems can read text in human language which helps the students to interact with an AI tutor or chatbot. Computer vision enables systems to analyse images and videos after which they can help in providing feedback on student work, or it can help in interactive learning experience. They provide individualized feedback to students based on analysis of massive amounts of data, automate teachers' administrative tasks, and can even predict which students are likely to succeed so that teachers can focus on kids who need help not to fall behind. They can be used to generate more engaging materials, so they can be implemented in personalized learning platforms, intelligent tutoring systems, automated assessment and feedback apps, and virtual learning environments. Interpretation and ISR of Be-Robots in Open World Robots can also function as teaching assistants, providing personalized support for students and delivering interactive and engaging learning (Yang et al., 2023).

2.2 Introduction to 5G Technology

5G is not just about faster mobile data, but also an entire system that promises massive speed, lower latency, and more reliable connectivity than its predecessors could ever dream of. What this means is that 5G will revolutionize education by facilitating seamless connectivity, immersive learning experiences, and real-time data analytics.

The quick evolvement of computing hardware is evidenced by miniaturization, integrating giant computing capacities into small devices; the technology also opens up new educational opportunities like VR field trips, studying in place, and students and teachers collaborating in real time.

- *Key features (low latency, high bandwidth, massive IoT)*: 5G is the next-generation cellular network technology, making it a high-speed, ultra-low latency, and massive technology. These features allow for new applications in education, such as VR and AR, remote learning, and real-time collaboration. 5G can accommodate many connected devices at the same time, which is perfect given classrooms usually have multiple students and devices. 5G networks provide increased throughput and lower latency which can help enable real-time data analytics that can enhance how educators can follow up on students' progress and adapt teaching patterns to suit them as needed. With 5G networks, a two-hour film can be downloaded in fewer than four seconds—whereas it takes over six minutes on 4G networks (Mohamed et al., 2021).
- *Comparison with previous generations (4G, LTE)*: 4G enables more immersive and interactive content along with real-time data analytics and feedback. 5G technology pulls together cell towers and allows for many more, tiny, low-power towers to cover a larger area with seamless connectivity, allowing access to a wider range of education applications, from VR field trips to real-time teacher collaboration, remote networked resources, and so on. The major challenges when dealing with backhaul requirements up until 4G cellular networks were to have enough capacity, availability, deployment cost, and long-distance reach. But as 5G network capabilities and services were being added to the 4G cellular networks, the backhaul network began presenting challenges in addition to those of 4G cellular networks, which include ultra-low latency requirements and the ultra-dense nature of the network. 5G technology is aimed at meeting new application requirements, such as ultra-low latency and massive connectivity that go beyond the traditional domain of wireless communication, which is improving data rates and capacity. The 5G networks can support higher frequency bands and broader channel bandwidths than those of LTE (Islam et al., 2021).

2.3 Synergistic Role of AI and 5G in Education

Education deserves transformative innovations, and the pairing of AI and 5G provides exposure to immersive, personalized, and accessible learning. There is a lot of headroom for education to change for the better and the intersection of AI and 5G opens up new frontiers for immersive, personalized, open educational technology. AI-driven learning solutions benefit from the high-speed, low-latency connectivity provided by 5G to create insightful and engaging experiences tailored to meet individual needs. By using AI and big data, 5G-enabled algorithms can optimize the huge amount of data that 5G-connected machines generate and deliver tailored feedback and support, automate administrative tasks for teachers, predict student performance, and identify students who risk falling behind. This confluence opens new avenues to create innovative solutions that tackle many of the biggest challenges in education including access, equity, and personalization. In addition, the distributed processing capability of AI will enable effective learning to improve efficiency and contribute to the solution of security issues, while enabling 5G systems to realize advanced network autonomous driving (NAD), optimizing system performance and user experience through online optimization (Wu et al., 2021). This enables educators to develop AI-based learning experiences that are tailored to meet the needs of every individual student.

- *Real-time feedback and assessment*: AI algorithms can assess students' performance and provide personalized feedback in real time to enable improvement. Extensive data that comes out from the algorithms can be transmitted with the power of 5G, and therefore we can receive feedback and assessment in real time. AI processes full-fledged student data—such as their learning styles, pace, and preferences—enabling personalized learning paths to be formed and adaptively altered to meet the unique requirements of every individual learner. Such tools include immediate feedback tools, or formative assessment learning systems, that can help keep students on track and provide the support they need to meet their learning objectives.
- *Immersive learning experiences (AR/VR/Metaverse)*: Remote and immersive learning experiences such as virtual field trips and augmented classes would be driven based on in-depth collaboration when the study material is delivered through high-resolution images and videos in real time. AI can tailor these experiences based on students' interests and learning patterns, enabling them to be collaborative and more impactful. Virtual assistants and other AI applications can provide personalized support and guidance to students, making it faster and easier to receive answers to their questions and to complete their assignments. Whether they are using data visualizations to make more immersive boss battles or generating whole worlds to run around in, they can also give students a more exciting learning experience that promotes deeper understanding of a subject and reinforces their ability to think critically about it. The latest revolution in higher education digital transformation is AI, which can change higher education and help students with personalized feedback, automate administrative tasks, take higher-quality education services, and much more. This is in contrast with traditional learning where students are treated as a homogenous entity irrespective of their individual learning and understanding abilities. AI can also help reduce scheduling and record-keeping, freeing up teachers' and administrators' time to focus on lesson planning and instructional improvement.
- *Data mining*: With the ability to analyse huge amounts of information, AI can also help drive instructional practices in the classroom. With its high-speed connectivity, 5G enables access to different educational resources, letting various resources be connected through the internet, reducing the disparity between urban and rural areas. 5G networks allow these tools to take advantage of AI in offering students personalized support to meet their unique learning needs.
- *Remote learning and smart classrooms*: 5G and AI can facilitate remote learning and smart classrooms, making education affordable and accessible. The 5G network ensures faster and stable connectivity which enables students to use AI-powered platforms to receive personalized learning experiences, irrespective of the geographical constraints. AI makes personalized education

possible through adaptive learning systems, appraising student data, and providing customized instruction, or meshes of education material, tailored to the needs and development level of each learner. AI-enabled adaptive learning systems can keep track of how students perform, and where their strengths and weaknesses lie, and create personalized learning pathways. AI-based tools can also be used by teachers to analyse behaviour and attitudes by studying facial expressions to know how focused the students are and how they understood the lesson (Saputra et al., 2023). When employed to analyse students' questions and give them appropriate activities, AI can return personalized feedback to students, enabling customized learning experiences and filling teacher shortages through intelligent tutoring systems.

3. Current Applications and use Cases

The emerging technology of AI applications in modern education enhance inclusive learning and interactive environment globally, as shown in Figure 14.3.

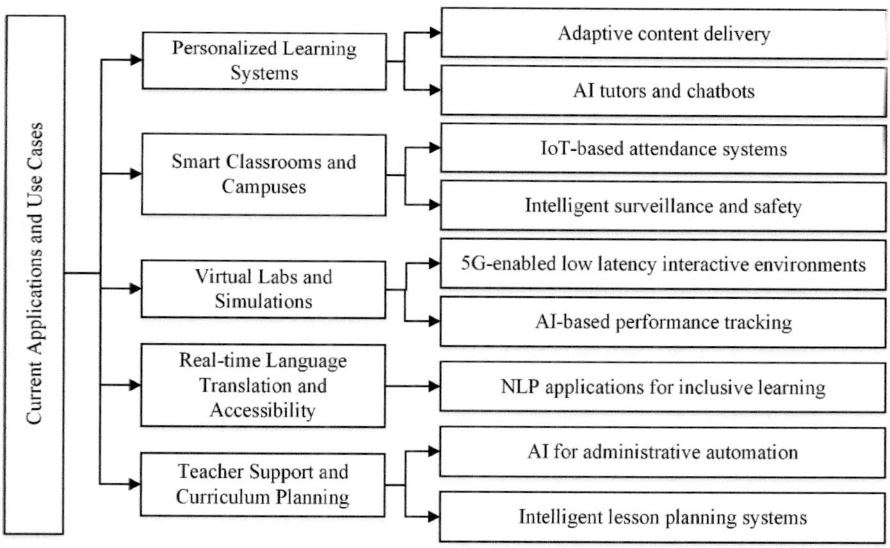

Fig. 14.3 Current AI applications.

3.1 Personalized Learning Systems

The AI algorithms can analyse and study a student's learning style, preference, and performance, and create personalized learning experiences for them. Personalized education applications driven by AI would learn from past experiences and study big data to identify the features and the learners' similarities and can provide the accurate traits of every student. These systems can personalize resources for each student according to their individual needs, providing customized content, feedback, and support (Ahmad et al., 2022). Additionally, personalized learning systems can help teachers distinguish among students, providing targeted assistance to students who may be struggling, with promise observed in applicable K-12 settings that can be used for both instruction and mixed-ability classes.

- *Adaptive content delivery*: AI algorithms can analyse and customize the difficulty level of the content based on students' performances. 5G enables the transfer of data in real time, which means adjustments can be made instantly.
- *Tutors and AI chatbots*: Both can provide students with personalized assistance, answering questions or assisting with assignments. The ability of AI systems to process and analyse large-scale datasets and extract meaningful patterns from them at scale provides a unique opportunity to understand the learners at a granular level. Using data obtained from an eLearning module, AI can assess an individual student's engagement levels and research trends, which indicate trends moving towards a certain mode of study or, in contrast, difficulty.

3.2 Smart Classrooms and Campuses

Artificial intelligence and 5G have the power to turn conventional classrooms into smart classrooms equipped with cameras, sensors, and other devices that record student behaviour and performance. This type of data can improve both teaching/learning outcomes and personalized learning environments (PLEs) (Gligorea et al., 2023). 5G ensures fast connections for learning materials, while AI simplifies administrative work, grading, and individual assessments.

- *IoT-based attendance systems*: With the help of IoT-based attendance systems, student attendance can be tracked automatically and real-time data can be provided to the teachers and the administrators. 5G also supports the stable and fast transmission of attendance data.
- *AI surveillance and safety*: AI-powered surveillance systems can monitor everything in schools, even teachers and students can be monitored from a distance. This allows for video footage to be transmitted in real time, which means the crew can respond to any incidents immediately, thanks to 5G.

3.3 Virtual Labs and Simulations

Virtual labs and simulations can be enabled by AI and 5G, as interactive classes, including hands-on experience, become available to students everywhere. These virtual labs and simulations can used for teaching all sorts of subjects, from science and engineering to medicine and healthcare. Concurrently, AI algorithms could be applied to design immersive and enjoyable learning environments, and 5G technologies would facilitate smooth connectivity with these learning resources.

- *5G-enabled low latency interactive environments*: 5G-enabled virtual labs offer students the experience of experimenting in a safe and controlled environment. The increased speed and lower latency of 5G allow students to join in on these simulations without lag, delivering a more accurate reflection of real-life events. Such virtual labs can recreate experiments that would be too risky or expensive to reproduce in a laboratory environment.
- *AI-based performance tracking*: Virtual laboratories allow for the creation of AI-based systems that track the performance of the students and provide

personalized feedback and guidance. These data can serve as a source of insights for instructors to personalize the learning experience and tailor support to reduce the learning curve in the area where students are struggling. AI technology can deliver instant feedback to students on their academic activities, thereby supporting enhanced performance through knowing immediately the strengths and weaknesses (Kaledio et al., 2024).

3.4 Real-time Language Translation and Accessibility

Language translation tools powered by AI can help break down language barriers and make education accessible to students from diverse backgrounds. With the advent of 5G, students who speak a different language may even have real-time translation of lectures and other educational materials, allowing them to participate in the learning process. AI is changing education by providing personalized learning experiences, automating administrative tasks, and enhancing accessibility for learners at all levels. AI-assisted approaches started to gain popularity, especially in English-as-a-second-language classes. With the internet and AI-based systems, online learning platforms have significantly transformed teaching and learning, assisted by painless and rapid access to learning content for educational innovations, which can be viewed in real time and connected through 5G for AI-based solutions (Yu & Nazir, 2021).

- *Inclusive learning with NLP applications*: Using NLP, educational software can facilitate real-time language translation, which allows non-English speaking students to understand educational material and participate in class discussions.

3.5 Teacher Support and Curriculum Planning

Artificial intelligence is also being used to assist teachers in doing their jobs. AI can automate grades, lesson plans, etc., so teachers can spend more time on engagement and mentorship, and not rote tasks. AI can also be used to analyse student data that can help identify areas of concern for struggling students and recommend tailored interventions. The power of AI can also be used to customize study materials and create tailored learning plans for every student. AI is used to create learning and assessment tools for some platforms that provide opportunities for dynamic and immersive learning (Godwin-Jones, 2024).

- *AI to automate administrative activities*: AI can help save teacher time by automating administrative activities, like grading, scheduling, or keeping records. By analysing data on how students perform, AI algorithms can also provide insights into learning patterns, which allows teachers to adjust instruction around the needs and preferences of every student. AI-driven algorithms can examine student performance, offering teachers guidance for adjusting curriculums to cover any gaps in students' knowledge.
- *Smart lesson design systems*: AI-based systems can take into account curriculum standards, learning objectives, and student performance data to create tailored lesson plans and resources for teachers. Tools that are empowered with AI can support educators with customized recommendations for instructional strategies,

activities, and assessments based on the learning styles and preferences of students, thus saving time and effort on lesson planning.

4. Synergistic Role for AI in 5G-Enabled Educational Ecosystems

This synergistic role encompasses crucial steps such as establishing the right infrastructure, effective data management, the development of AI models specific to educational needs, and the ongoing deployment and assessment of these AI solutions, as shown in Figure 14.4. There are specific challenges and opportunities at each stage that need to be addressed for successful implementation and sustainable growth time-frame as given in Table 14.1. A strong technology base that includes robust 5G networks and suitable hardware must be established to enable AI projects. These comprise deploying 5G base stations, edge computing devices, and high-performance computing resources in educational institutions. In parallel, educational institutions must implement robust data management frameworks that consider data privacy, safety, and ethical issues to enhance data collection and analysis (Tabuenca et al., 2024). This requires secure technologies for storing, processing, and sharing data, and rules for who is allowed to access and use the data.

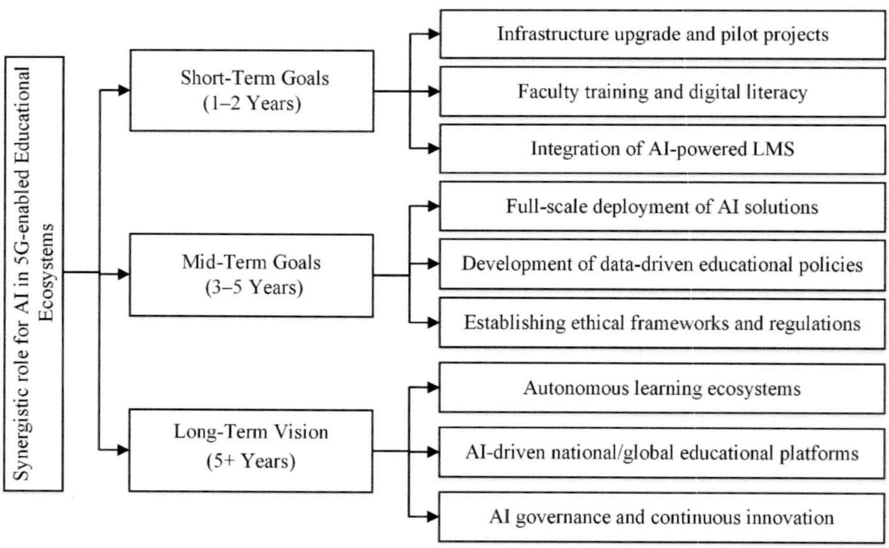

Fig. 14.4 Synergistic role of AI in 5G-enabled educational ecosystems.

4.1 Short-term Goals (1–3 Years)

At first, the priority needs to be building the fundamental elements and capacities for AI in education. Such initiatives include the rollout of 5G networks in schools, AI-based learning platforms, and training programmes for teachers and students. In the near term, AI-powered tools can help automate bureaucratic tasks like grading and scheduling, giving teachers more time for personalized attention.

Table 14.1 Synergistic role of AI in 5G-enabled educational ecosystems

Goal Category	*Timeframe*	*Objectives*	*Key Activities*
Short-term Goals	1–2 Years	This will include technical upgrades and pilot projects, faculty training and initiatives on digital literacy and learning, and integration of AI-powered "learning management systems" (LMS).	Evaluate current infrastructure and identify deficiencies, pilot test AI and 5G innovations, faculty training and upskilling covering digital literacy and AI bases, AI-delivered LMS integration.
Mid-term Goals	3–5 Years	Large-scale implementation of AI technologies, data-informed learning policies development, designing ethical policies	Scale up successful pilot initiatives; formulate evidence-based policies in education based on AI system-generated insights; develop ethical guidelines and regulations affecting AI in education, covering how algorithms are adopted, data privacy, and bias; establish cross-sectoral collaborations on solution building.
Long-term Vision	5+ Years	Educational innovation, ecosystems for teaching and learning in an autonomous age, national/global resourcing with an AI mechanism, governance of AI, and continuous innovation	Encourage independent learning ecosystems with adaptive AI; establish AI-led educational platforms on a national/global scale to ensure the availability of quality education; prepare AI regulations to ensure responsible and ethical use of AI in education; create continuous innovation models to adapt to changing technology and education needs.

- *Infrastructure upgrade and pilot projects*: These include working with telecom firms to develop 5G infrastructure for selected schools and universities, targeting areas with high student density with minimal access; and implementing pilot programmes to experiment with AI-powered tools for teaching (e.g., intelligent tutoring systems, automatic grading systems) in the classroom context. It helps to realize pilot projects for evaluating the effectiveness of AI-augmented solutions in different educational environments, collecting data on student progression and impact on authentic learning, as well as teacher acceptance and performance of the system.
- *Faculty training and digital literacy*: Critical elements include the integration of design-based professional development courses into the curriculum and the design of focused training plans enabling educators to inclusively implement AI technologies. These efforts should also prioritize data protection, ethical considerations, and deliverable steps that ensure 'smooth' integration of AI in teaching and learning environments. Awareness programmes for students should be implemented to educate them on AI concepts, responsible technology use, and

the potential benefits and risks of AI in education. Partnerships with technology companies and research organizations can help to create open-source AI tools and resources that are available to a broad network of educators and students.

- *Integration of AI-powered LMS*: Institutions should deploy AI-enabled learning management systems (LMS) that provide personalized content delivery, monitor student performance, and recommend personalized learning paths, enhancing the engagement and impact of educational experiences. With the introduction of AI-backed LMS, students will benefit from personalized learning experiences.

4.2 Mid-term Goals (3–5 Years)

The mid-term goals could be to increase AI deployments, improve analytics capabilities, and develop personalized learning experiences. This entails efforts to offer 5G access to more schools, creating new AI models for students to receive personalized learning experiences, and using data to make sound decisions. Moreover, such measures will also help to upscale the learners and prepare them for future job opportunities (Luan et al., 2020).

- *Full-scale deployment of AI solutions*: This involves xpanding 5G networks to new schools and universities, targeting features or areas that succeeded in pilot projects and where this type of advanced education technology is in high demand. In addition, tools and platforms enhanced by AI can be integrated at the instructor level as part of daily teaching routines, paving the way for personalized learning pathways, while also permitting adaptive assessment approaches that dynamically adapt to individual students' requirements. In K-12 education at least, we are not yet positioned to even really begin to harness the full power of AI (which can produce authentically personalized learning experiences that let students move at their own pace) and we need long-term research to help us accurately demarcate its positive or negative effects.
- *Development of data-driven educational policies (dashboards and reporting tools)*: Data analytics dashboards and reporting tools should be developed to provide information to educators, administrators, and policymakers about student performance, resource allocation, and programme efficiency. Data- and evidence-based decision-making should be encouraged across the educational system to increase student outcomes and close achievement gaps. Longitudinal studies are necessary for understanding the long-lasting effect of AI on student learning outcomes. Subsequent quarterly studies could also include new educational technologies in their repertoire in addition to AI, to take advantage of synergistic effects and contribute to innovation in teaching and learning based on evidence (Mundy, 2007).
- *Develop ethical guidelines and regulations*: There is a need to create guidelines and regulations that will address the ethical considerations of AI in education, such as data privacy, algorithmic bias, and transparency. Regulatory practices should be established to supervise the use of AI in education to block unintended negative consequences and ensure equal and just results for every student. The long-term effects need to be investigated further (AlTwijri & Alghizzi, 2024).

4.3 Long-term Vision (5+ Years)

In the long term, AI should be an enabler of an integrated education ecosystem for lifelong learning, innovation, and social mobility. These involve building AI-driven virtual learning spaces, establishing tailored training routes for students from all backgrounds, and cultivating a culture of educational innovation and trial.

- *Autonomous learning ecosystems*: Utilizing AI to build an AI-driven virtual learning environment to cater to individual needs can shift more students away from a one-size-fits-all model of learning. Each student will have a more engaging, responsive, and truly personalized experience that even extends beyond traditional classroom walls. Hence, continued education is promoted by providing AI-based services and tools to people to equip them with new capabilities and knowledge during their lives, helping in personal and professional enhancements.
- *AI-driven national/global educational platforms*: AI can be used to develop national or global online education initiatives that ensure access for all people to affordable and quality educational content, tailored learning experiences, collaborative learning, threby bridging the knowledge gap across vast geographical areas. Stakeholders should be educated about emerging technologies with the potential to transform traditional educational systems and pioneering countries should be encouraged to share their experiences to help mitigate the universal issues faced in technology adoption.
- *AI governance and continuous innovation*: Robust governance frameworks and regulatory structures should be established to oversee the ethical and responsible application of AI in education, ensuring compliance with data privacy regulations, promoting algorithmic transparency, and preventing bias and discrimination. There should be emphasis on a culture of continuous innovation and experimentation in education, promoting research and development initiatives that explore new AI applications and pedagogy, and supporting partnerships among educators, researchers, and industry actors to address evolving challenges and opportunities.

5. Uncertainties

The potential for AI in education is huge, but the challenges ahead are significant (see Table 14.2). To realize the gains that AI can bring to society, the challenges need to be addressed proactively. Other aspects of AI-based personalized education are as yet untried, such as overcoming the detrimental impacts of the lack of peers, generating and sustaining motivations for learning, enhancing diversity, and removing biases induced by data and algorithms. Such disparity is also due to infrastructural limitations in many regions that make it challenging for AI to be well adopted in education.

5.1 Technological Barriers

A solid technological infrastructure is essential for the success of AI in education. This entails consistent internet access, high-performance computational hardware, and adequately updated software and hardware.

- *Interoperability issues*: Wider deployment of AI technologies in education may be hindered by interoperability issues and lack of standardization among

Table 14.2 Challenges to implementation of AI in education

Key Barriers	*Description*	*Potential Impact*
Technological Barriers	*Interoperability issue*: Systems and platforms that do not integrate *Limitations in infrastructure*: This is especially pertinent about 5G and AI, as most systems depend on processing a significant amount of data; therefore, deploying these technologies with inadequate infrastructure, particularly in rural or underfunded areas, is difficult.	Stalls progress of AI-based solutions; permits academic inequities; and leads to increased complexity and cost in deployment.
Data Privacy and Security	*Data privacy and security*: The governance of data within AI systems *Responsible AI*: Business and data ethics for AI used in monitoring and assessment.	Undermines trust in AI-driven educational tools; ethical concerns regarding fairness and transparency; potential regulatory limitations; and compliance obstacles.
Digital Divide and Inequity	*Accessibility concerns*: Disparities in access to technology and digital resources among underserved populations could worsen pre-existing educational inequities.	It perpetuates existing inequities in learning and creates serious challenges to truly inclusive and equitable learning environments.
Resistance to Change	*Educator reluctance*: Concern over educator reluctance to adopt and implement new technologies in teaching. *Institutional inertia*: Overcome resistance to new technologies.	Restricts the use of AI in education; stifles innovations in teaching methods and serves as a roadblock to progress and modernization.
Lack of Policy and Standardization	*Deloitte found undefined regulatory frameworks*: Vague policies and regulations on the ethical use of AI in education. *Non-standardized implementation*: Failure to make things converge and be the same from one educational institution or region to another.	Generates ambiguity and confusion about best practices; disrupts the emergence of reproducibly good AI solutions; and could enable enchanted variable outcomes for students.

various AI systems and platforms, limiting the ability to share data and integrate tools, leading to disconnected learning experiences. Open standards and protocols that facilitate interoperability and data sharing among various AI systems and educational platforms need to be established so that these systems can integrate and work seamlessly with one another. There is a need for setting up testbeds and living labs in institutes of education to evaluate the performance, scalability, and interoperability of AI technologies in real-world environments, and then provide rich feedback for designing and deploying AI systems.

- *Rural or underfunded infrastructure gaps*: The digital divide is still here, with many unable to appropriately deploy AI-powered learning on account of unconnected schools or communities without sufficient devices. Closing infrastructure gaps in under-resourced areas requires coordinated investments in broadband infrastructure, device distribution programmes, and digital literacy initiatives by governments, organizations, and private sector partners, to ensure equitable access to AI-enabled learning opportunities for all students.

5.2 Data Privacy and Security

Protecting student privacy and data security is of utmost importance. The risks of surveillance and commodification of personal information are increasing, which highlights the need for effective data protection regimes to address data leaks and intrusions. Ethical concerns such as privacy violation and exploitation of the data in AI, especially those related to student data security, need to be resolved so that the personal information of the students can be secured and their trust in the educational institutions can be maintained (Huang, 2023).

- *Student data protection*: The use of a decentralized, peer-to-peer network means that no middleman is used for transfer, reducing the risk of student data exposure to central servers and networks vulnerable to attacks. There is a need to establish and implement comprehensive end-to-end data governance models that define the roles and responsibilities of stakeholders associated with each step within the data pipeline and align with applicable data protection/ethical standards.
- *Ethical consideration for AI usage in monitoring and assessment*: There is a need to design ethical guidelines about the use of AI in monitoring and assessment, and to undertake their effective implementation to ensure transparency, equity, and responsibility in educational contexts while factoring concerns related to algorithmic bias, data privacy, and risks to student well-being. As the integration of AI into the education space continues, it is especially important to ensure informed consent before the collection and use of student data and to prioritize transparency and accountability on the part of AI (Al-Zahrani, 2024).

5.3 Digital Divide and Inequity

This challenge encompasses the digital divide, or disparities in access to technology and digital literacy, which inhibits equitable AI deployment in education. There is a need to reduce the gap between those who have access to technology and digital literacy skills and those who do not, and to ensure that all individuals have the opportunity to benefit from AI in education. This may involve implementing initiatives that provide resources, support, and training to students, educators, and entire communities.

- *Accessibility concerns for underserved populations*: Some people (effectively, students from low-income backgrounds, rural regions, or with disabilities) may have no or limited access to the tools, infrastructure, and support required to take an interest in AI-powered learning situations. To help bridge the achievement gap in AI education and workforce opportunities, there is a need for fostering strategic partnerships with non-profit organizations, community groups, and educational institutions to develop outreach programmes, mentoring opportunities, and scholarship funding specifically for underrepresented populations in STEM and other disciplines.

5.4 Resistance to Change

A big challenge in adopting AI is getting on board the educators, students, and parents. The nature of AI education can lead many educators to be reluctant to implement new technology without training, fearing they will lose their jobs, or simply unsure how transformative AI can be in a classroom environment. So, support and training should be provided to the staff to address their concerns and to ensure the competent and effective integration of "AI into resource processes" to improve student performance.

- *The reluctance of educators to take on AI*: Training approaches need to be developed that help teachers learn how to incorporate AI in their instruction and be prepared to integrate those tools when they are ready to do so. Addressing educators' concerns and promoting a more positive attitude towards AI adoption can be done by emphasizing the potential of AI to augment their roles, ease administrative challenges, and personalize instruction.
- *Institutional inertia*: Existing processes and cultural barriers in educational institutions can make it difficult to adopt new technologies. To implement AI and create a culture of innovation and experimentation, educational institutions need to nurture an environment that emphasizes collaboration, taking risks, and continuous learning. This includes establishing cross-disciplinary teams to research and pilot AI solutions, rewarding instructors and staff members who promote innovation, and eliminating bureaucratic hurdles to prevent AI adoption.

5.5 Lack of Policy and Standardization

Finally, the current lack of comprehensive policies and standards for the integration of AI in education results in confusion, which can complicate responsible implementation. Stakeholders in education, government, and industry need to collaborate to create broad regulations and standards for the use of AI in education, which will pave the way for responsible AI innovation and ensure accountability and transparency in the practice. This involves setting ethical standards for the creation and use of AI, establishing norms for the privacy and security of information as well as establishing guidelines on accountability and transparency.

- *Undefined regulatory frameworks*: Without clear guidelines, educational institutions may find themselves uncertain about navigating the ethical and legal considerations associated with AI adoption, which has the potential to result in unintended consequences. By setting clear boundaries, on issues such as algorithmic bias, data privacy, and responsibility, stakeholders can ensure that AI is used ethically and responsibly to bolster learning and student success.
- *Fragmented implementation*: The diverse AI ecosystem lacks standardized protocols and formats, which hinders seamless integration across different platforms. There is a need to promote open standards and protocols that enable different AI tools and platforms to interoperate and work together effectively to foster innovative approaches to applying AI in education learning (Bura & Myakala, 2024).

6. Contingency

Through the marriage of 5G technology and AI in education, new frontiers can be realized, including personalized learning, greater accessibility, and new pedagogical approaches. Utilizing the fast connectivity, low latency, and high capacity of 5G networks, AI-based educational tools can provide personalized recommendations based on the student's specific requirements and preferences.

6.1 Enhanced Student Engagement and Motivation

Gamification, interactive simulations, and VR can be used in creating AI-powered educational platforms that engage students, keeping them motivated to learn by making them take part and thus ensuring knowledge retention. By integrating AI-powered adaptive feedback systems, we can offer students individualized information and assistance that adaptively strengthen their engagement and self-assurance in their learning process.

- *Gamified and immersive experiences*: The combination of AI with VR and AR technologies enables interactive learning experiences catered to the different learning preferences and styles of learners. VR and AR through devices allow for popular applications and experiences that could bring ancient and historical people, places, and moments to life in traditional classrooms in a way that allows for not just study or simulations, but also immersion through experiences, co-produced content, and projects.
- *AI-based real-time interactive platforms*: Students can receive real-time feedback as they progress through exercises and concepts, forcing them to complete unfamiliar concepts over and over until they gain the necessary confidence in their ability to learn. Machine learning algorithms allow these platforms to monitor student progress by detecting trends in learning patterns, and use those insights to evaluate the students' performance, to determine where that individual student is struggling, and to implement personalized interventions to guide them towards academic success.

6.2 Cost-effective and Scalable Education

Artificial intelligence is also capable of performing many administrative functions that optimize workflows and make better use of available resources, thereby saving educational institutions a significant amount of money. AI-based chatbots and virtual assistants can handle day-to-day questions and issues, freeing educators to focus on more complex tasks, such as creating course content and mentoring students. Moreover, AI-enabled personalized learning systems can be scaled up to support a large number of students, enabling underprivileged individuals to receive quality education.

- *AI for large-scale personalization*: AI holds the power to offer customized learning experiences at scale, enabling instructors to reach more students, mainly from deprived areas, with quality education. By tapping into data analysis, AI equips ed-tech to understand each student's unique learning style, pace and preferences, delivering personalized learning pathways that match their individual requirements, aspirations and academic objectives. This allows for individualized attention to ensure that all students have the support and resources they need to succeed academically.

- *Automation of administrative tasks*: AI can automate various administrative tasks, including grading, scheduling, and offering student support, thereby reducing the administrative load on educators and enabling them to concentrate on teaching and student engagement.

6.3 Lifelong and Inclusive Learning

Artificial intelligence will support lifelong learning by enabling personalized learning pathways, adaptive learning environments, and on-demand access to training and resources. Tools like AI-powered language translation and assistive technology can remove barriers to education for students with disabilities, ensuring everyone has access to learning and opportunities to succeed. AI-based career counselling systems are utilized to help people assess their abilities, discover career alternatives, and prepare the skills required to succeed in the job market.

- *Personalized learning assistance*: By using AI-powered technologies, personalized learning opportunities, support, and resources can be offered to people of any age or ability which can help them in lifelong learning and skill development. AI facilitates personalized learning opportunities, including adaptive assessments, tailored learning recommendations, and access to a wealth of learning materials on demand, letting students follow their interests and develop their knowledge and skills at their own pace.

6.4 Global Collaboration and Knowledge Sharing

Artificial intelligence can facilitate cross-cultural collaborations and knowledge sharing by connecting globally students and educators in virtual classrooms and online communities. AI-powered language translation systems can break free of the language seal, assist students in learning the material, and work with students speaking different languages.

- *Technology-driven translation and shared learning among regions*: AI-based device translation and shared learning among countries can promote international cooperation and cultural exchange by allowing people from different regions and backgrounds to communicate and share knowledge. AI-driven tools for translation can eliminate the language barrier and allow enterprising students from diverse language backgrounds to connect and work together, while virtual exchange programmes can facilitate the exchange of ideas, viewpoints, and best practices among peers across the globe.

7. Proposed Findings and Best Practices

Ensuring the potential of AI in education and learning advantaged by 5G ecosystems requires overcoming challenges to the wide and effective implementation of AI and integrating ethical best practices around its use. This section discusses the suggested approaches and action steps that educators, policymakers, and technical innovators can adopt to increase the advantages and decrease the perils of AI (see Table 14.3).

Table 14.3 Best practices for the implementation of AI in education

Finding/Best Practice	*Description*	*Key Components*
Policy and Governance Frameworks	A comprehensive investigation of ethical, pragmatic and societal implications is required for a responsible adoption of AI in education. To maintain transparency, accountability, and public trust, strong regulatory systems that support its ethical use in the learning environments should be developed	National strategies on AI specific to education; ethical frameworks governing data, algorithmic bias, transparency; policy guidelines regulating the usage of AI in institutions; focusing on data privacy and equal access to AI-powered resources.
Public-Private Partnerships	Enabling co-funded pilot programmes for the technology companies and the edtech startups to work together with the educational institutions. Such synergetic, cross-sectoral collaborations among academia, industry, and government sectors are vital.	Some potential areas of focus might include: collaboration on pilot programmes to test and validate proposed AI solutions; investment in funding partnerships to pilot and test AI research and development in education; industry partnerships to increase speed and efficiency.
Teacher Training & Capacity Building	Ensuring that educators receive continuous professional development and AI literacy programmes to effectively use AI tools in their teaching practices. For every recent negative press headline about advanced AI, there are educational opportunities for proactive investment in comprehensive teacher training programmes to ensure that educators develop the skills to successfully incorporate AI tools into their teaching practice.	Training courses on the use of AI in education; digital literacy and AI skill enhancement initiatives for educators; support and resources to educators for experimenting with AI-powered tools; teachers need to nurture a mindset for professional growth.
Infrastructure & Platform Development	Designing cloud-ready, secure, and scalable AI educational platforms using interoperability-ready standards and open-source infrastructure	Investment in robust and scalable infrastructure; development of secure and privacy-preserving cloud platforms; promotion of interoperability with open standards and tools.
Monitoring & Evaluation	With Key Performance Indicators (KPIs) and metrics to evaluate the success of each initiative, feedback loops, and continuous improvement models.	Identifying clear KPIs for AI in education; implementing feedback mechanisms to capture insights from educators and students; regularly assessing and refining AI solutions based on performance data.

7.1 Policy and Governance Frameworks

There is a need to devise appropriate policy and governance frameworks to hold stakeholders accountable for ethical standards, data usage, and algorithm transparency in AI-mediated learning. These types of frameworks must place firm guidelines around the ethical use of AI in education, to protect students' privacy and guard against inequitable opportunities and outcomes generated through biased AI algorithms.

- *Developing national education AI strategies*: Formulating country-specific education AI strategies would create a holistic framework for integrating AI with the education systems of the countries while keeping in mind ethical issues, data privacy concerns, and algorithm transparency. These strategies should also offer tangible principles to guide responsible AI application in education for the security of student data and the reduction of algorithmic bias. These policies should also ensure transparency at every stage in how data is processed and handled.
- *5G and AI ethical application directions*: The application and use of 5G and AI technologies in the education process must follow the values of transparency, fairness, and accountability. This will help to address issues like data privacy, algorithmic bias, and whether AI could take over the role of human teachers, thus ensuring responsible and ethical development while reducing any adverse effects.

7.2 Public-Private Partnerships

These partnerships promote collaborations between public and private sectors to harness expertise and resources for creating AI-powered educational solutions. They may enable the creation of cutting-edge AIED tools and platforms that address the varied requirements of students and educators.

- *Partnerships with tech companies and edtech startups*: There is a need to encourage partnerships among educational institutions, technology companies, and startups to foster innovation and develop AI-driven solutions that target specific pain points in education. This collaboration can result in developing tailored AIED tools and platforms that cater to the unique needs of learners and educators, ultimately improving educational outcomes and facilitating digital equity in the educational landscape.
- *Co-funded pilot programmes*: Partnerships among researchers, educators, and business partners must be developed to fund pilot programmes to test the effectiveness and scalability of AIED solutions across various educational settings. Such initiatives can also inform policy and resource allocation by providing valuable insights into the challenges that arise and opportunities that may exist in deploying AI in education.

7.3 Teacher Training and Capacity Building

Another step can be to encourage teacher training and professional development programmes that equip educators with the skills and knowledge needed to effectively integrate AI into their classroom practices. Such training programmes would promote a broad perspective—one that is based on digital literacy, pedagogical flourishing, and the ability to critically evaluate and manipulate AI-based learning resources.

- *Continuous professional development*: Educators should be provided with opportunities for continuing professional development so that they can keep up to date with the developments surrounding AIED and learn how to effectively integrate AI into their lessons. Such programmes should focus on nurturing teachers' digital literacy, pedagogical skills, and ability to evaluate and adapt AI-based educational tools.
- *AI literacy for educators*: AI literacy education programmes can be created for educators to support their understanding of the capabilities and limitations of AI, and to help them use AI tools with confidence to improve teaching and learning. Augmenting current practices with capabilities that enable AI in classroom spaces leads to innovative outcomes that benefit the students.

7.4 Infrastructure and Platform Development

There should be focus on developing strong 5G infrastructure and free educational platforms to run AI-driven educational apps. These platforms need to be built to scale, to interoperate, and be available to all students and educators regardless of their zip code or socio-economic status.

- *Secure cloud-based AI education platforms*: AI education platforms in the cloud should be designed for security, scalability, and interoperability and should provide students and teachers with intelligent tools and learning materials. These experiences are used as a cornerstone to build personalized, data-driven and social learning experiences. These platforms need to provide a safe and trusted environment for users by prioritizing security and privacy to enable innovation and improve learning outcomes.
- *Open-source tools and interoperability standards*: The use of open-source tools and interoperability standards for all AIED platforms should be advocated to facilitate innovation, minimize the cost of ownership, and ensure system compatibility. Now under consideration, these standards will make it easier to share data and resources, which will make collaboration easier and, hopefully, allow all students and teachers to access new tools being developed in this burgeoning ecosystem.

7.5 Monitoring and Evaluation Mechanisms

Such mechanisms involve creating very strong monitoring and evaluation plans to assess the impact of AI-driven educational interventions on students' learning outcomes, teachers' effectiveness, and educational inequity. Such mechanisms should incorporate both quantitative and qualitative data collection methods, along with feedback loops that allow for ongoing refinement and adjustment.

- *KPIs and metrics for success*: Key performance indicators and metrics can be established to monitor the efficacy and impact of AIED initiatives, considering factors such as student outcomes, educator satisfaction, and equitable access. They should provide measurable evidence of both the benefits and challenges of AI in education, helping to inform data-driven decisions about the value and distribution of resources.

- *Feedback loops and continuous improvement models*: Feedback loops and continuous improvement models can be developed to solicit input from students, teachers, and other stakeholders, using this information to iteratively improve and refine AIED solutions to ensure that they meet the evolving needs of the education community. These models should facilitate a more agile and responsive AI implementation approach, fostering innovation and ensuring that educational practices are always more effective.

8. Case Studies

The case studies in this section depict their personalized learning pathways, showing the potentiality for transformation that AI can bring to a 5G educational ecosystem. Using illustrations of successes in the field, they also expose ongoing gaps and challenges, all of which emphasize calls for proactive approaches to equitable and scalable implementation. The discursive and practical implications of studying these real-world applications are (hopefully) obvious as they highlight both the pitfalls and potential of embedding AI into education.

8.1 AI-powered Learning Platforms

The AI algorithms and 5G connectivity help these platforms customize educational content and track student progress. This does improve student engagement and academic results by allowing real-time feedback, adaptive learning modules, and individualized tutoring. AI algorithms collect and analyse data from various sources to compile comprehensive student profiles, enabling personalized learning experiences. By implementing machine learning processes to know patterns and trends in student performance, the AI-based platforms can derive the dynamic difficulty and rate of learning activities and also provide targeted and timely support and intervention (Egunjobi & Adeyeye, 2024).

8.2 5G-enabled Smart Campuses in Urban Areas

These campuses leverage AI-driven analytics to enhance security, optimize resource allocation, and deliver personalized learning experiences. The range of solutions implemented includes artificial intelligence (AI) surveillance systems, predictive maintenance solutions for campus infrastructure and intelligent energy management platforms—all to drive operational efficiency, safety and sustainability across educational institutions. Next, these campuses use AI-powered virtual assistants and chatbots to provide students with personalized support and guidance, so they can receive help with academic queries, course planning, or campus services at any given moment (Ojha et al., 2023).

8.3 AI in Remote and Rural Education: Pilot Projects

Through 5G connectivity, AI is being leveraged in these spheres to provide students with personalized learning experiences and quality educational material. Some of the technologies being used to bridge geographical barriers and resource shortfalls include AI-powered virtual tutors, language translation applications, and offline learning media. However, AI-enhanced platforms can provide students residing in remote areas with tailored feedback and assistance by analysing their learning practices as well as performance data, all without needing a typical classroom environment.

9. Future Trends and Research Directions

These educational systems, enabled by 5G, require an all-encompassing approach that should not only be focused on technological and pedagogical optimizations but also include moral aspects to kick-start the adoption of state-of-the-art techniques through AI in educational systems. Researchers working in this area need to be able to drive the development of personalized learning algorithms that will be adaptive, equitable, inclusive of, and responsive to all learners' individual needs, as shown in Figure 14.5.

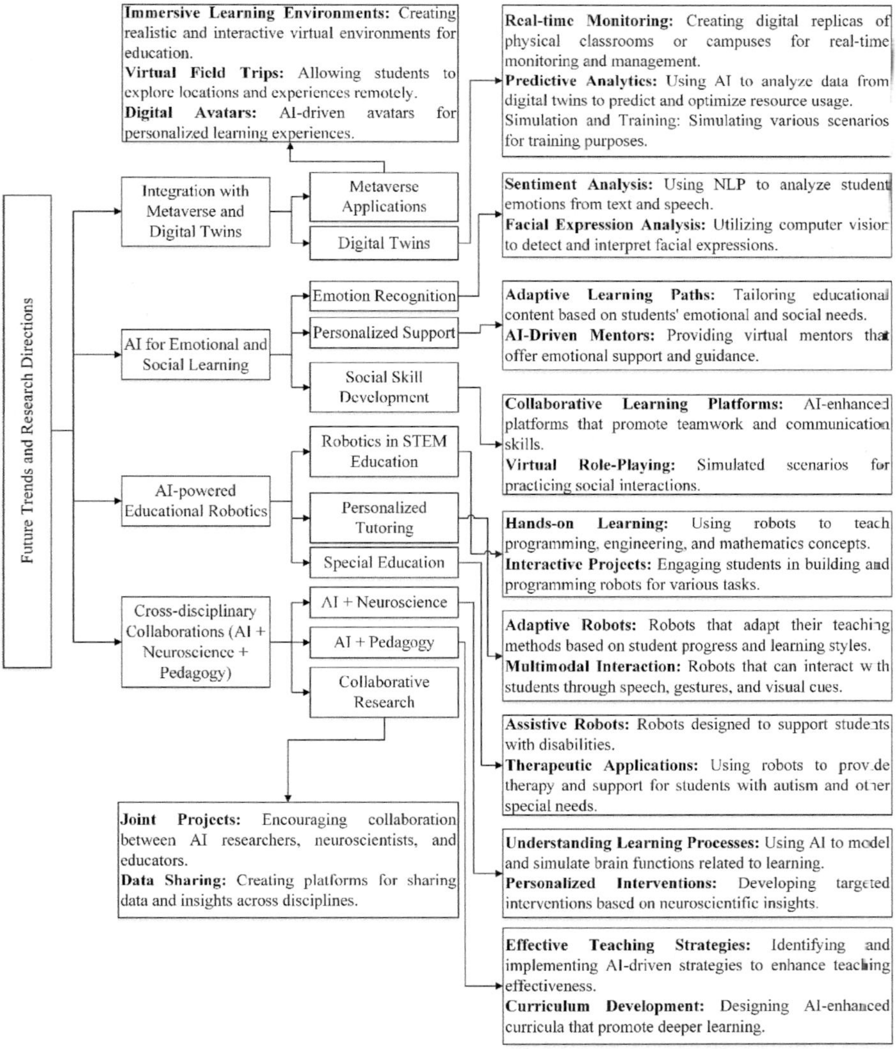

Fig. 14.5 Future trends and research directions.

- *Integration with metaverse and digital twins*: There is a need to investigate how AI is integrated with emerging technologies such as the metaverse and digital twins to create immersive and interactive learning environments to increase engagement and knowledge retention. Since the metaverse can simulate any contextual phenomenon through the creation of shared persistent worlds, it is one of the most significant innovations that provide us with an experience of collaborative learning, simulated experiences, and many gamified educational experiences. When combined with AI-powered personalization, these immersive settings can also adjust to a learner's unique style and preferences, offering tailored feedback and assistance to enhance educational results.
- *AI for emotional and social learning*: Research can be undertaken on using AI to facilitate emotional and social growth for students such as emotion recognition, empathy training, and conflict resolution. While students interact with AI-driven systems, the emphasis, therefore, must be placed on covering research around understanding emotional signals— helping these devices to offer responsive and empathetic responses. Simulation-based empathy training using AI simulators may have the potential to change the way empathy is trained and enable trainees to apply perspective-taking and resolution strategies in realistic settings. Not only does this promote emotional intelligence, but it helps to create more profound, more human learning experiences.
- *AI-powered educational robotics*: Emphasizing robotics and AI that are embedded in tools for education and child development is a good way to go. Teaching children how to code and solve problems, even using robots, helps to prepare them for future jobs. Educational robotics fosters a deep understanding of concepts while encouraging creativity, critical thinking, and collaboration through hands-on learning experiences for students. Machine learning enables AI algorithms to adjust robot behaviour as needs arise, tailoring support and teaching to the learner.
- *Cross-disciplinary collaborations (AI + neuroscience + pedagogy)*: Partnerships among educators, AI developers, policymakers, and industry stakeholders must be encouraged to ensure that AI solutions are aligned with educational objectives and ethical standards. The partnerships are irreplaceable in addressing the multifaceted challenges and opportunities AI presents in education, fostering the development of moral, pliable, and effective solutions that support student success and well-being. This chapter presents one of the first models to highlight how ongoing stakeholder dialogue and transparency contribute to individual awareness of the many implementation and usage decisions involved in the adoption of AI.

10. Conclusion

The convergence of AI and 5G technologies will transform education by allowing hyper-personalized learning experiences, intelligent campus infrastructure, and even higher-quality and more responsive educational delivery. Such synergy encourages real-time responses, immersive learning experiences, and remote learning opportunities to cater to the unique needs of modern education. Transforming this vision into reality involves overcoming numerous challenges, including technological limitations, data privacy, security issues, the digital divide, and resistance to change. Addressing these issues will require a concerted effort from educators, policymakers, technologists, and industry partners. Finally, AI in 5G for education can be harnessed with the help of stakeholders. The policymakers and public-private partners can make guidelines for AI in 5G for education and train teachers to make use of the features. They can also provide infrastructure for sustainable growth which would not allow the barriers of inequality be an issue in education. Finally, the introduction of AI and technology, such as 5G, in education will lead to better student engagement, more affordable education, and a chance to learn anytime around the world, even with foreign universities, so that whole generations reap their benefits.

References

Adewale, M. D., Azeta, A., Abayomi-Alli, A. and Sambo-Magaji, A. (2024). Impact of artificial intelligence adoption on students' academic performance in open and distance learning: A systematic literature review. *Heliyon*, *10*(22), e40025..

Ahmad, S., Umirzakova, S., Mujtaba, G., Amin, M.S. and Whangbo, T. (2023). Education 5.0: Requirements, enabling technologies, and future directions. arXiv preprint arXiv:2307.15846.

Ahmad, S.F., Alam, M.M., Rahmat, M.K., Mubarik, M.S. and Hyder, S.I. (2022). Academic and administrative role of artificial intelligence in education. *Sustainability*, *14*(3), 1101.

Akgun, S. and Greenhow, C. (2022). Artificial intelligence in education: Addressing ethical challenges in K-12 settings. *AI and Ethics*, *2*(3), 431–440.

AlTwijri, L. and Alghizzi, T.M. (2024). Investigating the integration of artificial intelligence in English as foreign language classes for enhancing learners' affective factors: A systematic review. *Heliyon*, *10*(10), e31053.

Al-Zahrani, A.M. (2024). Unveiling the shadows: Beyond the hype of AI in education. *Heliyon*, *10*(9), e30696.

Baker, R.S. (2021). Artificial intelligence in education: Bringing it all together. In *Digital Education Outlook: Pushing the Frontiers with AI, Blockchain, and Robots* (pp. 43–54). Paris: OECD Publishing.

Bura, C. and Myakala, P.K. (2024). Advancing transformative education: Generative AI as a catalyst for equity and innovation. arXiv preprint arXiv:2411.15971.

Dey, D.N.C. (2024). Enhancing educational tools through artificial intelligence in perspective of need of AI. Available at https://papers.ssrn.com/sol3/papers.cfm?abstract_id=5031275 (accessed on 5 August 2025).

Dimitriadou, E. and Lanitis, A. (2023). A critical evaluation, challenges, and future perspectives of using artificial intelligence and emerging technologies in smart classrooms. *Smart Learning Environments*, *10*(1), 12.

Egunjobi, D. and Adeyeye, O.J. (2024). Revolutionizing learning: The impact of augmented reality (AR) and artificial intelligence (AI) on education. *International Journal of Research Publication and Reviews*, *5*(10), 1157–1170.

Gligorea, I., Cioca, M., Oancea, R., Gorski, A.T., Gorski, H. and Tudorache, P. (2023). Adaptive learning using artificial intelligence in e-learning: A literature review. *Education Sciences*, *13*(12), 1216.

Godwin-Jones, R. (2024). Distributed agency in second language learning and teaching through generative AI. arXiv preprint arXiv:2403.20216.

Huang, L. (2023). Ethics of artificial intelligence in education: Student privacy and data protection. *Science Insights Education Frontiers*, *16*(2), 2577–2587.

Islam, N., Rashid, M.M., Pasandideh, F., Ray, B., Moore, S. and Kadel, R. (2021). A review of applications and communication technologies for internet of things (IoT) and unmanned aerial vehicle (UAV) based sustainable smart farming. *Sustainability*, *13*(4), 1821.

Jian, M.J.K.O. (2023). Personalized learning through AI. *Advances in Engineering Innovation*, *5* (1).

Jiang, H., Li, J., Zhao, P., Zeng, F., Xiao, Z. and Iyengar, A. (2021). Location privacy-preserving mechanisms in location-based services: A comprehensive survey. *ACM Computing Surveys (CSUR)*, *54*(1), 1–36.

Kaledio, P., Robert A. and Frank L.A. (2024). The impact of artificial intelligence on students' learning experience. Available at https://papers.ssrn.com/sol3/papers.cfm?abstract_id=4716747 (accessed on 5 August 2025).

Kohnke, S. and Zaugg, T. (2025). Artificial intelligence: An untapped opportunity for equity and access in STEM education. *Education Sciences*, *15*(1), 68.

Lampou, R. (2023). The integration of artificial intelligence in education: Opportunities and challenges. *Review of Artificial Intelligence in Education*, *4*, e15.

Luan, H., Geczy, P., Lai, H., Gobert, J., Yang, S.J., Ogata, H. ... and Tsai, C.C. (2020). Challenges and future directions of big data and artificial intelligence in education. *Frontiers in Psychology*, *11*, 580820.

Mahmoud, C.F. and Sørensen, J.T. (2024). Artificial intelligence in personalized learning with a focus on current developments and future prospects. *Research and Advances in Education*, *3*(8), 25–31.

Mohamed, E.S., Belal, A.A., Abd-Elmabod, S.K., El-Shirbeny, M.A., Gad, A. and Zahran, M.B. (2021). Smart farming for improving agricultural management. *The Egyptian Journal of Remote Sensing and Space Science*, *24*(3), 971–981.

Mundy, K. (2007). Global governance, educational change. *Comparative Education*, *43*(3), 339–357.

Ojha, S., Mohapatra, S., Narendra, A. and Misra, I. (2023). From robots to books: An introduction to smart applications of AI in education (AIEd). In: Yashwant Singh, Paulo J. Sequeira Gonçalves, Pradeep Kumar Singh, Maheshkumar H. Kolekar. *The International Conference on Recent Innovations in Computing* (pp. 311–322). Singapore: Springer Nature Singapore.

Owoc, M.L., Sawicka, A. and Weichbroth, P. (2019). Artificial intelligence technologies in education: Benefits, challenges and strategies of implementation. In: Mieczysław Lech Owoc, Maciej Pondel. *IFIP International Workshop on Artificial Intelligence for Knowledge Management* (pp. 37–58). Cham: Springer International Publishing.

Perez-Ortiz, M., Novak, E., Bulathwela, S. and Shawe-Taylor, J. (2021). An AI-based learning companion promoting lifelong learning opportunities for all. arXiv preprint arXiv:2112.01242.

Saputra, I., Astuti, M., Sayuti, M. and Kusumastuti, D. (2023). Integration of artificial intelligence in education: Opportunities, challenges, threats and obstacles. A literature review. *The Indonesian Journal of Computer Science*, *12*(4), 1590–1600.

Tabuenca, B., Uche-Soria, M., Greller, W., Hernández-Leo, D., Balcells-Falgueras, P., Gloor, P. and Garbajosa, J. (2024). Greening smart learning environments with artificial intelligence of things. *Internet of Things*, *25*, 101051.

Wu, J., Li, R., An, X., Peng, C., Liu, Z., Crowcroft, J. and Zhang, H. (2021). Toward native artificial intelligence in 6G networks: System design, architectures, and paradigms. arXiv preprint arXiv:2103.02823.

Yang, Q.F., Lian, L.W. and Zhao, J.H. (2023). Developing a gamified artificial intelligence educational robot to promote learning effectiveness and behavior in laboratory safety courses for undergraduate students. *International Journal of Educational Technology in Higher Education*, *20*(1), 18.

Yu, H. and Nazir, S. (2021). Role of 5G and artificial intelligence for research and transformation of English situational teaching in higher studies. *Mobile Information Systems*, *2021*(1), 3773414.

Chapter 15

Integrating Generative AI into the Curriculum and Learning Design

Rajalakshmi Nagarnaidu Rajaperumal[1*], *Jeevanantham Vellaichamy*[2], *Saravanan Krishnan*[3] and *Saravanan Srinivasan*[4]

1. Introduction

The application of artificial intelligence (AI) in education has entered a pivotal moment with the introduction of generative AI (gen AI), a class of models that generates text, images, code, and other media. Large language models (LLMs) like Open AI's GPT, Google's Gemini, and Meta's LLaMA are creating new opportunities for optimization in content creation, personalization, and innovation in teaching. With educational institutions facing the challenges of rapidly evolving technologies, growing diversity in learners, and changes in industry expectations, gen AI provides a never-before-seen opportunity to change the thinking of how curriculum and learning design is being conceptualized. Traditional curriculum development is often a labour-intensive and linear process. In addition to static resources and time constraints, curriculum projects have a limited scope. In contrast, gen AI allows practitioners to co-create live and customized content, automate assessments, and support learners in real time. Gen AI enables the development of personalized, inclusive, interdisciplinary curriculum that develops twenty-first-century competencies, including critical thinking, creativity, digital literacy, and ethical reasoning. In addition, gen AI can serve as both a teaching tool and a design assistant that can support teachers in designing instructional programmes, connecting learning objectives to programme goals, and enhancing student engagement through multimodal delivery of content. As generative

[1] Department of Computer Science and Engineering, Vel Tech Rangarajan Dr. Sagunthala R&D Institute of Science and Technology, Tamil Nadu, India. https://orcid.org/0000-0001-9705-0917

[2] Department of Computer Science and Engineering, Vel Tech Rangarajan Dr. Sagunthala R&D Institute of Science and Technology, Tamil Nadu, India. https://orcid.org/0000-0003-3787-4855

[3] Department of Computer Science and Engineering, Anna University, Tamil Nadu, India. https://orcid.org/0000-0001-9517-9792

[4] Department of Computer Science and Engineering, SRM Institute of Science and Technology, Tamil Nadu, India. https://orcid.org/0000-0001-7884-3532

* Corresponding author: rajirajasekaran@gmail.com

DOI: 10.1201/9781003639367-15

technologies are built into the curriculum frameworks, educators will find pedagogical ecosystems to become nimbler, more responsive, and more student-centred.

This chapter focuses on the potential uses of generative AI in curriculum and learning design, and particularly its potential to speed up the development of curriculum, support differentiated learners, enable innovative practices, and allow educators to uphold pedagogical integrity. This chapter will outline some ways Gen-AI has been used and offer some considerations to support educators, administrators, and policymakers in using gen AI for educational transformation.

2. Incorporating Generative AI into Curriculum and Learning Design

Artificial intelligence is revolutionizing education as it increases engagement, personalized instruction, and productivity for administrators and educators. Generative AI, as a distinct area of AI focused on content production, is emerging as a transformative tool for curriculum creation and for instructional designers. LLMs can create course outlines, learning objectives, assessments, even lessons—all in less time, and all aligned with learning standards (Q. Lu et al. 2024). Zhao et al. (2022) argued that generative AI can provide real-time multilingual support for adaptive instructional materials to support different learning levels and cultural contexts. Open AI's ChatGPT, Google's Gemini, and Anthropic's Claude are some of the platforms that can create more equitable and accessible learning environments. Nonetheless, there are significant concerns surrounding misinformation, use of AI for ethical purposes, and those surrounding bias in data. However, generative AI, described in a paper by the United Nations Educational, Scientific and Cultural Organization (UNESCO, 2023), supports multilingual and multimodal delivery of content in order to democratize education even in contexts where resources are limited. Generative AI significantly adds value as a co-creative collaborator, enhancing efficiencies in curriculum design and increasing learner engagement. Generative AI is not simply an innovation in teaching technology but also an innovation in pedagogy and is changing how curricula are designed, delivered, and experienced. (Yeh, 2024) discovered that ChatGPT can help EFL (English as a Foreign Language) instructors to develop interactive and adaptive learning resources, including picture books and karaoke practice, to support more effective language instruction. (Yuan, 2023) investigated the implementation of an AI chatbot in English language teaching at a Chinese primary school. Findings showed that students who learned with the AI chatbot performed better than their counterparts in regular classrooms as far as oral English ability and communication willingness were concerned. Kusuma et al. (2024) investigated the application of ChatGPT in the professional growth of English language instruction, with a view to strengthening pre-service EFL teachers' teaching methodologies. Through providing a secure and non-critical environment, chatbots not only enhanced students' language ability but also alleviated anxiety and boosted their English-speaking confidence. Evmenova et al. (2024) explored the quality and scope of AI in offering feedback to elementary school pupils with diverse levels of writing proficiency. The research revealed that ChatGPT was capable of reading purposeful prompts and grading student essays according to a criterion-based rubric provided. Nonetheless, the quality of feedback was inconsistent across various ChatGPT models and prompts, and there was an apparent lack of correspondence between AI-drafted feedback and personal student attributes, e.g., proficiency level or special needs. In a similar manner, (Yuxiu, 2024) developed an AI translation instructional system module that decreased the distance between translation systems using algorithms from neural machine translation (NMT) and statistical machine translation (SMT) to increase translation ability and instructional efficiency. The module was implemented in high school locations for students using the system's real time translation tools, translation assessments, translation feedback in a way that successfully enhanced students' translation skill acquisition, and other means of learning. Jia et al. (2024) examined the difference between advice on geometry and algebra problems given by ChatGPT and human instructors using the Turing test. They found that the majority of human subjects had trouble differentiating between the AI and human-generated tips for students, indicating that ChatGPT could be utilized to assist teachers in minimizing the cognitive load associated with producing mathematical tips for students. Tassoti (2024) further presented the "Five S" prompt framework to help students make their prompts for chemistry related queries more effective, leading to increased satisfaction among students using AI responses from a Discord chatbot based on ChatGPT. Bai et al. (2024) proposed a three-step process for prompt engineering to facilitate scenario-based learning with ChatGPT, leading to increased intrinsic motivation, improved learning performance, and more positive attitudes towards coding and computational thinking among students. ChatGPT is generating customized content and helping to develop communication skills of the student to enhance a more dynamic and student-focussed learning environment.

3. Methodology for Integrating Generative AI into Curriculum and Learning Process

This section outlines a practical methodology for integrating generative AI into curriculum and learning design processes. The proposed framework is structured into six iterative phases: Needs Analysis, Model Selection, Content Co-creation, Validation and Review, Implementation, and Continuous Improvement.

3.1 Evaluating the Alignment of Learner Needs and Industry Demands in Curriculum Design

Evaluating *alignment* is the most critical step in the design of the curriculum, where content is aligned with learners' needs and industry demands. This process is initiated by the teachers, who will declare with precision the specific learning outcomes to be attained, and the learners (students), and the current gaps in the curriculum. An inclusive approach is taken, conducting consultations among major stakeholders like the faculty members, students, and industry players to gain varied opinions. The stage also involves a rigorous review of existing curricula and assessment data to analyse relevance and effectiveness. Finally, the learning outcomes determined are aligned with critical skills and competencies so that the curriculum is outcome based and in accordance with real-world requirements. The systematic analysis thus guarantees that the following stages of curriculum development are responsive, inclusive, and future-oriented.

3.2 Expected Course Outcomes for Data Science Course

The *data science* programme aims at providing the learners with both fundamental and advanced skills along the data science pipeline. The programme starts by presenting the fundamental ideas, domain, and data science lifecycle so that students can grasp the essential phases of data handling such as collection, pre-processing, analysis, and visualization. Focus is given to the practical implementation of statistical techniques and probability principles to derive useful insights from unstructured and structured data. The course combines hands-on experience with commonly used programming libraries and tools like R, Python, NumPy, Pandas, Scikit-learn, and Matplotlib to carry out data wrangling and data transformation operations. The students will be taught to construct and test machine learning models with supervised and unsupervised methods, promoting analytical reasoning to solve real-world issues. Further, students are guided to build comprehensive, evidence-based solutions by combining algorithmic techniques with domain knowledge and effective visualization methods. The course also instils ethical considerations, data privacy awareness, and proper communication, resulting in ethical data-driven decision-making. Through this holistic approach, the course will provide students with training for technical roles along with interdisciplinary roles in the emerging field of data science.

3.3 Selection of Large Language Model

After determination of curriculum requirements in the form of *course outcomes*, the subsequent crucial process is strategic selection of appropriate gen AI tools to enhance teaching and learning objectives. The selection of appropriate tools relies on certain needs such as the modality of content required like text creation, code development, or image content generation along with language assistance, accessibility functionality, and an uninterrupted integration with current learning management systems (LMS). Tools are evaluated for their pedagogical alignment and technical compatibility. Prominent examples include OpenAI's ChatGPT and DALL·E, Google Gemini, Microsoft Copilot, and open-source models from Hugging Face. Wangsa et al. (2024) compared five leading chatbot models: ChatGPT, Bard, Llama, Ernie, and Grok. The review highlights that each model excels in different contexts, with no single model emerging as superior. This phase ensures that the chosen gen AI tools are both contextually appropriate and technologically robust, laying the groundwork for effective content co-creation and instructional enhancement.

3.4 Structured and Pedagogically Sound Course Content Generation

In current curriculum planning, instructors are increasingly employing genAI tools to work in tandem with the development of structured and pedagogically sound course content. The process begins by creating thorough course charts such as in Table 15.1, unit plans such as in Table 15.2, daily calendars that are integrated with properly defined intended learning outcomes (ILOs). Teachers begin by typing prompts into gen AI systems, specifying such things as course objectives, prerequisite knowledge, target learner profiles, desired skills, and preferred pedagogical practices or delivery modes (e.g., flipped classroom or blended learning). In reaction to these inputs, the

AI generates preliminary drafts of course outlines, including modular course plans, in-depth weekly unit plans with sub-topics, topics, learning activities, and suggested readings, and weekly timetables that map out learning goals into instructional time and tests.

Table 15.1 Course chart

Course Outcomes	*Mapped ILO*	*Bloom's Level*	*Assessment Method*
Describe the data science lifecycle and the role of data in analytical problem-solving in all fields	Knowledge, Understanding	Understand	Quiz, descriptive text
To impart statistical methods and machine learning algorithms for data analysis and prediction	Application, Analysis	Apply	Assignment, lab exercise

Table 15.2 Unit plan

Unit	*Title*	*Topics Covered*	*Mapped Course Outcomes*
1	Introduction to Data Science	Definition, Lifecycle, Applications, Data Types, Data Sources, Structured vs Unstructured Data	CO1
2	Statistical Foundations	Descriptive Stats, Probability, Hypothesis Testing, Inferential Statistics	CO2

While these drafts are a good starting point, instructors play an important role in the process as they go through and refine the content on a critical level to ensure academic sophistication, contextual and cultural sensitivity, and institutional compliance. Refining also ensures accessibility, promotes interdisciplinary connections, and adds real-world application. Moreover, prompt engineering becomes an essential ability in which instructors can lead the AI step by step towards responses that still have disciplinary correctness and pedagogical nuance.

3.5 Prompt Engineering

Prompt engineering at this stage is imperative. Quality, pertinence, and originality of the outputs of gen AI to a great extent depend on how prompts are presented. Teachers must utilize specific, context-based cues to guide the AI and gradually optimize inputs in order to maximize pedagogical value of responses and integrate course-specific vocabulary, exemplars, and learner factors into prompting.

This round-the-clock, interactive cycle of human knowledge and gen AI capacity assures that the ultimate product in the form of curriculum courses and course content are not merely effective and engaging but also pedagogically correct and inclusive.

An example of prompt is as follows:

prompt = """

I am designing an undergraduate Data Science programme. The programme aims to equip learners...

```
"""
To generate response
    messages=[
      {"role": "system", "content": "You are a curriculum expert in Data Science."},
      {"role": "user", "content": prompt}
      ],
```

3.6 Results and Implementation of Generative AI in Curriculum Design

The proposed work is implemented to support the rapid development of a new undergraduate data science course. The faculty members can use gen AI to create first drafts of course outlines, weekly lesson plans, and example assignments for data science subjects. The faculty panels need to review and revise the AI-created content. This process shaved more than 40% off curriculum planning time while ensuring continued compliance with accreditation guidelines. It further enabled quick iteration based on industry insights. In order to deploy an efficient gen AI solution, the process starts by selecting a proper base model like an open-source model from Hugging Face. The domain-specific data like course outcomes for data science courses must be pre-processed which involves gathering, cleaning, and preparing relevant datasets for the particular problem scenario. Prompt engineering is crucial in defining model behaviour by designing inputs that elicit the right and context-specific responses. For high performance and contextual relevance, retrieval-augmented generation (RAG) can be used to combine external knowledge sources dynamically at inference time. Once the system has been set up, it is important to test the output for correctness, relevance, and consistency, both through automated metrics and human judgement. Once validated, the model is implemented as a solution, with the delivery ensuring that it conforms to performance and scalability requirements. Finally, integration into current applications, workflows, or platforms facilitates efficient user interaction and optimizes the value of the AI-based solution.

3.7 Templating the Prompt to Receive Expected Response from Trained LLM

Prompt engineering is a technique that extends the capabilities of LLMs and vision-language models by using task-specific instructions, prompts. These prompts allow for seamless integration of pre-trained models into downstream tasks, eliciting desired behaviours based on the given prompt. These prompts are given as input to the LLM to design curriculum course content and learning materials. Good prompts are clear, focus on a single goal, and leave no room for wild guessing. How you word your ask tone, word order, and plain clarity shape the model's reply so it matches what you pictured. Because the request gives precise limits to follow, giving the learning material for the data science course always yields correct response in technical queries. So, the lessons, quick summaries, step-by-step explanations, or fresh creative pieces can be created. Prompt engineering is the art of crafting tidy, purposeful questions that matter a lot while using an open-source model. A prompt is only the text you enter into the system to guide what appears when you interact with instruction-tuned models like MiniMax on Hugging Face. The model has specific boundaries to adhere to when using the request. For this reason, whether you require lessons, summaries, course content, learning material explanations, or a new piece of creative writing, prompt engineering is the skill of creating neat, intentional requests. Prompt engineering is crucial when working with MiniMax model, that one line or a few lines usually determine how specific, readable, and well-formatted the response will be generated. These models watch for indications that indicate, "This is work I should do" because they are trained to follow instructions that appear to be brief activities. Typically, a successful prompt begins with a clear directive and, when beneficial, provides just enough context to ensure the model understands the entire narrative. The model is informed of its precise role by saying something like, "Explain course content based on course objectives".

3.8 The Role of Prompt Template in Curriculum and Learning Design

As shown in Figure 15.1, the prompt template plays a major role in generating the response from trained LLMs.

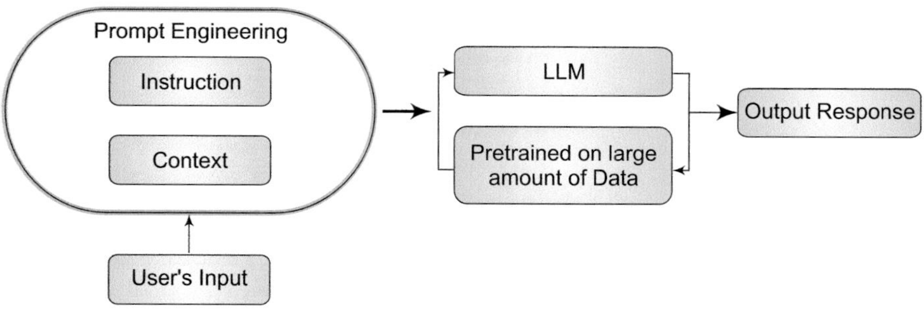

Fig. 15.1 Role of prompt in curriculum and learning design.

The crucial role that prompt engineering plays in utilizing LLMs for curriculum and learning design is depicted in Figure 15.1. Usually created by educators or instructional designers, user input starts the process and is subsequently improved through prompt engineering. This entails organizing the contextual background and the instruction required for the model to produce insightful answers. The LLM receives the prompt after being pre-trained on a large amount of domain-specific and educational data. After processing this input, the LLM produces an output response that is appropriate for the educational setting. Course material, learning goals, test questions, and individualized learning pathways are a few examples of this output. Figure 15.1 illustrates how carefully considered prompts and pre-trained LLMs work together to produce scalable and contextually appropriate content. Hence, five prompts are being described here about the generation of data science course and learning materials.

3.8.1 Prompt 1: Course Overview Generator

prompt = PromptTemplate(
input_variables=["Data Science Course Outcome"],
template="Design an exhaustive course overview for a Data Science course given the following input parameters:\nCourse Title: Data Science\nAcademic Level: Under Graduate\nCredits: 3\nTarget Audience: Students\nPrerequisites: python programming, Statistics\nTools/Technologies to be used: Python, Scikit Learn\n{Data Science Course Outcome}",)

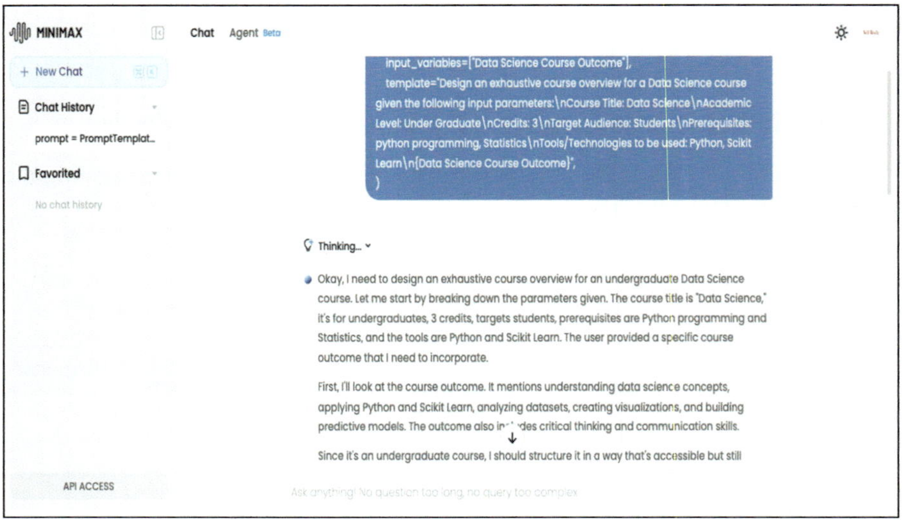

Fig. 15.2 Prompt input for course overview generation.

Provide: Course Description (100–150 words), Course Objectives (3–5 bullet points).

A key component of starting the curriculum design process with AI is prompt input for course overview generation. This input is organized using prompt engineering in Figure 15.2, which blends contextual information with instructional intent. The purpose of this prompt is to give the LLM enough context and clear instructions so that it can produce an insightful and pertinent course overview. The LLM will respond with pedagogically sound and domain-specific content thanks to the meticulous input construction, which the educator can then refine and validate for eventual use in curriculum design and pathways for learning. Figure 15.2 illustrates how skilled prompts and pre-trained LLMs work together to produce scalable and contextually relevant content, which serves as the basis for curriculum development that is dynamic, responsive, and learner-centred.

Figure 15.3 illustrates the AI-generated course description and learning objectives derived from a well-structured prompt input using MiniMax LLM model.

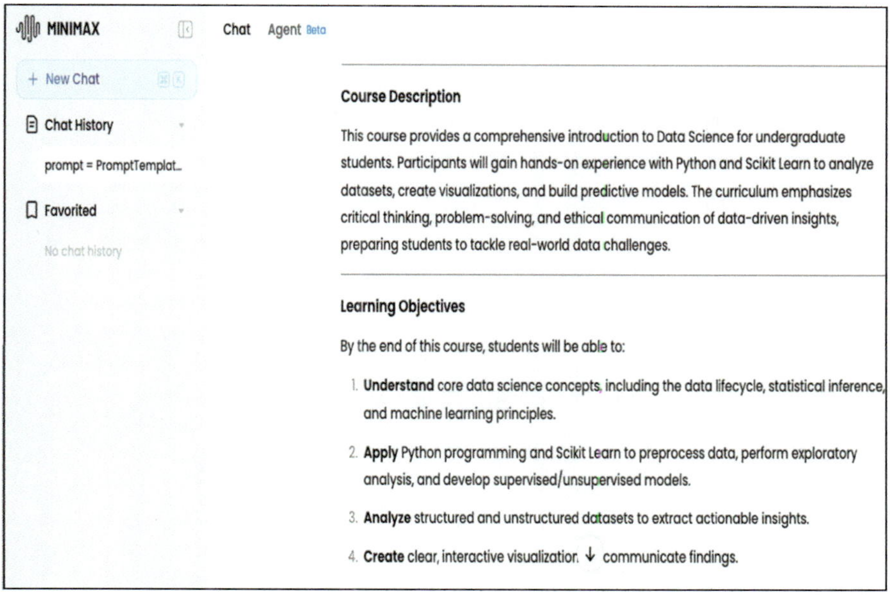

Fig. 15.3 Generation of course description and learning objectives.

3.8.2 Prompt 2: Generation of ILOss

You are an instructional designer applying Bloom's Taxonomy. Based on the course description and objectives, create 5–7 ILOs that indicate what the students should be able to do by the end of the course.

> course_info = """"Course Description:
> { This course provides a comprehensive introduction to Data Science, teaching students to analyse, clean, and interpret data using real-world datasets and Python libraries. It covers data preprocessing, visualization, statistical inference, machine learning, and results communication, promoting ethical data use and problem-solving. }

Course Objectives:

> {To familiarize students with the data science process and the importance of data in analytical decision-making.
> To acquire proficiency in data wrangling, cleansing, and exploration through Python techniques.
> To impart statistical methods and machine learning algorithms for data prediction and analysis.
> To enable students to effectively visualize and communicate insights.
> To instill sensitivity towards ethical implications and best practices in data use.}
> Every ILO should employ a Bloom's verb (e.g., explain, analyze, design, evaluate, implement, create) and include measurable learning outcomes.""""

Figure 15.4 shows that ILOs are being created using open source models and every ILO should employ a Bloom's verb of explain, implement, and create, and include measurable learning outcomes.

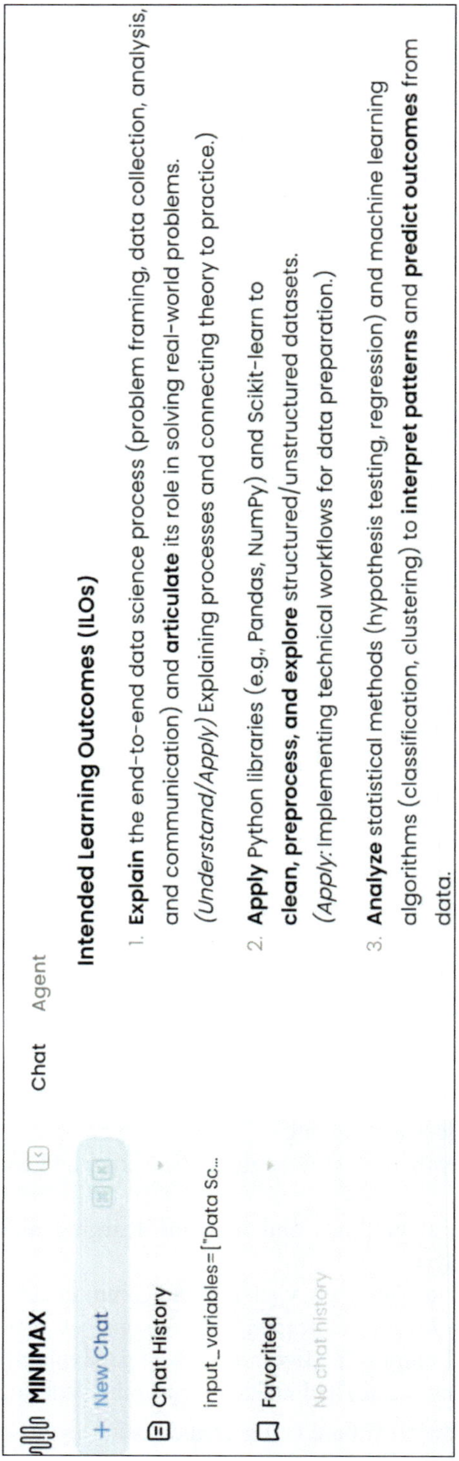

Fig. 15.4 Generation of intended learning outcomes (ILOs).

3.8.3 Prompt 3: Weekly Schedule Planner

prompt_template = """"""
You are an expert curriculum designer.

Based on the following intended learning outcomes (ILOs) and a total duration of {weeks} weeks, create a structured, week-by-week course schedule for a Data Science course.

ILOs:

1. Describe the data science lifecycle and the role of data in analytical problem-solving in all fields.
2. Apply Python programming techniques to prepare, refine, and explore datasets, identifying significant insights and underlying trends.
3. Implement statistical analysis and machine learning methods to interpret data and develop accurate forecasting models.
4. Create effective data representations and articulate analytical conclusions using industry-standard visualization platforms and clear communication strategies.
5. Evaluate ethical considerations in data management and formulate fair, accountable policies for data collection and utilization.

For each week, provide the following details:

- Week Number
- Unit Title
- Brief Description of Topics to be Covered
- Key Outcomes or Skills Developed

Format the output clearly as a structured list or table.""""""

3.8.4 Prompt 4: Unit Plan Generator

course_prompt_template = """"""
You are an expert curriculum designer.

Your task is to generate a detailed, structured, week-by-week course schedule for a Data Science course based on the provided intended learning outcomes (ILOs) and the course duration of {weeks} weeks.

Intended learning outcomes (ILOs):

1. Describe the data science lifecycle and the role of data in analytical problem-solving in all fields.
2. Apply Python programming techniques to prepare, refine, and explore datasets, identifying significant insights and underlying trends.
3. Implement statistical analysis and machine learning methods to interpret data and develop accurate forecasting models.
4. Create effective data representations and articulate analytical conclusions using industry-standard visualization platforms and clear communication strategies.
5. Evaluate ethical considerations in data management and formulate fair, accountable policies for data collection and utilization.

For each week, provide the following information in a clear and structured format:

Week Number
Unit Title
Brief Description of Topics to be Covered
Key Outcomes or Skills Developed
Format the output as a structured list or markdown table. Ensure content is aligned with the ILOs and follows a logical progression from foundational to advanced topics"""".

3.8.5 Prompt 5: Assessment Strategy Generator

Based on the intended learning outcomes, develop an assessment strategy for the entire course. Include both formative and summative assessments.

ILOs:{

1. Describe the data science lifecycle and the role of data in analytical problem-solving in all fields.
2. Apply Python programming techniques to prepare, refine, and explore datasets, identifying significant insights and underlying trends.
3. Implement statistical analysis and machine learning methods to interpret data and develop accurate forecasting models.
4. Create effective data representations and articulate analytical conclusions using industry-standard visualization platforms and clear communication strategies.
5. Evaluate ethical considerations in data management and formulate fair, accountable policies for data collection and utilization.

}

Design:

Formative assessments (quizzes, in-class exercises, labs)

Summative assessments (midterm, final exam, project)

Weightage or grading strategy

Alignment of each assessment type with relevant ILOs

To effectively generate a week-by-week course schedule for a data science programme using open-source model, a structured and code-ready prompt template is designed. This prompt guides the model to create a modular curriculum aligned with well-defined ILOs. The template includes a clear task description for the AI, outlines five ILOs covering understanding of data life cycle, Python-based data analysis, machine learning algorithm, data visualization, and ethical practices. The model was told that the arrangement of the course should be in weekly parts, with each part containing the unit title, topics to be discussed, and the key outcomes or skills gained. The prompt highlights the importance of a logical sequence, starting from basic concepts and progressing to advanced topics. Intended to be used in applications like Lang Chain through the LLM chain component, this method implies that educators and developers can still automate curriculum design but at the same time maintaining the pedagogical integrity and outcome alignment.

4. Curriculum and Learning Design Using Fine Tuning and Embedding Approaches

A structured approach fine-tunes an LLM like LLaMA 2 for curriculum preparation, using domain-specific data of course outlines, unit plans, learning outcomes, assessment templates, and parameter efficient techniques. This adapts the model to pedagogical structures, academic terminology, and instructional patterns.

4.1 Model Training and Embedding

An advanced LLM is trained to understand and generate text. The ability of LLaMA 2 in assisting the curriculum development is enhanced by using fine tuning approach. LLaMa is fine-tuned using the carefully prepared datasets of curriculum components such as formatted input and output pairs to enhance its ability to provide relevant course and learning design information. In parallel, embedding approach can also be employed to convert textual data into dense vector representations, capturing semantic relationships and contextual relationships between curriculum components. These vectors allow for the organization of educational content into a structure compatible with embedding based retrieval systems. For example, course outcomes or ILOs, course objectives are converted into dense vector representations to represent the semantic meaning in high-dimensional space, this approach being called as embedding. This vector-based representation enables the mapping of course outcomes to broader programme outcomes through similarity measures such as cosine distance. Although LLaMA 2 is primarily generative model, designed for text generation, it can represent curriculum elements in a high-dimensional vector space through internal hidden state representations, allowing semantic relationships between different elements to be quantified using distance metrics. As a result, this dual approach, which uses embeddings for semantic analysis and fine-tuning for content generation, allows for sophisticated applications in data-driven instructional design, automated curriculum planning, and semantic search.

4.2 Data Science Course Generation through Fine Tuning the LLaMA Model

This process begins with collecting or constructing a high-quality, domain-related dataset of input–output pairs, such as prompts for developing curriculum dataset like course outcomes, weekly plans, unit plans, lab sessions, assessment rubrics, project guidelines, programme outcomes mapping, and Bloom's Taxonomy. For example, an input prompt like "Provide a unit plan for an introductory Data Science course" can be paired with structured output on topics of data collection, cleaning, analysis, and data visualization. These prompt response pairs collectively form the curriculum dataset. The LLaMA 2 model processes these text inputs to represent curriculum elements in a high-dimensional vector space. Each text input is preprocessed using tokenization and embedding techniques to analyse the similarity distribution of educational intentions. So, course outcomes, unit titles, and necessary industry competencies are all densely

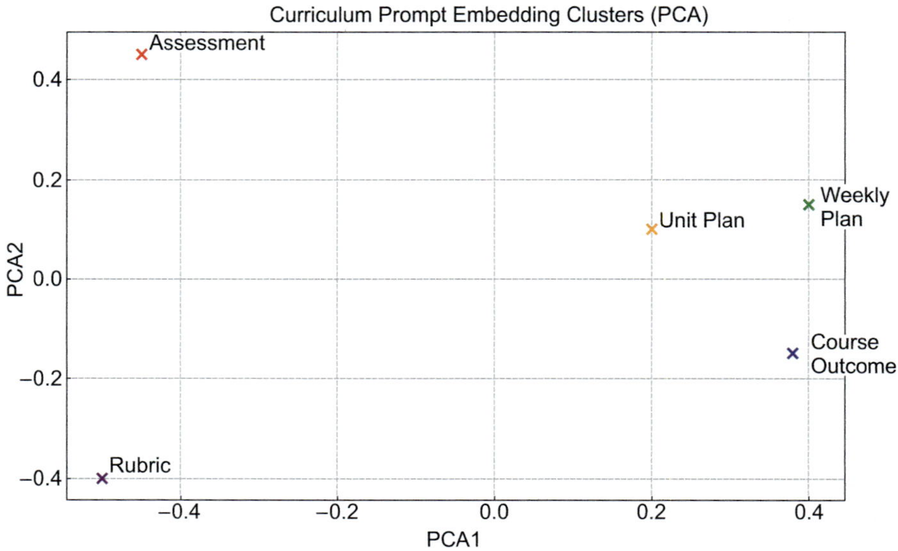

Fig. 15.5 Curriculum prompt embedding clusters

represented as vectors using embeddings. The embeddings make it easier to align the generated course material with external baselines, like accreditation or occupation standards. To align and become relevant, for example, embeddings of data science occupation descriptions can be compared to embeddings of learning outcomes from the refined model. Furthermore, embeddings enable the clustering of related topics, as shown in Figure 15.5, which facilitates logical content organization for instructors by grouping ideas like data wrangling and data cleaning into cohesive units. This technique enables curriculum designers to compare, cluster, or retrieve pedagogically related content, supporting coherent course structures.

The integration of embedding and fine-tuning approaches enable the automatic creation of scalable, responsive, comprehensive, and industry-relevant data science course to meet the educational needs. Once prepared, the dataset is utilized to fine tune the LLaMA model with the help of Hugging Face Transformers and Parameter Efficient Fine Tuning (PEFT) tool to fine tune only low rank adaptation layers to generate the pedagogically sound and context-sensitive content appropriate for the teaching of data science course. The model aids educators in creating data science–specific course outcomes, schedules, and assessment plans, enhancing efficiency, scalability, and academic alignment through AI integration.

5. Continuous Improvement

After the course is launched, the responses of the learners and teachers are collected and analysed to figure out the strong points and the areas where the course can be improved. The use of gen AI is very important in this optimization period as it enables the user to quickly modify the existing one, to develop new material according to student performance evaluations, and to receive alternative suggestions. Repetition is intended to ensure that the assimilation of gen AI, in addition to being the fittest, is moral and consistent with the educational objectives. By combining human creativity with AI's agile capabilities, this approach is an evolutionary and scalable curriculum design model which is able to adjust to the changing learner needs and the educational environments.

6. Conclusion

Gen AI can re-imagine curriculum design as a more dynamic, data-driven, inclusive, and needs-responsive form. Automating content creation, improving instructional alignment, and enabling continuous updates, gen AI can make the curriculum development process more effective. The following process minimizes design time and effort, improves inclusivity, scalability, and responsiveness to trends, and advances the transition to more agile, data-driven, and learner-focused education ecosystems. Gen AI can generate content in various formats such as text, images, video; it can translate and localize the content; and it can offer differentiated and inclusive instruction and improve participation, understanding, and retention. The process of developing a successful gen AI solution for designing a data science course begins with selecting an appropriate base model of LLM. Then there is prompt engineering or fine tuning the LLM used to get the response from the model. Therefore, designing prompts that simulate the course plan, unit plan, and week plan to obtain correct and context-specific responses or integrating LLAMA 2, an advanced LLM, with a sophisticated embedding model and vector database offers a robust solution for delivering accurate and contextually relevant curriculum and learning design.

References

Bai, S., Gonda, D.E. and Hew, K.F. (2024). Write-curate-verify: A case study of leveraging generative AI for scenario writing in scenario-based learning. *IEEE Transactions on Learning Technologies*, *17*, 1313–1324.

Evmenova, A.S., Regan, K., Mergen, R. and Hrisseh, R. (2024). Improving writing feedback for struggling writers: Generative AI to the rescue? *TechTrends*, *68*, 790–802.

Jia, J., Wang, T., Zhang, Y. and Wang, G. (2024). The comparison of general tips for mathematical problem solving generated by generative AI with those generated by human teachers. *Asia Pacific Journal of Education*, *44*(1), 8–28.

Kusuma, I.P.I., Roni, M., Dewi, K.S. and Mahendrayana, G. (2024). Revealing the potential of ChatGPT for English language teaching: EFL preservice teachers' teaching practicum experience. *Studies in English Language and Education*, *11*(2), 650–670.

Lu, J., Zheng, R., Gong, Z. and Xu, H. (2024). Supporting teachers' professional development with generative AI: The effects on higher order thinking and selfefficacy. *IEEE Transactions on Learning Technologies*, *17*, 1279–1289.

Lu, Q., Yao, Y., Xiao, L., Yuan, M., Wang, J. and Zhu, X. (2024). Can ChatGPT effectively complement teacher assessment of undergraduate students' academic writing? *Assessment & Evaluation in Higher Education*, *49*(5), 616–633.

UNESCO. (2023). Guidance for generative AI in education and research. UNESCO. https://www.unesco.org/en/articles/guidance-generative-ai-education-and-research

Wangsa, K., Karim, S., Gide, E. and Elkhodr, M. (2024). A systematic review and comprehensive analysis of pioneering AI chatbot models from education to healthcare: ChatGPT, Bard, Llama, Ernie and Grok. *Future Internet*, *16*(7), 219.

Yeh, H.-C. (2024). The synergy of generative AI and inquiry-based learning: Transforming the landscape of English teaching and learning. *Interactive Learning Environments*, *33*(1), 88–102. https://doi.org/10.1080/10494820.2024.2294509..

Yuan, Y. (2023). An empirical study of the efficacy of AI chatbots for English as a foreign language learning in primary education. *Interactive Learning Environments*, *32*(10), 6774–6789.

Yuxiu, Y. (2024). Application of translation technology based on AI in translation teaching. *Systems and Soft Computing*, *6*, 200072.

Zhao, L., Wu, X. and Luo, H. (2022). Developing AI literacy for primary and middle school teachers in China: Based on a structural equation modelling analysis. *Sustainability*, *14*(21), 14549.

Chapter 16

A Comprehensive Review of Automated Learning Systems to Monitor the Pervasiveness of Musculoskeletal Pain Among Students in the Classroom

Dilliraj Ekambaram[1], *Vijayakumar Ponnusamy*[2*] and *Nemanja Zdravkovic*[3]

1. Introduction

The emergence of AI-based exercise analysis for musculoskeletal disorders can be attributed to the development of motion analysis systems, which identified and assessed movement patterns using machine learning algorithms (Roggio et al., 2024). The system for musculoskeletal disorder (MSD) was initially designed for monitoring sportspersons and athletes to give feedback on performance on the track by the individual players. Dr Scott Delp from Stanford University used the machine learning and motion analysis system that aids the rehabilitation process for patients who were suffering from musculoskeletal pains. He used OpenSim system to simulate the movement of a human to produce the feedback on rehabilitating the patients (Delp et al., 2007). The prevalence of MSD among working population in India ranges from 20% to 70%, reported by Mahajan et al. (2023). We acknowledge this correction. Since the name of the journal and the details of the work are already provided in the References, we agree that the deleted text was redundant. Thank you for pointing this out. Most physicians around the globe suggest that exercise therapy is the best

[1] Department of Electronics and Communication Engineering, SRM Institute of Science and Technology, Tamil Nadu, India. https://orcid.org/0000-0002-9245-2746

[2] Department of Electronics and Communication Engineering, SRM Institute of Science and Technology, Tamil Nadu, India. https://orcid.org/0000-0002-3929-8495

[3] Faculty of information Technology, Metropolitan University, Serbia. https://orcid.org/0000-0002-2631-6308

* Corresponding author: vijayakp@srmist.edu.in

DOI: 10.1201/9781003639367-16

solution to avoid the prevalence of MSD. The economic burden of MSDs in India is also significant, with estimates suggesting that it accounts for 1% to 2% of the country's gross domestic product (GDP) (Mishra et al., 2024; Nayak & Anitha, 2024; Jebaraj et al., 2022).

The prevalence of MSD is not to be considered only among the working professionals. It is also essential to monitor it in the education profession and among students (Tiwari et al., 2024; Cieza et al., 2020; Prasetya et al., 2024). This chapter reveals the prevalence of MSD among students and teaching professionals. High levels of presenteeism, absenteeism, and sick leave among teachers have also been linked to MSDs. One study found that 81% of the 230 students studied may have a high frequency of musculoskeletal problems. Researchers in disciplines like pedagogy and psychology are paying more and more attention to evaluations of educational quality (Tahernejad et al., 2024). But still, the health problems experienced by educators remain unrevealed. One of the research studies noted that the prevalence of MSD ranged between 39% and 95% (Erick & Smith, 2015). These high prevalence rates in educators are linked to personal, professional, and psychosocial issues.

The following subsections discuss the studies that have looked into the connection between teachers' working circumstances and MSDs. Recent studies have shown that artificial intelligence (AI) plays an essential part in automatically monitoring students' conduct in the classroom, preventing and managing classroom-based MSDs through wearable devices, and enabling visual inspection predictive systems for students' sitting and standing postures to avoid the prevalence of MSD (Chan et al., 2022). This prevention management system of classroom-based musculoskeletal disorder (CMSD) provides effective solutions to maintain the students' spinal bones, leg bones, joints, and muscles. Various pose estimation frameworks like Google MediaPipe (google-ai-edge, 2024) and OpenPose (*CMU-Perceptual-Computing-Lab/Openpose*, 2020) were widely utilized to ensure the analysis of human postures. Figure 16.1 shows the pose estimation frameworks and details.

MediaPipe contains thirty-three key points to address the challenges of interference or perception models for visual-based activity recognition applications. This framework is used to maintain low interference in different lighting conditions targeting single human motion action recognition (Lugaresi et al., 2019). OpenPose is a multi-person pose estimation framework that has twenty-five key points. Interference may cause loss of key points of information due to occlusion of body parts on multi-person interaction (Rookba et al., 2020). The following subsections deliberate the risk factors of MSD, learning management systems, ergonomics for CMSD, and common deep learning systems for rehabilitation systems through exercise therapy.

1.1 Risk Factors of MSD

Most of the learning communities around the globe are affected by musculoskeletal pains due to work-related risk factors such as prolonged sitting (spending long periods sitting in the same place with poor sitting posture), poor ergonomics (not having proper resources like chairs and desk), repetitive motions, unfavourable environmental conditions, and stress. On the other hand, students suffer from these pains due to improper sitting position, resting their heads on tables, being in the

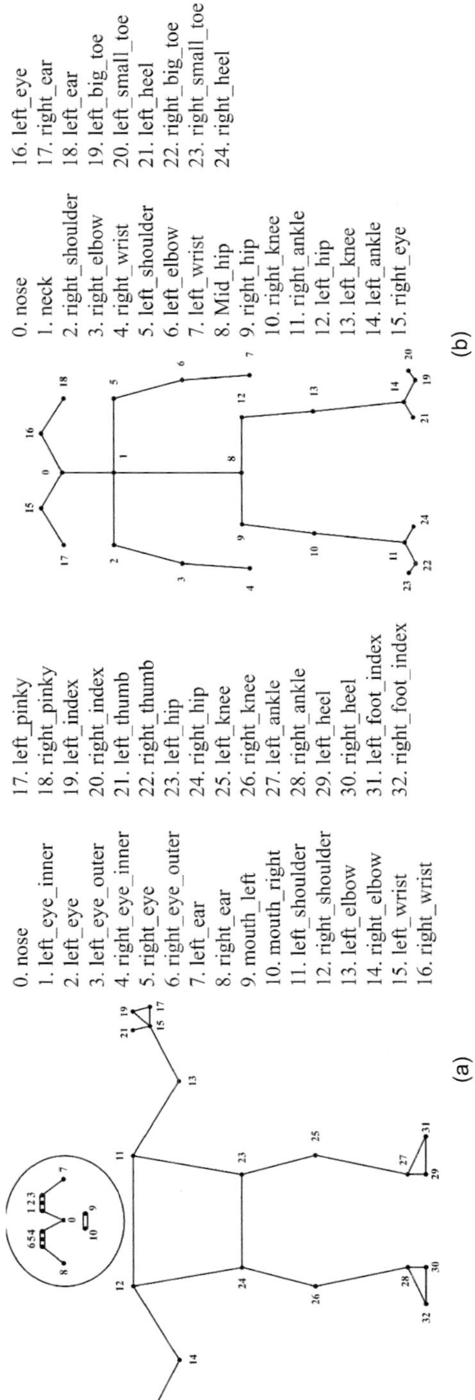

Fig. 16.1 Human pose estimation frameworks (Yoon et al., n.d.).

same posture for a long duration, and inadequate infrastructure. Various medical and mechanical methods, such as massage, heat, cold, exercise, and surgery have been used to treat musculoskeletal diseases.

1.2 Learning Management System (LMS)

Usually, learning management system (LMS) helps the educationalist to provide the educational content, track the progress of the students, engage in parent–teacher communication forums, and so on (Aljaloud et al., 2022). Providing education details in LMS gives information about students' progress to their parents. With LMS, the inclusion of a report notifying of prevalence of MSD in a student also helps the parents to monitor and prevent the severity of the disorder in the student. LMS offers detailed reports and analytics on student progress, performance, and course usage, allowing instructors to make informed decisions about their teaching approach. LMSs are widely used in education, corporate training, and professional development, and can be customized to meet the specific needs of different organizations and educational institutions (Harithasan et al., 2022). By use of an LMS, the learning process's efficiency and effectiveness can be raised besides giving teachers and students a flexible and easily available platform for evaluation and learning (Srinivasa et al., 2022).

1.3 Ergonomics

Ergonomics help to lower the MSD risk in a classroom setting. Those who maintain prolonged static postures or engage in repeated tasks—like sitting in a classroom—may find MSDs a frequent issue. Several ergonomic strategies in the classroom can help lower MSD risk: adjustable desks and chairs can lower the risk of MSDs (Abdul Rahim et al., 2022); excellent posture, including keeping the back straight and the feet flat on the floor, can also lower stress on the muscles and joints. MSDs can be avoided by letting students stand up, stretch, and move around during break time (Harithasan et al., 2022). Maintaining suitable lighting and temperature conditions can assist in lowering eye strain and muscle tension. Using ergonomic writing tools—pencils with a comfortable grip, for example—helps lower the risk of MSDs in the hands and wrists. Reducing the risk of CMSDs (Cento et al., 2022) by limiting technology use and encouraging the use of paper and pencil will help. Regular exercise at home also helps to prevent MSD in the future.

1.4 Deep Learning Models for Rehabilitation Exercises

The intensity level of MSD prevalence among students can be assessed through the anatomical structure of sitting and standing postures (Boesch, 2021; Gu et al., 2020). The factors mentioned earlier might lead people to experience low-back pain, and neck, knee, shoulder, and elbow joint pains, which lead to MSDs (Suri et al., 2010; National Academies of Sciences, 2020; Romero-Morales et al., 2023). Usually, LMS helps the educationalist to provide the educational content, track the progress of the students, engage in parent–teacher communication forums, and so on (Veluvali & Surisetti, 2021). Providing education details in LMS gives information about students' progress to their parents. With LMS, the inclusion of a report notifying of prevalence of MSD in a student also helps the parents to monitor and prevent the severity of the disorder in the student (Alturki & Aldraiweesh, 2021; Ahmad et al., 2022). Using an LMS can increase the efficiency and effectiveness of the learning process and provide students and instructors with a flexible and accessible platform for learning and assessment (Akpen et al., 2024; Strakos et al., 2023).

A study by Cai (2022) investigated the analysis of fitness motions in real time using deep learning based on convolutional neural networks (CNNs), which offer a robust and efficient real-time image diagnosis process along with a reasonably high detection accuracy (97.8%). The 8GB NVIDIA GeForce 1060 graphics processing unit (GPU) is used to implement this system, and the real-time analysis processing time is 2000 seconds.

Another study by Qiu et al. (2022) utilized gradient-weighted class activation mapping (Grad-CAM), and the visual result of matching the right and wrong poses was shown for the traditional Chinese rehabilitation exercises (eight section brocade). The deep learning architecture of a pair-based Siamese convolutional neural network (ST-AMCNNN) allows this framework to achieve an overall accuracy of 70.02%. Grad-CAM is able to generate dynamic feedback by using heat maps to represent student poses, utilizing a GPU (RTX 2080 Ti).

Compared to a direct image-based method, Moran et al. (2022) use three-dimensional coordinates for significant human body landmarks, which necessitates a model with significantly fewer input parameters and a smaller input size. Since the classification model uses the blaze pose estimation technique to accept a temporal series of key point positions, it can take into account the sequential position dependencies of an exercise (Kaur et al., 2024). This framework's accuracy for four fitness exercise poses is greater than 90%. The Intel i5-10200U Central Processing Unit (CPU) is used to implement the model of this framework. To model and identify body sensor-based activity, Uddin and Soylu (2018) suggested a time-sequential information-based deep neural structured learning (NSL), a promising deep learning method. The dependable time-sequential features are combined with NSL based on long short-term memory (LSTM) for activity modelling. The proposed method has a recall rate of approximately 99% for a publicly available dataset.

Using a virtual reality (VR) rehabilitation system with a Kinect sensor, Xiao et al. (2022) treat patients with movement disorders who are at or above Brunnstrom Stage III and possess a particular motor ability. It takes around 0.4 seconds to display the movements in an immersive environment that mimics real-time user action. Accurate rehabilitation is made easier for patients by audio feedback. The Dynamic Time Warping-K Nearest Neighbour (DTW-KNN), LSTM, and Gated Recurrent Unit (GRU) neural networks are used by Shen et al. (2022). The experimental findings show that the classification accuracy of DTW-KNN is 72.2%, 47.2%, and 69.4% for the three upper limb rehabilitation actions (elbow flexion, forearm) under different completion scenarios, while the corresponding classification accuracy of LSTM is 97.2%, 94.4%, and 91.7%. Using one of several pose estimate techniques, Hussein et al. (2020) interpret the fundamental idea of determining the body's key points. Dynamic temporal warping is then used to compare the patient's key points with those of the pickle files. The system, which is a desktop application with an easy-to-use interface, uses a 3D avatar to help patients complete the exercises correctly. In addition to rating them on the task, the technology indicates which parts of their bodies exhibit bad posture. Hussain et al., proposed technology provides high-accuracy pose comparison and posture estimation with a minimum accuracy of 80%.

Anand Thoutam et al. (2022) proposed deep learning–based detection of correct yoga exercise poses and provided feedback on incorrect postures to the users. This system uses an NVIDIA DGX V-100 GPU device to analyse the correctness of yoga pose through the following steps: feature extraction (Keras multi-person pose estimation with twelve body joints), classification (Multilayer Perceptron), and feedback generation. This approach provides an accuracy of 99.58% with less computational complexity. Rangari et al. (2022) utilize the RGB stands for Red, Green, and Blue camera to record exercisers' poses while in motion. To categorize the various activities of the human body, they use eighteen human body joints in 2D coordinates without imparting any wearable devices.

Sharma et al. (2022) developed the iYogacare system for self-correction of errors in performing yoga poses for users. The iYogacare system tracks postures over time by capturing feature points and generating user activity reports on a daily and weekly basis, following standard protocols. This task portion is primarily considered for accurate posture detection with XGBoost with a RandomSearch CV model of 99.2%—the pre-processing and data training of this framework on Ubuntu 18.04.4 LTS terminal and Google Colab. Nair and Sakthivel (2022) proposed deep learning with fifteen layers of CNN-based method to help the patient determine whether each of the twelve hand and arm rehabilitation exercises has been completed. The usefulness of such systems in post-stroke rehabilitation therapy is demonstrated by the fact that the suggested deep learning–based system can successfully detect the completion of rehabilitation activities with an accuracy of more than 98%.

This chapter comprises of four sections: the introduction section describes the prevalence of MSD among learning communities and AI-based MSD analysis systems, the risk factors of MSD, the LMS, and the importance of ergonomics in MSD; the second section describes the essential pieces of information about existing AI-based student behaviour monitoring; the third section presents the discussion; and the final section concludes.

2. Related Works

The term "classroom-based musculoskeletal disorders" refers to conditions or injuries related to the bones, muscles, and joints that develop as a direct result of prolonged sitting, repetitive motions, and other activities that are associated with school or work that takes place in a classroom setting (Yin Albert et al., 2022). Face detection, identification, and assessment of students' gazes can measure their attentiveness in the classroom (Zhang et al., 2023). Because it helps students become more focused, developing an automated learning feedback system is essential to modern educational practices (Kaewkaisorn et al., 2024). Proctoring an examination means keeping an eye on what students are doing while they are in the testing room (Naicker et al., 2020). Table 16.1 describes the purpose, methods, the dataset used, and significant results of the existing studies.

Table 16.1 Summary of the existing system's methods, results, and dataset used

Reference	*Purpose of the Work*	*Methods*	*Dataset Used*	*Major Result*
Yin Albert et al. (2022)	For multi-student targets in classroom teaching scenarios, real-time classroom behaviour recognition	2D-CNN expanded 3D-CNN with spatiotemporal domain	The raw data utilized in this work with over 900 mins of video in the classroom	The accuracy achieved through this framework is 93.43%.
Ngoc Anh et al. (2019)	Accurately provide the progress of a student	Random forest with synthetic minority oversampling technique	Real-time feed dataset collected at FPT University	Random forest with 100 estimators—produces an accuracy of 93.33% (previous study: 92.82%).
Lin et al. (2021)	Monitoring the behaviour of students in the classroom	OpenPose framework to collect 3D skeleton through Microsoft Kinect sensor for pose estimation. Students' behaviour is classified using a deep neural network (DNN) and skeleton-based bounding box created to recognize the person.	Datasets used in this work were collected from 459 high school students in Iran.	A 15.15% higher precision accuracy compared to skeleton-based and 12.15% higher recall metrics with skeleton-based.
Aljaloud et al. (2022)	To identify student blackboard interaction-based KPI - Key Performance Indicators that predicted learning outcomes	Combined CNN-LSTM approach	The dataset consists of seven courses and seven features from the 35,000 students in an institution. The total number of components used in this work is 1.715 million features.	Using all seven features, the CNN-LSTM approach yielded a precision score of 94.2%.
Su et al. (2021)	Developing an automatic learning feedback system is crucial in today's educational systems as it helps to increase students' focus.	RGB image-based CNN approach	OpenPose framework to collect 3D skeleton through RGB colour images. The behaviour of students is classified using a CNN created to recognize the activity using skeleton information.	The average recognition accuracy of this system is 89.37%.

The methodology for locating current literature on "automated learning systems to monitor the pervasiveness of musculoskeletal pain among students in classroom" encompassed three essential elements. Initially, pertinent academic databases such as PubMed, IEEE Xplore, Google Scholar, Scopus, and ScienceDirect were chosen for their healthcare, technological, and educational research coverage. A search for pertinent papers was conducted using a mix of keywords including "automated learning systems", "musculoskeletal pain", "students", and "classroom". The inclusion criteria targeted studies of student populations and AI-based monitoring systems, excluding those beyond these parameters. The titles and abstracts of identified articles were evaluated, followed by comprehensive reviews of pertinent studies. The snowballing technique was utilized to trace references and citations in significant publications, facilitating a more extensive identification of relevant works.

2.1 Evaluation of Students' Activity in the Classroom Premises Using Various Machine Learning Algorithms

Random forest (RF) (Ngoc Anh et al., 2019), linear support vector machines (LSVM) (Naicker et al., 2020), logistic regression, and artificial neural networks are among the classifiers used for the various systems. The architectural representation of machine learning-based analysis of students' behaviour in the classroom is shown in Figure 16.2.

Recent decades have seen the discovery of a number of factors that impact the prevalence of scoliosis, such as a fully developed skeleton, body mass index (BMI), genes, and age. Two to three per cent of the population may be affected with adolescent idiopathic scoliosis (AIS), the most common form of the condition (Liang et al., 2022). It is critical to track the progress of stroke survivors in order to assess their level of recovery, as stroke is recognized as a critical worldwide health concern (Pollock et al., 2014). Although clinical diagnosis is a popular and prevalent traditional method for evaluating stroke rehabilitation (Igwesi-Chidobe, 2012; Alizadehsani et al., 2013), it has the drawbacks of being subjective, expensive (Chen et al., 2021), and inconvenient for patients to have to attend health centres multiple times.

2.2 Student Activity Monitoring through OpenPose Estimation Technique

In a classroom, various students behaviours such as reading, writing, listening the lecture, taking notes, or using digital gadgets are may appear visually similar. The problems of automatically identifying, detecting, and annotating disengagement of listening, using mobile phones, and roaming around classroom are the present significant challenges to most conventional techniques. To advance educational research, it is vital to describe students' behaviours, for which spotting subjects and their behaviour is always helpful. Classroom video recordings typically contain these diverse behaviors, making precise observation and analysis of each learners behaviour inside the classroom. Before recognizing multi-subject behaviour, the spatial identification of subjects and the temporal detection of activities are required. Therefore, the student class behaviour dataset suggested in this chapter is divided into a detection section, a recognition section, and a caption section.

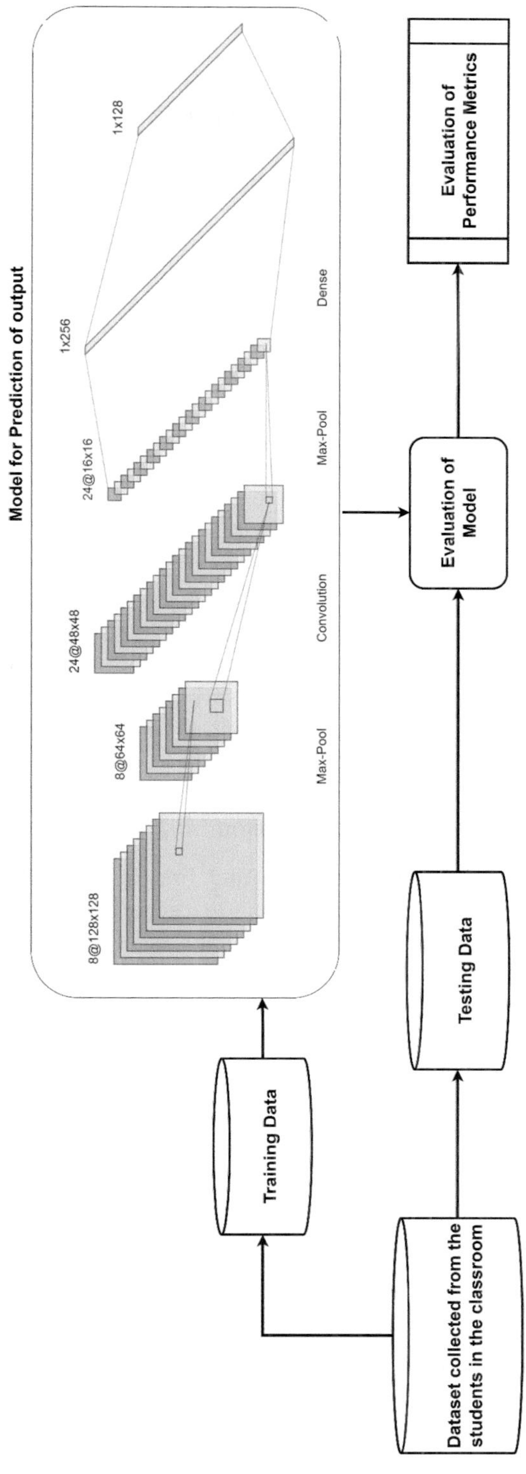

Fig. 16.2 ML-based student behaviour analysis architecture.

Several challenges may arise by changing the video recording camera's field of view, resolution, and lighting in the classroom environment which may significantly affect the accuracy of student's pose recognition. The OpenPose framework estimates two-dimensional (2D) human poses to solve this problem (Boesch, 2021; Gu et al., 2020). This framework is highly customizable, allowing developers to fine-tune the parameters to meet the specific needs of their application. The drawback in this framework is that faulty connections in pose estimation happen in congested spaces like classrooms. There are two basic flows in OpenPose, a multistage CNN (Cao et al., 2018). Both the first flow and the second flow forecast confidence maps (St) expressed as:

$$S^t = \rho^t(F, L^{T_P}, S^{t-1}), \forall T_P + T_C \tag{16.1}$$

where, ρ^t-Stage t, ($T_P < t \leq T_P + T_C$), Part Affinity Fields (PAFs) that represent the connections between different body parts (L^t) expressed as:

$$L^t = \phi^t(F, L^{t-1}), \forall 2 \leq t \leq T_P \tag{16.2}$$

where, ϕ^t-Stage t, ($t \leq T_P$). The greedy inference approach is utilized to create 2D key points (joints) for humans present in the image. Figure 16.3 displays the OpenPose architecture to detect the skeleton information of humans.

The architecture of OpenPose consists of several components such as image pre-processing, pose estimation, PAF generation, non-maximum suppression, and pose clustering.

2.3 Deep Learning–based Student Behaviour Analysis

The technology can examine student data and find behavioural patterns and correlations that may not be immediately obvious to teachers. This can assist teachers in personalizing their approach to instruction and enable pupils to perform by revealing insights into students' strengths and limitations. Deep learning algorithms that analyse student behaviour are trained on a sizable dataset containing data from several students. This may include test scores, participation in class, and attendance records. These data are fed into the algorithms, which look for trends and connections between behavioural characteristics and academic performance. The development of the recurrent neural network is the LSTM model (Ekambaram & Ponnusamy, 2024; Nandakumar et al., 2023). It is frequently used in speech recognition and text analysis. It has a memory cell to help with precise predictions. L2-GraftNet adds additional layers to the underlying architecture in a controlled way by using a process known as "grafting" (Saba et al., 2021). For analyzing motion in sequences of images or videos in a time-series format, Long Short-Term Memory (LSTM) networks are required. An LSTM is composed of three main components: as shown in Figure 16.4 the forget gate, the input gate, and the output gate, which together regulate the flow of information through the network.

Research on recovery training recognition has mostly focused on practical, real-time uses of the technology, including smart home technology, situational awareness,

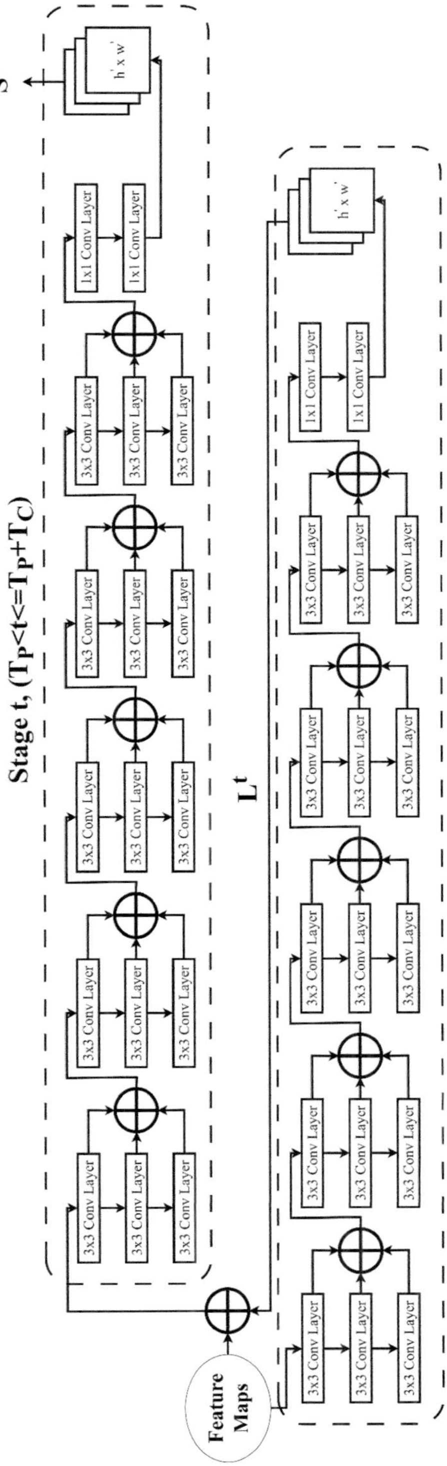

Fig. 16.3 OpenPose architecture for human pose detection (Gu et al., 2020).

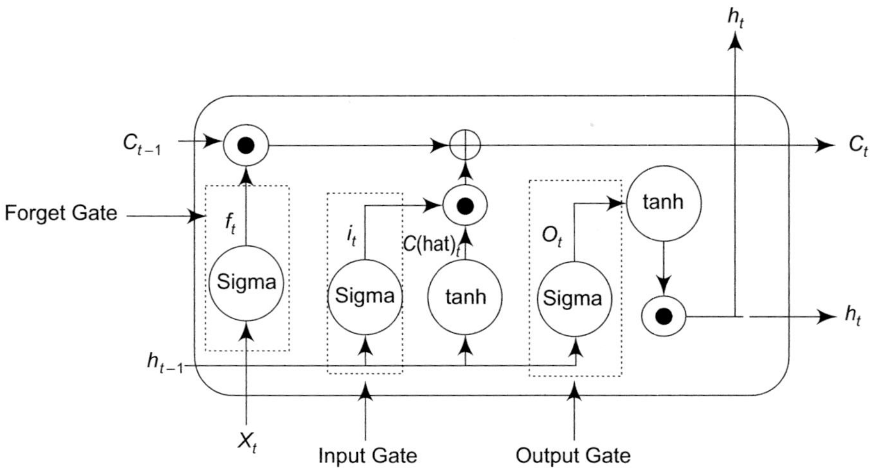

Fig. 16.4 Structure of CNN-LSTM (Ekambaram & Ponnusamy, 2024).

and other healthcare and military-related domains (Shalaby et al., 2022). From this vantage point, a deep learning system needs to keep tabs on their everyday activities, and there has been a rise in the use of affordable fitness trackers for medical purposes (Solongontuya et al., 2021). Due to the developments in deep learning and their higher quality, CNN models are commonly used in human posture evaluation (Vijayakumar & Dilliraj, 2022).

3. Discussion

The combination of real-time monitoring and machine learning algorithms can provide a comprehensive and accurate assessment of the risk factors for musculoskeletal pain in the classroom. This section explores the various interpretations of the existing AI systems and discusses the research findings with a focus on the future.

3.1 Interpretations of the Existing System

Significant interpretations of existing systems have been discussed. Ngoc Anh et al. (2019) proposed a 3D position estimation algorithm. A reliable way to gauge the estimated user attention is to combine estimated locations and eye gaze. It enables to respond in the actual world where classroom layouts differ (configuration, large space, etc.) by combining several cameras (Lin et al., 2021). Using an error correction approach, issues like an erroneous person and pose detection were fixed in the suggested method. This approach reduces the number of broken links in the skeleton data. Su et al. (2021) proposed that without requiring the storage or transmission of visual data to a distant cloud, analytic algorithms can operate in real time using standard RGB cameras on portable edge devices. This online, edge-computing concept guarantees simple deployment and can significantly allay privacy worries.

3.2 Research Findings

The research findings from the existing systems are:

- They are inclusive of more behaviour predictions in the system like standing, chatting, and other actions in any environment.
- The processing and maintenance costs of the current system are very high.
- Depth-wise video analysis is required to find the smartphone action by the students in the classroom.
- Enhancement of lightweight deep learning architectures helps reduce the system's computational complexity.
- The occlusion of body parts in skeleton-based pose estimation may cause a severe loss of accuracy in recognizing students' behaviour.

4. Conclusion

According to the review's results, the learning/learned communities are at risk of MSD. Educators are frequently affected by MSD, which most often affects the wrist, neck, lower back, and shoulder (Tahernejad et al., 2024; Bhatia et al., 2024; Abdul Rahim et al., 2022). Ergonomics plays an important role in reducing the risk of CMSDs. This includes proper furniture design, encouraging good posture, providing break time, maintaining appropriate lighting and temperature levels, encouraging the use of ergonomic writing tools, and limiting the use of technology. Educators and school administrators can work together to create an ergonomic environment that promotes student health and well-being. In addition to ergonomic practices, regular exercise, stretching, and breaks can help reduce the risk of CMSDs (Mahmud et al., 2011). Overall, CMSDs can hurt students' academic performance and quality of life. Educators and school administrators must take proactive measures to reduce the risk of CMSDs and promote student health and well-being.

A comprehensive evaluation in this chapter focuses on how automated learning system technologies analyse the posture movement of students in real time to prevent chronic MSD. These technologies help to identify risk factors like poor ergonomics and avert long-term health issues (Chan et al., 2022; Hoe et al., 2018). However, data protection and technology implementation in the classroom remains difficult. Constant surveillance can improve behaviour but sometimes causes psychological issues like uneasiness (Markova et al., 2024). The review emphasizes multidisciplinary methodologies and long-term health studies to improve these systems. These findings have policy significance since ergonomic norms and technology-driven health treatments in the classroom may improve students' health and academic performance (Safari et al., 2025; Khamaisi et al., 2024).

Students, particularly those who spend extended periods sitting, are prone to developing MSDs as improper seating and static postures can lead to chronic pain (Demissie et al., 2024; Daneshmandi et al., 2017). In fact, research has shown that prolonged sedentary behaviour among students is associated with an increased prevalence of neck, back, and shoulder pain. Addressing this issue early through automated learning systems in educational environments can help prevent the development of these conditions, which can impact students' academic performance and overall well-being. By focusing on students, we aim to tackle an often-overlooked population that stands to benefit significantly from such interventions, just as athletes do in their respective contexts.

The research can be taken forward in the future by using cutting-edge deep learning models that offer lesser training time with reasonable computing power, recognition of actions like hand rising, standing, and chatting in various environments, and processing of incomplete skeleton data that is caused by the occlusion of different human parts. Future research on the work can make use of state-of-the-art techniques such as manifold learning, LSTMs, quantum deep learning, brain-like computing approaches, and lightweight shallow deep learning models.

References

Abdul Rahim, A.A., Jeffree, M.S., Ag Daud, D.M., Pang, N. and Sazali, M. F. (2022). Factors associated with musculoskeletal disorders among regular and special education teachers: A narrative review. *International Journal of Environmental Research and Public Health*, *19*(18), 11704.

Ahmad, S., El-Affendi, M. A., Anwar, M. S. and Iqbal, R. (2022). Potential future directions in optimization of students' performance prediction system. *Computational Intelligence and Neuroscience*, *2022*, 1–26.

Akpen, C.N., Asaolu, S., Atobatele, S. et al. Impact of online learning on student's performance and engagement: a systematic review. Discov Educ 3, 205 (2024). https://doi.org/10.1007/s44217-024-00253-0.

Alizadehsani, R., Habibi, J., Hosseini, M.J., Mashayekhi, H., Boghrati, R., Ghandeharioun, A., Bahadorian, B. and Sani, Z. A. (2013). A data mining approach for diagnosis of coronary artery disease. *Computer Methods and Programs in Biomedicine*, *111*(1), 52–61.

Aljaloud, A.S., Uliyan, D.M., Alkhalil, A., Elrhman, M.A., Alogali, A.F.M., Altameemi, Y.M., Altamimi, M. and Kwan, P. (2022). A deep learning model to predict student learning outcomes in LMS using CNN and LSTM. *IEEE Access*, *10*, 85255–85265.

Alturki, U. and Aldraiweesh, A. (2021). Application of learning management system (LMS) during the Covid-19 pandemic: A sustainable acceptance model of the expansion technology approach. *Sustainability*, *13*(19), 10991.

Anand Thoutam, V., Srivastava, A., Badal, T., Kumar Mishra, V., Sinha, G.R., Sakalle, A., Bhardwaj, H. and Raj, M. (2022). Yoga pose estimation and feedback generation using deep learning. *Computational Intelligence and Neuroscience*, *2022*, 1–12.

Bhatia, V., Vaishya, R.O., Jain, A., Grover, V., Arora, S., Das, G., Abdulla, A.M., Sainudeen, S., Babiker, A. and Saluja, P. (2024). Identification of prevalence of musculoskeletal disorders and various risk factors in dentists. *Heliyon*, *10*(1), e23780–e23780.

Boesch, G. (2021). *A Guide to OpenPose in 2021*. Viso.ai, 20 September. Available at https://viso.ai/deep-learning/openpose/. (Accessed on 5 June 2025)

Cai, H. (2022). Application of intelligent real-time image processing in fitness motion detection under internet of things. *The Journal of Supercomputing*, *78*(6), 7788–7804.

Cao, Z., Hidalgo, G., Simon, T., Wei, S.-E. and Sheikh, Y. (2018). *OpenPose: Realtime multi-person 2D pose estimation using part affinity fields*. Available at https://doi.org/10.48550/arxiv.1812.08008. (Accessed on 8 June 2025).

Cento, A.S., Leigheb, M., Caretti, G. and Penna, F. (2022). Exercise and exercise mimetics for the treatment of musculoskeletal disorders. *Current Osteoporosis Reports*, *20*, 249–259.

Chen, X., Guan, Y., Shi, J.-Q., Du, X.-L. and Eyre, J. (2021). Automated stroke rehabilitation assessment using wearable accelerometers in free-living environments. *ArXiv:2009.08798 [Cs, Eess, Stat]*. available at https://arxiv.org/abs/2009.08798. (Accessed on 10 June 2025).

Cieza, A., Causey, K., Kamenov, K., Hanson, S.W., Chatterji, S. and Vos, T. (2020). Global estimates of the need for rehabilitation based on the Global Burden of Disease study 2019: A systematic analysis for the Global Burden of Disease Study 2019. *The Lancet*, *396*(10267), 2006–2017.

CMU-Perceptual-Computing-Lab/openpose. (2020). GitHub. 5 November. Available at https://github.com/CMU-Perceptual-Computing-Lab/openpose. (Last Accessed on 5 June 2025)

Daneshmandi, H., Choobineh, A., Ghaem, H. and Karimi, M. (2017). Adverse effects of prolonged sitting behavior on the general health of office workers. *Journal of Lifestyle Medicine*, *7*(2), 69–75.

Delp, S.L., Anderson, F.C., Arnold, A.S., Loan, P., Habib, A., John, C.T., Guendelman, E. and Thelen, D. G. (2007). OpenSim: Open-source software to create and analyze dynamic simulations of movement. *IEEE Transactions on Biomedical Engineering*, *54*(11), 1940–1950.

Demissie, B., Bayih, E.T. and Demmelash, A.A. (2024). A systematic review of work-related musculoskeletal disorders and risk factors among computer users. *Heliyon*, *10*(3), e25075.

Ekambaram, D. and Ponnusamy, V. (2024). Real-time monitoring and assessment of rehabilitation exercises for low back pain through interactive dashboard pose analysis using Streamlit—A pilot study. *Electronics*, *13*(18), 3782.

Erick, P.N. and Smith, D.R. (2015). Musculoskeletal disorders in the teaching profession: An emerging workplace hazard with significant repercussions for developing countries. *Industrial Health*, *53*(4), 385–386.

google-ai-edge. (2024). *GitHub - google-ai-edge/mediapipe: Cross-platform, customizable ML solutions for live and streaming media.* GitHub, 18 December. Available at https://github.com/google-ai-edge/mediapipe/tree/master. (Last Accessed on 5 June 2025).

Gu, Y., Zhang, H. and Kamijo, S. (2020). Multi-person pose estimation using an orientation and occlusion aware deep learning network. *Sensors*, *20*(6), 1593.

Harithasan, D., Singh, D.K.A., Abd Razak, N.A.B. and Baharom, N.B. (2022). Personal, academic stressors and environmental factors contributing to musculoskeletal pain among undergraduates due to online learning: A mixed method study with data integration. *International Journal of Environmental Research and Public Health*, *19*(21), 14513.

Hoe, V.C., Urquhart, D.M., Kelsall, H.L., Zamri, E.N. and Sim, M.R. (2018). Ergonomic interventions for preventing work-related musculoskeletal disorders of the upper limb and neck among office workers. *Cochrane Database of Systematic Reviews*, Issue No. 10, Art. No. CD008570.

Igwesi-Chidobe, C. (2012). Obstacles to obtaining optimal physiotherapy services in a rural community in southeastern Nigeria. *Rehabilitation Research and Practice*, *2012*, 1–8.

Jebaraj, P., Vellaiputhiyavan, K., Mohan, V. and Alex, R.G. (2022). Prevalence and risk factors for work-related musculoskeletal disorder among female domestic workers employed in private homes in south India: A cross-sectional study. *Indian Journal of Medical Sciences*, *74*, 72–78.

Kaewkaisorn, K., Pintong, K., Bunyang, S., Tansawat, T. and Siriborvornratanakul, T. (2024). Student attentiveness analysis in virtual classroom using distraction, drowsiness and emotion detection. *Discover Education*, *3*(1), 56.

Kaur, H., Kaur, T., Bhardwaj, V., *et al.* (2024). An ensemble deep learning model for classification of students as weak and strong learners via multiparametric analysis. *Discover Applied Sciences*, *6*, 595.

Khamaisi, R.K., Perini, M., Morganti, A., Placci, M., Grandi, F., Peruzzini, M. and Botti, L. (2024). An innovative integrated solution to support digital postural assessment using the TACOs methodology. *Computers & Industrial Engineering*, *194*, 110376.

Liang, R., Yip, J., Fan, Y., Cheung, J.P.Y. and To, K.-T.M. (2022). Electromyographic analysis of paraspinal muscles of scoliosis patients using machine learning approaches. *International Journal of Environmental Research and Public Health*, *19*(3), 1177.

Lin, F.-C., Ngo, H.-H., Dow, C.-R., Lam, K.-H. and Le, H. L. (2021). Student behavior recognition system for the classroom environment based on skeleton pose estimation and person detection. *Sensors*, *21*(16), 5314.

Lugaresi, C., Tang, J., Nash, H., McClanahan, C., Uboweja, E., Hays, M., Zhang, F., Chang, C.-L., Yong, M. G., Lee, J., Chang, W.-T., Hua, W., Georg, M. and Grundmann, M. (2019). MediaPipe: A framework for building perception pipelines. *ArXiv*. Available at https://doi.org/10.48550/arxiv.1906.08172. (Last accessed on 6 June 2025).

Mahajan, D., Gupta, M.K., Mantri, N., Joshi, N.K., Gnanasekar, S., Goel, A.D., Srinivasan, S., Gonade, N.M., Sharma, S.K., Garg, M.K. and Bhardwaj, P. (2023). Musculoskeletal disorders among doctors and nursing officers: An occupational hazard of overstrained healthcare delivery system in western Rajasthan, India. *BMC Musculoskeletal Disorders*, *24*(1), 349.

Mahmud, N., Kenny, D.T., Md Zein, R. and Hassan, S.N. (2011). Ergonomic training reduces musculoskeletal disorders among office workers: Results from the 6-month follow-up. *The Malaysian Journal of Medical Sciences*, *18*(2), 16–26.

Markova, V., Markov, M., Petrova, Z. and Filkova, S. (2024). Assessing the impact of prolonged sitting and poor posture on lower back pain: A photogrammetric and machine learning approach. *Computers*, *13*(9), 231.

Mishra, S., Avinash, G., Kundu, M.G., Verma, J., Sheth, A. and Dutta, A. (2024). Work-related musculoskeletal disorders among various occupational workers in India: A systematic review and meta-analysis. *Journal of Occupational Health*, *67*(1).

Moran, A., Gebka, B., Goldshteyn, J., Beyer, A., Johnson, N. and Neuwirth, A. (2022). Muscle vision: Real time keypoint based pose classification of physical exercises. *ArXiv (Cornell University)*. Available at https://doi.org/10.48550/arxiv.2203.12111 (Last accessed 4 June 2025).

Naicker, N., Adeliyi, T. and Wing, J. (2020). Linear support vector machines for prediction of student performance in school-based education. *Mathematical Problems in Engineering*, *2020*(5), 1–7.

Nair, B.B. and Sakthivel, N.R. (2022). A deep learning-based upper limb rehabilitation exercise status identification system. *Arabian Journal for Science and Engineering*, *48*, 1237–1271.

Nandakumar, R., Ponnusamy, V. and Kumar Mishra, A. (2023). LSTM based spectrum prediction for real-time spectrum access for IoT applications. *Intelligent Automation & Soft Computing*, *35*(3), 2805–2819.

National Academies of Sciences. (2020). Musculoskeletal disorders. In *Selected Health Conditions and Likelihood of Improvement with Treatment*. National Academies Press (US). Available at https://www.ncbi.nlm.nih.gov/books/NBK559512/. (Last Accessed 5 June 2025).

Nayak, A. and Anitha, C.T. (2024). Prevalence of musculoskeletal disorders and healthcare utilization among female homemakers in rural Odisha: A cross-sectional study. *The Indian Journal of Occupational Therapy*, *56*(2), 42–47.

Ngoc Anh, B., Tung Son, N., Truong Lam, P., Phuong Chi, L., Huu Tuan, N., Cong Dat, N., Huu Trung, N., Umar Aftab, M. and Van Dinh, T. (2019). A computer-vision based application for student behavior monitoring in classroom. *Applied Sciences*, *9*(22), 4729.

Pollock, A., Baer, G., Campbell, P., Choo, P.L., Forster, A., Morris, J., Pomeroy, V.M. and Langhorne, P. (2014). Physical rehabilitation approaches for the recovery of function and mobility following stroke. *Cochrane Database of Systematic Reviews*, Issue no. 4, Art. No. CD001920.

Prasetya, T. AE., Mamun, A.A., Rahmania, A., Ahmed, M., Uddin, A. SMS., Nilamsari, N. and Wardani, R. WK. (2024). Prevalence and associated risk factors of musculoskeletal disorders among information technology (IT) professionals: A systematic review. *Narra J*, *4*(3), e1100.

Qiu, Y., Wang, J., Jin, Z., Chen, H., Zhang, M. and Guo, L. (2022). Pose-guided matching based on deep learning for assessing quality of action on rehabilitation training. *Biomedical Signal Processing and Control*, *72*, 103323.

Rangari, T., Kumar, S., Roy, P.P., Dogra, D.P. and Kim, B.-G. (2022). Video based exercise recognition and correct pose detection. *Multimedia Tools and Applications*, *81*, 30267–30282.

Roggio, F., Trovato, B., Sortino, M. and Musumeci, G. (2024). A comprehensive analysis of the machine learning pose estimation models used in human movement and posture analyses: A narrative review. *Heliyon*, *10*(21), e39977.

Romero-Morales, C., López-López, D., Almazán-Polo, J., Mogedano-Cruz, S., Sosa-Reina, M.D., García-Pérez-de-Sevilla, G., Martín-Pérez, S. and González-de-la-Flor, Á. (2023). Prevalence, diagnosis and management of musculoskeletal disorders in elite athletes: A mini-review. *Disease-a-Month*, *70*(1), 101629.

Rookba, H., Hussein Abdallah, B., Ahmed Abdallah, A., Osama Abdel-Aal, R., Reda Numan, R., Khaled Darwish, A. and El-Behaidy, W. (2020). Automatic feedback for physiotherapy exercises based on PoseNet. *FCI-H Informatics Bulletin*, *2*(2), 10–14.

Saba, T., Rehman, A., Jamail, N.S.M., Marie-Sainte, S.L., Raza, M. and Sharif, M. (2021). Categorizing the students' activities for automated exam proctoring using proposed deep L2-GraftNet CNN network and ASO based feature selection approach. *IEEE Access*, *9*, 47639–47656.

Safari, M., Naserbakht, A.H., Badri Kouhi, A. and Varmazyar, S. (2025). Artificial intelligence and emerging technologies in assessing ergonomic risk factors in the workplace: A systematic review. *Work*, 10519815251349793.

Sezer, B., Kartal, S., Sıddıkoğlu, D. and Kargül, B. (2022). Association between work-related musculoskeletal symptoms and quality of life among dental students: A cross-sectional study. *BMC Musculoskeletal Disorders*, *23*(1), 41.

Shalaby, E., ElShennawy, N. and Sarhan, A. (2022). Utilizing deep learning models in CSI-based human activity recognition. *Neural Computing and Applications*, *34*, 5993–6010.

Sharma, A., Shah, Y., Agrawal, Y. and Jain, P. (2022). Real-time recognition of yoga poses using computer vision for smart health care. *ArXiv (Cornell University)*. Available at https://doi.org/10.48550/arxiv.2201.07594. (Last accessed on 7 Jun 2025).

Shen, C., Ning, X., Zhu, Q., Miao, S. and Lv, H. (2022). Application and comparison of deep learning approaches for upper limb functionality evaluation based on multi-modal inertial data. *Sustainable Computing: Informatics and Systems*, *33*, 100624.

Solongontuya, B., Cheoi, K.J. and Kim, M.-H. (2021). Novel side pose classification model of stretching gestures using three-layer LSTM. *The Journal of Supercomputing*, *77*(9), 10424–10440.

Srinivasa, K.G., Kurni, M. and Saritha, K. (2022). Learning, teaching, and assessment methods for contemporary learners. In *Springer Texts in Education*. Springer Nature Singapore.

Strakos, J.K., Douglas, M.A., McCormick, B. and Wright, M. (2023). A learning management system-based approach to assess learning outcomes in operations management courses. *The International Journal of Management Education*, *21*(2), 100802.

Su, M.-C., Cheng, C.-T., Chang, M.-C. and Hsieh, Y.-Z. (2021). A video analytic in-class student concentration monitoring system. *IEEE Transactions on Consumer Electronics*, *67*(4), 294–304.

Suri, P., Morgenroth, D.C., Kwoh, C.K., Bean, J.F., Kalichman, L. and Hunter, D.J. (2010). Low back pain and other musculoskeletal pain comorbidities in individuals with symptomatic osteoarthritis of the knee: Data from the osteoarthritis initiative. *Arthritis Care & Research*, *62*(12), 1715–1723.

Tahernejad, S., Hejazi, A., Rezaei, E., Makki, F., Sahebi, A. and Zangiabadi, Z. (2024). Musculoskeletal disorders among teachers: A systematic review and meta-analysis. *Frontiers in Public Health*, *12*.

Tiwari, J., Halder, P., Sharma, D., Saini, U.C., Rajagopal, V. and Kiran, T. (2024). Prevalence and association of musculoskeletal disorders with various risk factors among older Indian adults: Insights from a nationally representative survey. *PLoS ONE*, *19*(10), e0299415–e0299415.

Uddin, M.Z. and Soylu, A. (2021). Human activity recognition using wearable sensors, discriminant analysis, and long short-term memory-based neural structured learning. *Scientific Reports*, *11*(1), 16455.

Veluvali, P. and Surisetti, J. (2021). Learning management system for greater learner engagement in higher education—A review. *Higher Education for the Future*, *9*(1), 234763112110498.

Vijayakumar, P. and Dilliraj, E. (2022). A comparative review on image analysis with machine learning for extended reality (XR) applications. In *Smart Innovation, Systems and Technologies* (pp. 307–328). Springer Nature.

Xiao, B., Chen, L., Zhang, X., Li, Z., Liu, X., Wu, X. and Hou, W. (2022). Design of a virtual reality rehabilitation system for upper limbs that inhibits compensatory movement. *Medicine in Novel Technology and Devices*, *13*, 100110.

Yin Albert, C.C., Sun, Y., Li, G., Peng, J., Ran, F., Wang, Z. and Zhou, J. (2022). Identifying and monitoring students' classroom learning behavior based on multisource information. *Mobile Information Systems*, *2022*, 1–8.

Yoon, H., Jo, E., Ryu, S., Yoo, J.-I. and Kim, J. (n.d.). *MediaPipe-based LSTM-Autoencoder Sarcopenia Anomaly Detection and Requirements for Improving Detection Accuracy*. Available at https://ceur-ws.org/Vol-3362/ISE2022_short01_Yoon_MediaPipe.pdf (accessed on 18 June, 2025).

Zhang, M., Li, H. and Tian, S. (2023). Visual analysis of machine learning methods in the field of ergonomics—Based on Cite Space V. *International Journal of Industrial Ergonomics*, *93*, 103395.

Index

For Product Safety Concerns and Information please contact our EU representative GPSR@taylorandfrancis.com Taylor & Francis Verlag GmbH, Kaufingerstraße 24, 80331 München, Germany